Manual of First and
Second Fixing Carpentry

Les Goring is a member of the Chartered Institute of Building and a former Senior Lecturer in Wood Trades at Hastings College of Arts & Technology. In addition to being a Member of the Institute of Wood Science, he is also a Fellow of the Institute of Carpenters and a Licentiate of the City & Guilds Institute. All drawings in Les Goring's books were drawn by the author.

To Mary, Penny, Jon and Jenny, and other Gorings
of whom there are many

Books by the same author:

First-Fixing Carpentry Manual
Manual of Purpose-Made Woodworking Joinery

Manual of First and Second Fixing Carpentry

Fourth Edition

Les Goring

Routledge
Taylor & Francis Group

LONDON AND NEW YORK

Fourth edition published 2018
by Routledge
2 Park Square, Milton Park, Abingdon, Oxon, OX14 4RN

and by Routledge
711 Third Avenue, New York, NY 10017

Routledge is an imprint of the Taylor & Francis Group, an informa business

First edition published by Edward Arnold Publishers 1998
Third edition published by Butterworth-Heinemann 2010

British Library Cataloguing-in-Publication Data
A catalogue record for this book is available from the British Library

Library of Congress Cataloging-in-Publication Data
Names: Goring, L. J., author.
Title: Manual of first and second fixing carpentry / Les Goring.
Other titles: Manual of first & second fixing carpentry
Description: Fourth edition. | New York, NY : Routledge, 2018. | Revised edition of: Manual of first & second fixing carpentry / Les Goring. 2007. | Includes index.
Identifiers: LCCN 2017045944| ISBN 9781138295995 (pbk) | ISBN 9781138296008 (hbk) | ISBN 9781315099804 (ebk) | ISBN 9781351584203 (Adobe Reader) | ISBN 97813515841997 (ePub) | ISBN 9781351584180 (Mobipocket)
Subjects: LCSH: Carpentry--Handbooks, manuals, etc. | Finish carpentry--Handbooks, manuals, etc.
Classification: LCC TH5608 .G6142 2018 | DDC 694/.6--dc23
LC record available at https://lccn.loc.gov/2017045944

ISBN: 978-1-138-29600-8 (hbk)
ISBN: 978-1-138-29599-5 (pbk)
ISBN: 978-1-315-09980-4 (ebk)

Typeset in Adobe Caslon and Futura
by Servis Filmsetting Ltd, Stockport, Cheshire

Contents

Preface   xi
Acknowledgements   xii
Health and Safety Awareness   xiii
Abbreviations   xiv
Technical Data   xv

1 Reading Construction Drawings   1

   1.1 Introduction   1
   1.2 Orthographic Projection   3
   1.3 Oblique Projections   5

2 Tools Required: their Care and Proper Use   8

   2.1 Introduction   8
   2.2 Marking and Measuring   8
   2.3 Handsaws   12
   2.4 Hammers   15
   2.5 Screwdrivers   16
   2.6 Marking Gauges   16
   2.7 Chisels   17
   2.8 Oilstones and Diamond Whetstones   17
   2.9 Hand Planes   19
   2.10 Ratchet Brace   20
   2.11 Bits and Drills   20
   2.12 Individual Handtools   22
   2.13 Portable Powered and Cordless Circular Saws   24
   2.14 Powered and Cordless Drills and Screwdrivers   25
   2.15 Powered and Cordless Planers   26
   2.16 Powered and Cordless Jigsaws   27
   2.17 Powered and Cordless SDS Rotary Hammer Drills   28
   2.18 Powered (Portable) Routers   28

2.19 Jigs   29
2.20 Nailing Guns   30

3 Carpentry Fixing-Devices   33

3.1   Introduction   33
3.2   Nails and Pins   33
3.3   Screws and Plugs   35
3.4   Cavity Fixings   39

4 Making and Fixing Shelving Arrangements   40

4.1   Introduction   40
4.2   Shelf Material   40
4.3   Beams or Joists Likened to Shelves   40
4.4   Understanding Basic Mechanics   40
4.5   Shelf Supports   41
4.6   Loads on Shelves Likened to Loads on Floor Joists   43
4.7   Blockboard or Plywood as Shelves   44
4.8   Mid-area Shelf-supports   45
4.9   Ladder-frame Shelf Unit   45

5 Making Site-Equipment Items   50

5.1   Saw Stool   50
5.2   Hop-ups   58
5.3   Board-and-stand   59
5.4   Straightedges, Concrete-levelling Boards and Plumb Rules   61

6 Fixing Doorframes, Linings and Doorsets   63

6.1   Introduction   63
6.2   Fixing Doorframes   65
6.3   Frame Details   67
6.4   Fixing Door Linings   68
6.5   Setting Up Internal Frames Prior to Building Block-partitions   71
6.6   Storey Frames   72
6.7   Subframes   72
6.8   Built-up Linings   73
6.9   Moisture Effect from Wet Plastering   73
6.10 Doorsets   74
6.11 Fire-resisting Doorsets   75

7 Fixing Wooden and uPVC Windows   77

7.1   Introduction   77
7.2   Casement Windows   77

7.3   Glazing   78
7.4   Window Boards   80
7.5   Boxframe Windows   81
7.6   Fixing uPVC Casement Windows into Existing Wooden Boxframes   83

8   Fixing Traditional and Modern Floor Joists and Flooring   86

8.1   Introduction   86
8.2   Ground Floors   87
8.3   Laying T&G Timber Boarding   90
8.4   Floating Floor (with Continuous Support)   91
8.5   Floating Floor (with Discontinuous Support)   91
8.6   Fillet or Battened Floors   92
8.7   Beam-and-Block Floors   93
8.8   Engineered-Timber Floors   93
8.9   Upper Floors   94
8.10 Strutting   97
8.11 Fitting and Fixing Timber Joists   100
8.12 Fixing Flooring Panels on Joists   100
8.13 Fitting and Fixing Engineered Joists   101
8.14 Posi-Joist™ Steel-Web System   105
8.15 Overlay Flooring   107
8.16 Notching and Drilling Traditional Floor/Ceiling Joists   109

9   Fixing Interior and Exterior Timber Grounds   111

9.1   Introduction   111
9.2   Skirting Grounds   111
9.3   Architrave Grounds   112
9.4   Apron-lining Grounds   112
9.5   Wall-panelling Grounds   112
9.6   Framed Grounds   113
9.7   External Grounds   113

10 Fixing Stairs and Balustrades   114

10.1 Introduction   114
10.2 Installation Procedure   115
10.3 Fixing Tapered Steps   123
10.4 Fixing Balustrades   125

11 Stair Regulations Guide to Design and Construction   129

11.1 The Building Regulations 2010   129

12 Making and Fixing Formwork for *In Situ* Concrete Stairs

12 Making and Fixing Formwork for *In Situ* Concrete Stairs   139

   12.1   Introduction   139
   12.2   Details of Design   139
   12.3   Designing Stairs   140
   12.4   Formwork for a Half-turn Stair   143

13 Constructing Traditional and Modern Roofs   151

   13.1   Introduction   151
   13.2   Basic Roof Designs   152
   13.3   Roof Components and Terminology   152
   13.4   Basic Setting-out Terms   158
   13.5   Geometrical Setting-out of a Hipped Roof   160
   13.6   Roofing Ready Reckoner   161
   13.7   Metric Rafter Square   163
   13.8   Alternative Method for the Use of the Metric Rafter Square 164
   13.9   Bevel-formulas for the Roofing Square   166
   13.10 Roofmaster Square   167
   13.11 Setting Out a Common (Pattern) Rafter   170
   13.12 Setting Out a Crown (or Pin) Rafter   171
   13.13 Setting Out a Hip Rafter   172
   13.14 Setting Out Jack Rafters   175
   13.15 Pitching Details and Sequence   176
   13.16 Pitching a Hipped Roof (Double-ended)   179
   13.17 Flat Roofs   180
   13.18 Dormer Windows and Skylights   184
   13.19 Skylights (Roof Windows)   188
   13.20 Eyebrow Windows   188
   13.21 Lean-to Roofs   194
   13.22 Chimney-trimming and Back Gutters   194
   13.23 Trussed Rafters   195
   13.24 Erection Details and Sequence for Gable Roofs   197
   13.25 Hipped Roofs Under 6m Span   197
   13.26 Hipped Roofs Over 6m Span   198
   13.27 Alternative Hipped Roof up to 11m Span   198
   13.28 Valley Junctions   199
   13.29 Gable Ladders   199
   13.30 Roof-trap Hatch   200
   13.31 Chimney Trimming   200
   13.32 Water-tank Supports   201
   13.33 Work at Height Regulations 2005 (Amended April 2007)   202

14 Erecting Timber-stud and Metal-stud Partitions 203

14.1	Introduction	203
14.2	Traditional Braced Partition	203
14.3	Traditional Trussed Partition	203
14.4	Modern Timber-stud Partitions	204
14.5	Door-stud and Door-head Joints	206
14.6	Stud Joints to Sill or Head Plate	207
14.7	Door-stud and Sill-plate Joints	208
14.8	Corner and Doorway Junctions	209
14.9	Floor and Ceiling Junctions	210
14.10	Erecting Metal-stud Partitions	212

15 Geometry for Arch Shapes 218

15.1	Introduction	218
15.2	Basic Definitions	218
15.3	Basic Techniques	219
15.4	True Semi-elliptical Arches	221
15.5	Approximate Semi-elliptical Arches	223
15.6	Gothic Arches	224
15.7	Tudor Arches	224
15.8	Parabolic Arches	226
15.9	Hyperbolic Arch	226

16 Making and Fixing Arch Centres 228

16.1	Introduction	228
16.2	Solid Turning Piece (Single-rib)	228
16.3	Single-rib Centres	229
16.4	Twin-rib Centres	230
16.5	Four-rib Centres	231
16.6	Multi-rib Centres	233

17 Fixing Architraves, Skirting, Dado and Picture Rails 235

17.1	Architraves	235
17.2	Skirting	240
17.3	Dado Rails and Picture Rails	242

18 Fitting and Hanging Doors 243

18.1	Introduction	243
18.2	Fitting Procedure	243
18.3	Hanging Procedure	248

19 Fitting Locks, Latches and Door Furniture — 249

19.1 Locks and Latches — 249
19.2 Mortice Locks — 249
19.3 Mortice Latches — 250
19.4 Mortice Dead Locks — 251
19.5 Cylinder Night Latches — 251
19.6 Fitting a Letter Plate — 251
19.7 Fitting a Mortice Lock — 252
19.8 Fitting Door Furniture — 253

20 Fixing Pipe Casings and Framed Ducts — 254

20.1 Introduction — 254
20.2 Pipe Casings — 254
20.3 Framed Ducts — 254

21 Designing and Installing a Fitted Kitchen — 256

21.1 Introduction — 256
21.2 Ergonomic Design Considerations — 256
21.3 Planning the Layout — 257
21.4 Dismantling the Old Kitchen — 258
21.5 Pre-fitting Preparation — 259
21.6 Fitting and Fixing Base Units — 260
21.7 Cutting, Jointing and Fitting Worktops — 261
21.8 Fixing the Wall Units — 263
21.9 Adding Finishing Items — 264

22 Site Levelling and Setting Out — 265

22.1 Introduction — 265
22.2 Establishing a Datum Level — 265
22.3 Setting Out the Shape and Position of the Building — 266

Appendix: Glossary of Terms — 271
Index — 277

Preface

This book was written because there is a need for trade books with a strong practical bias, using a DIY step-by-step approach – and not because there was any desire to add yet another book to the long list of carpentry books already on the market. Although many of these do their authors credit, the bias is mainly from a technical viewpoint with wide general coverage and I believe there is a potential market for books (manuals) that deal with the sequence and techniques of performing the various, unmixed specialisms of the trade. Such is the aim of this book, to present a practical guide through the first two of these subjects, namely first-fixing and second-fixing carpentry.

These definitions mean that any work required to be done **before** plastering takes place – such as roofing and floor joisting – is referred to as first-fixing carpentry; second-fixing carpentry, therefore, refers to any work that takes place **after** plastering – such as fixing skirting boards, architraves and door-hanging.

Most carpenters cover both areas of this work, although some specialize in either one or the other. To clarify the mix up between *carpentry* and *joinery*, items of joinery – such as staircases and wooden windows – are manufactured in workshops and factories and should be regarded as a separate specialism.

The book, hopefully, will be of interest to many people, but it was written primarily for craft apprentices (a rare breed in this present-day economy), trainees and building students, established trades-people seeking to reinforce certain weak or sketchy areas in their knowledge and, as works of reference, the book may also be of value to vocational teachers, lecturers and instructors. Finally, the sequential, detailed treatment of the work should appeal to the keen DIY enthusiast.

Les Goring, Hastings, East Sussex, UK

Acknowledgements

The author would like to thank the following people and companies for their co-operation in supplying technical literature or other assistance used in the first, second and third editions, or in this, the fourth edition of the book. Special thanks go to Jenny Goring, Jonathan Goring, Kevin Hodger, Peter Shaw and Tony Moon, who assisted on various occasions:

Alpha Pneumatic Supplies Ltd, Unit 7, Hatfield Business Park, Hatfield, Hertfordshire AL10 9EW, UK (www.nailers.co.uk) for information on air nailers and compressors; Andrew Thomas and Ian Harris of ITW (Illinois Tool Works) Construction Products (www.itwcp.co.uk); Brian Redfearn of Hastings College; CSC Forest Products Limited (OSB flooring panels); Dave Aspinall and Ronni Boss of Hilti GB Ltd (www.hilti.co.uk); Doris Funke, Commissioning Editor, Elsevier Limited; Tony Moore, Senior Editor at Routledge; Scott Oakley and Gabriella Williams, Assistant Editors at Routledge; Eric Brown and Bradley Cameron of DuPont™ Tyvek®, Hither Green Trading Estate, Clevedon, North Somerset BS21 6XU (www.tyvekhome.com); Gang-Nail Systems Ltd, a member company of the International Truss Plate Association; Health and Safety Executive (HSE), for the Work at Height Regulations 2005 (as amended 2007); Helen Eaton, Editorial Assistant, Elsevier Limited; Jenny, my daughter, for meticulously collating and compiling the original Index and the amended Index to the second edition; Kevin Hodger, ex colleague (inventor of the Roofmaster Square); Kingsview Optical Ltd (suppliers of the Roofmaster Square) Harbour Road, Rye, East Sussex TN31 7TE, UK, telephone +44 (0) 1797 226202; Laybond Products Ltd; Mark Kenward, MRICS, extraordinary carpentry student, now a chartered surveyor; McArthur Group Ltd, for their DVD catalogue on nails and screws, etc. (www.mcarthur-group.com); Melvyn Batehup (for his suggested revision-areas); Mike Owst, Programme Area Leader in Construction Studies at Hastings College; Mike Willard of Bexhill Locksmiths & Alarms; Pace Timber Engineering Ltd, Bleak Hall, Milton Keynes MK6 1LA (www.pacetimber.co.uk); Percy Eldridge, former lecturer at Hastings College; Peter Bullock of MiTeK Industries Ltd (re MiTeK Posi-Joist™ Steel Web System), Grazebrook Industrial Park, Peartree Lane, Dudley DY2 OXW (www.mitek. co.uk); Peter Oldfield and Peter Shaw of Hastings College; Rachel Hudson, Commissioning Editor (Newnes), Butterworth-Heinemann (for prompting a definition of 1st- and 2nd-fixing carpentry in the preface); Schauman (UK) Ltd (plywood flooring panels); South Coast Roofing Supplies Ltd, St Leonards-on-Sea; Steve Pearce of South Coastal Windows and Doors (uPVC) Ltd; Sydney Clarke of Helifix Ltd; Tony Fleming, former Head of Construction Studies at Hastings College; Tony Moon of A & M Architectural Design Consultants, Hastings, East Sussex (01424 200222); Toolbank of Dartford, Kent, for information from their 'Big Blue book' on tools and accessories (www.toolbank.com); Trus Joist™, East Barn, Perry Mill Farm, Birmingham Road, (A441), Hopwood, Worcestershire B48 7AJ, UK (www. trustjoist.com); York Survey Supply Centre (www. YorkSurvey.co.uk); Freyan Eames, C&J lecturer at Sussex Coast College, Hastings; Richard Evans of Pears Stairs' technical department; and the last two important contributors: Jonathan, my son, for patiently posing with his hands for the original illustration at Figure 2.26 in Chapter 2 and for turning me on to word processing prior to rewriting the second edition, via his gift of a laptop – and Darren Eglington, my son-out-of-law, for his time and patience in teaching me how to use it – and the subsequent PCs that followed.

Health and Safety Awareness

Although carpentry operations involving hand tools and power tools etc. used on construction sites, other work places or in one's home are obviously the main subjects in this written work, this book does not purposely set out to instruct readers in the safe use of tools and equipment – or in site-safety awareness. But, make no mistake about it, once you start using tools and handling materials and equipment on a construction site or in your home or workshop you enter into a work-world full of potential danger.

So, you need to learn about safety awareness and work-place hazards, especially if you have not undergone a Site Carpentry or Construction Course at a vocational College – or have not worked with tools or in a construction-site environment before.

In my experience, hazard-recognition always seems to lie just beneath the surface of conscious awareness – and yet potential dangers need to be thought about and identified *before* we start work and *as we work*. More easily said than done, of course, as we all have other issues on our minds. We also need to realize that serious injuries can be caused by misusing the simplest of tools. I know of two such accidents that occurred recently with a retractable-bladed craft knife (often referred to as a *Stanley knife*), caused by an over-applied force related to the proximity of the worker's other hand; and another potentially fatal accident many years ago involving a young worker who tripped and fell through an unguarded mid-area opening in a recently built flat roof over a single-storey building. A carpenter was working there, forming a raised curb around the opening as an upstand for a manufactured lantern light to be fitted later. There

was no scaffold platform beneath the square-opening to break his fall. Of course, this accident happened long before the Health & Safety Executive (HSE) brought in stringent legislation, involving on-site risk-assessment (with heavy penalties to contractors for non-conformity), and imposed penalties related to a *duty of care*, making *everyone* involved in the workplace (the employer, the workers, subcontractors, etc.) responsible for their actions and for how *their* actions might affect (endanger) others.

Miraculously, the young worker who fell through the roof-aperture got up from the concrete ground-slab after the fall and walked away, embarrassed, with only bruises! But even nowadays, with much improved safety regulations and procedures in place (controlled by HSE penalties), the construction industry still has the highest number of fatalities per year – recorded at 30 fatal injuries so far in the current, ongoing 2016/2017 period.

Finally, I would recommend that all readers of this manual should check out the following online publications:

1. Health and Safety at Work Act (HASAWA) 1974
2. Reporting of Injuries, Diseases and Dangerous Occurrences Regulations (RIDDOR) 2013
3. Provision and Use of Work Equipment Regulations (PUWER) 1998
4. Personal Protective Equipment (PPE) at Work Regulations 1992
5. Work at Height Regulations 2005 (as amended 2007).

Abbreviations

bdg	boarding	GL	ground level
bldg	building	hdb	hardboard
BMA	bronze metal antique	hwd	hardwood
BS	British Standards (Institution)	ms	mild steel
c/c	centre to centre (measurement)	O/A	over all (measurement)
℄	centre line	par	planed all round
cpd	cupboard	ppd	prepared (timber planed all round)
DPC	damp-proof course	PVA	polyvinyl acetate (adhesive)
DPM	damp-proof membrane	swd	softwood
dia, ø	diameter	T&G	tongued and grooved
EML	expanded metal lathing	TRADA	Timber Research and Development
ex.	prefix to material size before being worked		Association
ffl	finished floor level	vh	vertical height

Technical Data

STANDARD TIMBER-SIZES

Table 1 shows the basic sectional sizes for sawn softwood recommended by the British Standards to be available to the industry – it should be borne in mind that any non-standard requirement represents a special order and is likely to cost more.

Standard metric lengths are based on a 300 mm module, starting at 1.8 m and increasing by 0.3 m to 2.1 m, 2.4 m, 2.7 m and so on, up to 6.3 m. Non-standard lengths above this, usually from North American species, may be obtained up to about 7.2 m.

For anybody more used to working in imperial sizes rather than metric, it is worth bearing in mind that measured pieces of timber required for a particular job, must be divisible by 0.3 to meet the modular sizes available. For instance, 3.468 m ÷ 0.3 = 11.56 modules. Therefore, this would have to be increased commercially to 12 modules, i.e., 12 × 0.3 = 3.6 m, which is a commercially available size. If you prefer, you can think of this little sum as 12 × 3 = 36, then insert the decimal point. This simple mental arithmetic, based on the three-times-table, can be used for all the commercially available sizes between 6 × 3 = 18 (1.8 m) and 21 × 3 = 63 (6.3 m).

Table 1 Sawn structural timber sizes (From BS 336: 1995)

mm	inches	75 / 3	100 / 4	125 / 5	150 / 6	175 / 7	200 / 8	225 / 9	250 / 10	275 / 11	300 / 12
22	$\frac{7}{8}$		✓	✓	✓	✓	✓	✓			
25	1	✓	✓	✓	✓	✓	✓	✓			
38	$1\frac{1}{2}$	✓	✓	✓	✓	✓	✓	✓			
47	$1\frac{7}{8}$	✓	✓	✓	✓	✓	✓	✓	✓		✓
63	$2\frac{1}{2}$		✓	✓	✓	✓	✓	✓			
75	3		✓	✓	✓	✓	✓	✓	✓	✓	✓
100	4		✓		✓		✓	✓	✓		✓
150	6				✓		✓				✓
250	10								✓		
300	12										✓

STANDARD DOOR-SIZES

Door frames and linings may vary in their opening sizes, but are normally made to accommodate standard doors. Again, it must be realized that special doors, made to fit non-standard frames or linings, would considerably increase the cost of the job. The locations given to the groups of standard door-sizes in Table 2, below, are only a guide, not a fixed rule.

Table 2 Standard doors and their usual location

	Metric			Imperial		
	Height (m)	Width (mm)	Thickness (mm)	Height (ft/in)	Width (ft/in)	Thickness (in)
Main entrance	2.134	914	45	7'0"	3'0"	$1\frac{3}{4}$
doors	2.083	864	45	6'10"	2'10"	$1\frac{3}{4}$
	2.032	813	45	6'8"	2'8"	$1\frac{3}{4}$
	1.981	838	45	6'6"	2'9"	$1\frac{3}{4}$
	1.981	762	45	6'6"	2'6"	$1\frac{3}{4}$
Room doors	2.032	813	45 or 35	6'8"	2'8"	$1\frac{3}{4}$ or $1\frac{3}{8}$
	1.981	762	45 or 35	6'6"	2'6"	$1\frac{3}{4}$ or $1\frac{3}{8}$
Bathroom/toilet	1.981	762 or 711	35	6'6"	2'6" or 2'4"	$1\frac{3}{8}$
doors	1.981	686	35	6'6"	2'3"	$1\frac{3}{8}$
Cupboard doors	1.981	610	35	6'6"	2'0"	$1\frac{3}{8}$
	1.981	533	35	6'6"	1'9"	$1\frac{3}{8}$
	1.981	457	35	6'6"	1'6"	$1\frac{3}{8}$

Note: When doorframes and doors are required to be fire-resisting, special criteria laid down by the British Standards Institution and the Building Regulations must be adhered to – a detailed reference to this is given in Chapter 6.

1

Reading Construction Drawings

1.1 INTRODUCTION

Construction drawings are necessary in most spheres of the building industry, as being the best means of conveying detailed and often complex information from the designer to all those concerned with the job. Building tradespeople, especially carpenters and joiners, should be familiar with the basic principles involved in understanding and reading drawings correctly. Mistakes on either side – in design or interpretation of the design – can be costly, as drawings form a legal part of the contract between architect/client and builder. This applies even on small jobs, where only goodwill may suffer; for this reason, if a non-contractual drawing or sketch is supplied, it should be kept for a period of time after completion of the job, in case any queries should arise.

1.1.1 Retention of Drawings or Sketches

A simple sketch supplied by a client in good faith to a builder or joinery shop for the production of a replacement casement-type window, is shown in Figure 1.1(a). The client's mistake in measuring between plastered reveals is illustrated in Figure 1.1(b). Retention of the sketch protects the firm from the possibility of the client's wrongful accusation.

Another important rule is to study the whole drawing carefully and be reasonably familiar with the details before starting work.

The details given in this chapter are based on the recommendations laid down by the British Standards Institution, in their latest available publications entitled *Construction drawing practice*, BS 1192: Part 1: 1984, and BS 1192: Part 3: 1987. BS 1192: Part 5: 1990, which is not referred to here, is a guide for the structuring of computer graphic information.

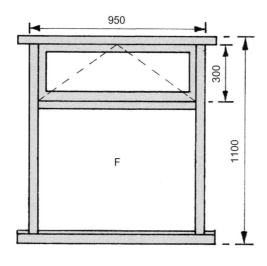

(a)

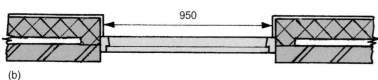

(b)

Figure 1.1 (a) Client's sketch drawing. **(b)** Horizontal section showing client's mistake

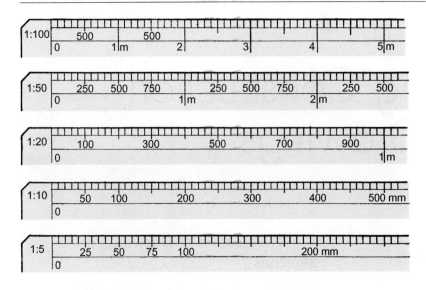

Figure 1.2 Common metric scales

1.1.2 Scales Used on Drawings

Parts of metric scale rules, graduated in millimetres, are illustrated in Figure 1.2. Each scale represents a ratio of given units (millimetres) to one unit (one millimetre). Common scales are 1:100, 1:50, 1:20, 1:10, 1:5 and 1:1 (full size). For example, scale 1:5 = one-fifth ($\frac{1}{5}$) full size, or 1 mm on the drawing equals 5 mm in reality.

Although a scale rule is useful when reading drawings, because of the dimensional instability of paper, preference should always be given to written dimensions found on the drawing.

1.1.3 Correct Expressions of Dimensions

The abbreviated expression, or unit symbol, for metres is a *lower* case (small) letter m, and letters mm for millimetres. Symbols are not finalized by a full stop and do not use a letter 's' for the plural. Confusion occurs when, for example, $3\frac{1}{2}$ metres is written as 3.500 mm – which means, by virtue of the decimal point in relation to the unit symbol, $3\frac{1}{2}$ millimetres! To express $3\frac{1}{2}$ metres, it should have been written as 3500 mm, 3.5 m, 3.50 m, or 3.500 m. Either one symbol or the other should be used throughout on drawings; they should not be mixed. Normally, whole numbers should indicate millimetres, and decimalized numbers, to three places of decimals, should indicate metres. Contrary to what seems to be taught in schools, the construction and engineering industry in the UK does not communicate in centimetres. All references to measurement are made in millimetres and/or metres, i.e. 2 cm should be expressed as 20 mm.

1.1.4 Sequence of Dimensioning

The recommended dimensioning sequence is illustrated in Figure 1.3. Length should always be given

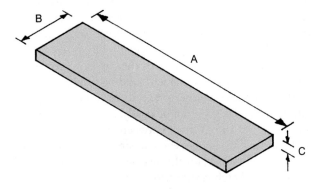

Figure 1.3 Dimensioning sequence = $A \times B \times C$

first, width second and thickness third, for example $900 \times 200 \times 25$ mm. However, if a different sequence is used, it should be consistent throughout.

1.1.5 Dimension Lines and Figures

A dimension line with open arrowheads for basic/modular (unfinished) distances, spaces or components is indicated in Figure 1.4(a). Figure 1.4(b) indicates the more common, preferred dimension lines, with solid arrowheads, for general use in finished work sizes.

All dimension figures should be written above and along the line; figures on vertical lines should be written, as shown, to be read from the right-hand side.

1.1.6 Special-purpose Lines

Figure 1.5: Section lines seen on drawings indicate imaginary cutting planes, at a particular point through the drawn object, to be exposed to view. The view is called the section and is lettered A–A, B–B and so on, according to the number of sections to be exposed. It is important to bear in mind that the arrows indicate the direction of view to be seen on a separate section drawing.

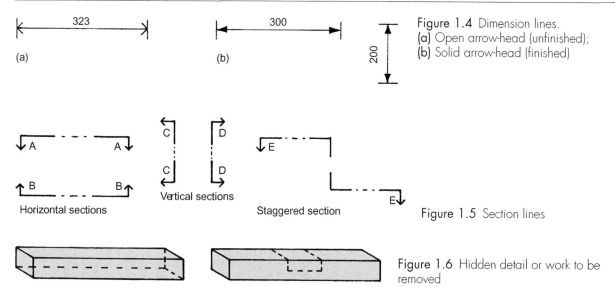

Figure 1.4 Dimension lines.
(a) Open arrow-head (unfinished);
(b) Solid arrow-head (finished)

Vertical sections

Horizontal sections

Staggered section

Figure 1.5 Section lines

Figure 1.6 Hidden detail or work to be removed

Figure 1.6: Hidden detail or work to be removed, is indicated by a broken line.

Figure 1.7 Break lines

Figure 1.7(a): End break-lines (zig-zag pattern) indicate that the object is not fully drawn.

Figure 1.7(b): Central break-lines (zig-zag pattern) indicate that the object is not drawn to scale in length.

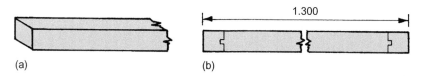

Figure 1.8 Centre or axial line

Figure 1.8: Centre or axial lines are indicated by a thin dot-dash chain.

1.2 ORTHOGRAPHIC PROJECTION

1.2.1 Introduction

Orthography is a Latin/Greek-derived word meaning 'correct spelling' or 'writing'. In technical drawing it is used to mean 'correct drawing'; orthographic projection, therefore, refers to a conventional drawing method used to display the three-dimensional views (length, width and height) of objects or arrangements as they will be seen on one plane – namely the drawing surface.

The recommended methods are known as *first-angle* (or European) *projection* for construction drawings, and *third-angle* (or American) *projection* for engineering drawings.

1.2.2 First-angle Projection

The box in Figure 1.9(a) is used here as a means of explaining first-angle projection (F.A.P.). If you can imagine the object shown in Figure 1.9(b) to be suspended in the box, with enough room left for you to walk around it, then by looking squarely at the object from all sides and from above, the views seen would be the ones shown on the surfaces in the background.

1.2.3 Opening the Topless Box

In Figure 1.9(c) the topless box is opened out to give the views as you saw them in the box and as they should be laid out on a drawing. Figure 1.9(d) shows the BS symbol recommended for display on drawings to indicate that first-angle projection (F.A.P.) has been used.

Note that when views are separated onto different drawings, becoming unrelated orthographically, descriptive captions should be used such as 'plan', 'front elevation', 'side elevation', etc.

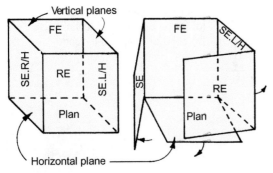

Figure 1.9 (a) Theory of first-angle orthographic projection (SE = side elevation, FE = front elevation, RE = rear elevation, R/H = right-hand side, L/H = left-hand side)

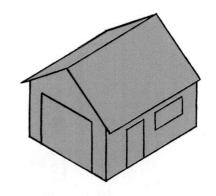

Figure 1.9 (b) Example object

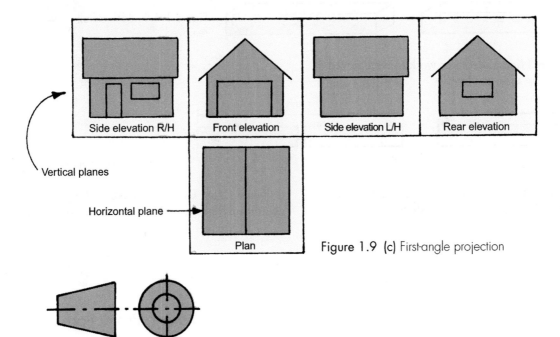

Figure 1.9 (c) First-angle projection

Figure 1.9 (d) F.A.P. symbol

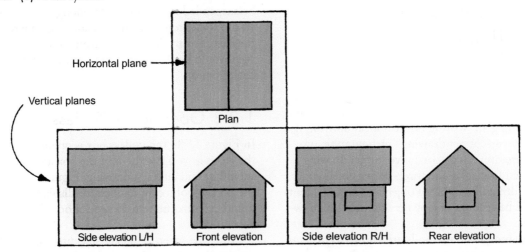

Figure 1.9 (e) Third-angle projection

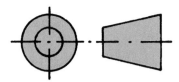

Figure 1.9 (f) T.A.P. symbol

1.2.4 Third-angle Projection

This is shown in Figure 1.9(e) for comparison only. This time the box has a top instead of a bottom; the views from the front and rear would be shown on the surface in the background, as before, but the views seen on the sides would be turned around and seen on the surfaces in the foreground; the view from above (plan) would be turned and seen on the surface above. Figure 1.9(f) shows the BS symbol for third-angle projection (T.A.P.).

1.2.5 Pictorial Projections

Figure 1.10: Another form of orthographic projection produces what is known as pictorial projections, which preserve the three-dimensional view of the object. Such views have a limited value in the make-up of actual working drawings, but serve well graphically to illustrate technical notes and explanations.

1.2.6 Isometric Projection

This is probably the most popular pictorial projection used, because of the balanced, three-dimensional effect. Isometric projections consist of vertical lines and base lines drawn at 30°, as shown in Figure 1.10(a). The length, width and height of an object thus drawn are to true scale, expressed as the ratio 1:1:1.

1.3 OBLIQUE PROJECTIONS

There are three variations of oblique projections.

1.3.1 Cavalier Projection

Shown in Figure 1.10(b) with front (F) drawn true to shape, and side (S) elevations and plan (P) drawn at 45°, to a ratio of 1:1:1. Drawn true to scale by this method, the object tends to look misshapen.

1.3.2 Cabinet Projection

Shown in Figure 1.10(c), this is similar to cavalier except that the side and plan projections are only drawn to half scale, i.e. to a ratio of $1:1:\frac{1}{2}$, making the object look more natural.

1.3.3 Planometric Projection

Shown in Figure 1.10(d), this has the plan drawn true to shape, instead of the front view. This comprises verticals, lines on the front at 30° and lines on the side elevation at 60°. It is often wrongly referred to as axonometric.

1.3.4 Perspective Projections

Figure 1.11: Parallel perspective, shown in Figure 1.11(a) refers to objects drawn to diminish in depth to a vanishing point.

Angular perspective, shown in Figure 1.11(b) refers to an object whose elevations are drawn to diminish to two vanishing points. This is of no value in pure technical drawing.

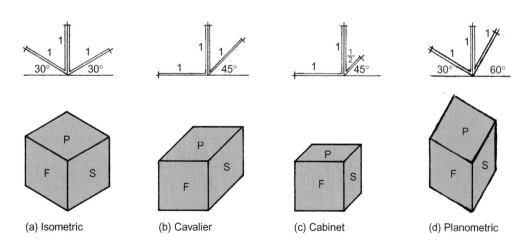

(a) Isometric   (b) Cavalier   (c) Cabinet   (d) Planometric

Figure 1.10 Pictorial projections (F = front, P = plan, S = side elevation)

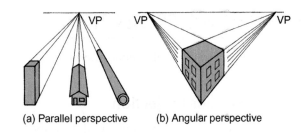

(a) Parallel perspective   (b) Angular perspective

Figure 1.11 Perspective projections (VP = vanishing point)

1.3.5 Graphical Symbols and Representation

Figure 1.12: Illustrated here are a selection of graphical symbols and representations used on building drawings.

Figure 1.13: On more detailed drawings, various materials and elements are identified by such sectional representation as shown here.

To help reduce the amount of written information on working drawings, abbreviations are often used. A selection are shown here:

BMA = bronze metal antique
DPC = damp-proof course
DPM = damp-proof membrane

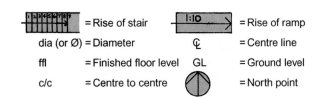

= Rise of stair
dia (or Ø) = Diameter
ffl = Finished floor level
c/c = Centre to centre
= Rise of ramp
= Centre line
GL = Ground level
= North point

Figure 1.12 Graphical symbols and representations

EML = expanded metal lathing
par = planed all round
PVA = polyvinyl acetate
T&G = tongue and groove
bdg = boarding
bldg = building
cpd = cupboard
hbd = hardboard
hwd = hardwood
ms   = mild steel
swd = softwood

1.3.6 Window Indication

Figure 1.14: Windows shown on elevational drawings usually display indications as to whether a window is fixed (meaning without any opening window or vent)

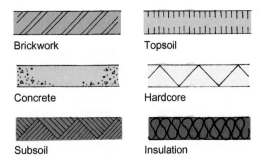

Brickwork   Topsoil

Concrete   Hardcore

Subsoil   Insulation

Figure 1.13 Sectional representation of materials

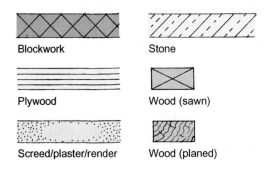

Blockwork   Stone

Plywood   Wood (sawn)

Screed/plaster/render   Wood (planed)

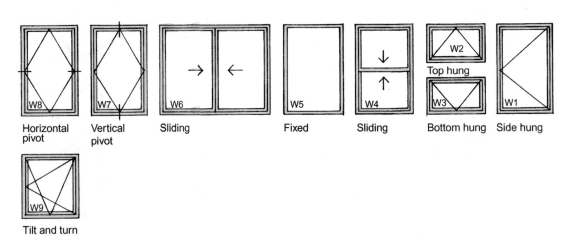

Horizontal pivot   Vertical pivot   Sliding   Fixed   Sliding   Bottom hung   Side hung

Tilt and turn

Figure 1.14 Opening/fixed window indication – numbered clockwise round the exterior of the building

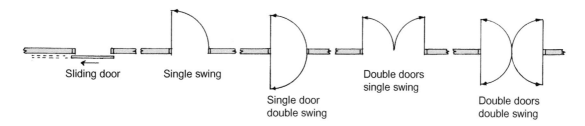

Figure 1.15 Plan view of door indication

or opening (meaning that the window is to open in a particular way, according to the BS indication drawn on the glass area).

1.3.7 Door Indication

Figures 1.15 and 1.16: Doors shown on plan-view drawings are usually shown as a single line with an arrowed arc indicating their opening-direction, as illustrated. Alternatively, the 90° arrowed arc may be replaced by a 45° diagonal line, from the door-jamb's edge to the door's leading edge. Figure 1.16 is the indication for revolving doors.

1.3.8 Block Plans

Figure 1.17: Block plans shown on construction drawings, usually taken from Ordnance Survey maps, are to identify the site (e.g. No. 1 Woodman Road, as illustrated) and to locate the outline of the building in relation to its surroundings.

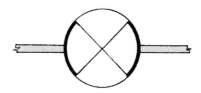

Figure 1.16 Revolving doors

1.3.9 Site Plans

Figure 1.18: Site plans locate the position of buildings in relation to setting-out points, means of access, and the general layout of the site; they also give information on services and drainage, etc.

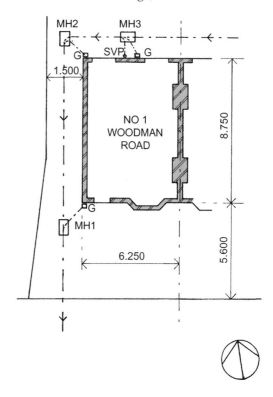

Figure 1.18 Site plan (scale 1:200)

1.3.10 Location Drawings

These are usually drawn to a scale of 1:50 and are used to portray the basic, general construction of buildings. Other, more detailed, drawings cover all other aspects.

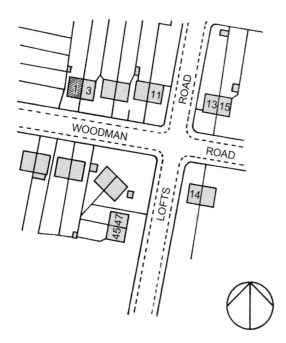

Figure 1.17 Block plan (scale 1:1250)

2

Tools Required: their Care and Proper Use

2.1 INTRODUCTION

The whole range of tools for first- and second-fixing carpentry is quite extensive and includes power and battery-operated (cordless) tools in the essential list. The following details, therefore, do not cover all the tools that you *could* have, rather all the tools that you *should* have.

2.2 MARKING AND MEASURING

2.2.1 Pencils

Figure 2.1: These must be kept sharp for accurate marking. Although sharpening to a pin-point is quite common, for more accurate marking and a longer-lasting point, they can easily be sharpened to a chisel-point, similar to the sharpening illustrated in Figure 2.1(c). Stumpy sharpening (Figure 2.1(a)) should be avoided; sharpen at an angle of about 10° (Figure 2.1(b)). Use grade HB for soft, black lines on unplaned timber and – if you prefer – grade 2H on planed timber. Choose a hexagon shape for better grip and anti-roll action, and a bright colour to detect easily when left lying amongst shavings, etc. Oval or

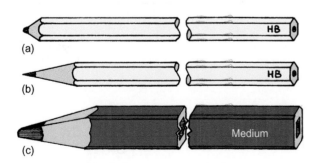

Figure 2.1 Pencils

rectangular-shaped carpenters' pencils (Figure 2.1(c)), of a soft or medium grade lead, are better for heavy work such as roofing, joisting, marking unplaned timber, etc. – although one disadvantage is that they cannot be put behind the ear for quick availability, as is the usual practice with ordinary pencils.

2.2.2 Tape Rule

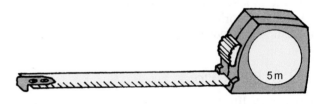

Figure 2.2 Tape rule

Figure 2.2: This is essential for fast, efficient measuring on site work. For this type of carrying-rule, sizes vary between 2 m and 10 m. Models with lockable, power-return blades and belt clips, one of 3.5 m and one of 8 m length are recommended. When retracting these power-return rules, slow down the last part of the blade with the sliding lock to avoid damaging the riveted metal hook at the end or nipping your fingers. To reduce the risk of kinking the sprung-steel blade, do not leave extended after use.

2.2.3 Folding Rule

Figure 2.3: This rule is optional, having been superseded by the tape rule. However, it is sometimes preferred for measuring/marking small sizes. Its unfolded length is 1 m and it is 250 mm folded. It is marked in single millimetre, 5 mm, 10 mm (centimetre) and 100 mm (decimetre) graduations. It is still available in boxwood or – better still – in virtually unbreakable white or grey engineering plastic with tipped ends, permanently tensioned joints and bevelled edges for easier, more accurate reading/marking when the rule is laid out flat. These rules, although tough, should always be

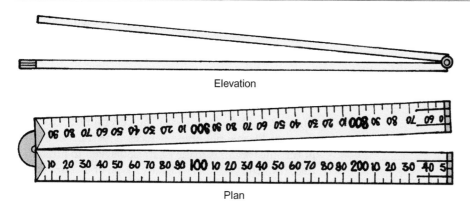

Elevation

Plan

Figure 2.3 Folding rule

folded after use to avoid possible hinge damage, especially from underfoot if left on the floor.

2.2.4 Chalk Line Reel

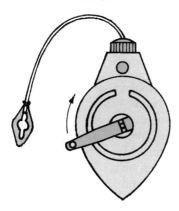

Figure 2.4 Chalk line reel

Figure 2.4: This tool is very useful for marking straight lines by holding the line taut between two extremes, lifting at any mid point with finger and thumb and flicking onto the surface to leave a straight chalk line. The line is retractable by winding a hinged handle housed in the die-cast aluminium case, that folds back after use. Powdered chalk is available in colours of red, white, blue, orange, green and yellow. The reel has a subsidiary use as a plumb bob – but it is not ideal for this purpose.

2.2.5 Spirit Level

Figure 2.5: This is an essential tool for plumbing and levelling operations. Sizes vary between 200 mm and 2 m long, but a level of 800 mm length is recommended for general usage and easy accommodation in the tool kit. A small level of 200 mm to 300 mm length, traditionally called a boat level or torpedo level, is also recommended for use in restricted areas. Heavy-duty levels of aluminium alloy die-cast, or lightweight models of extruded aluminium, with clear,

Figure 2.5 Spirit level

tough plexiglass vials (containing spirit and trapped bubble), epoxy-bonded into their housings to give lasting accuracy, are the most popular levels nowadays.

Even though these levels are shockproof, they should not be treated roughly, as body damage can affect accuracy. After use, avoid leaving levels lying on the floor or ground to be trodden on, especially when partly suspended, resting on other objects such as scrap timber. When checking or setting up a level or plumb position, be sure that the bubble is precisely positioned between the lines on the vial for accurate readings.

2.2.6 Straightedges

Figure 2.6 Straightedge

Figure 2.6: In the absence of very long spirit levels, straightedges may be used. These are parallel, straight softwood boards of various lengths, for setting out or (with a smaller spirit level held against one edge) for plumbing and levelling. If transferring a datum point in excess of the straightedge length, the risk of a cumulative error is reduced by reversing the straightedge end-for-end at each move. Traditionally, large holes were drilled along the centre axis to prevent the board from being claimed for other building uses.

2.2.7 Plumb Bob

Figure 2.7: There is still a use, however limited, for these traditional plumbing devices. The short one in the illustration is made of steel, blacked to inhibit rust; the other, which is a heavier type of $4\frac{1}{2}$ ounces (128 g), has a red plastic body filled with steel shot and a 3 m length of nylon line. They should, as illustrated, always be suspended away from the surface being checked and measured for equal readings at top and bottom. If in possession of a tarnishable steel plumb bob, wipe with an oily rag occasionally. Although commonly called plumb bobs, if they are pointed on the underside, they are really *centre bobs*. The point is very useful for plumbing to a mark on the floor.

2.2.8 Combination Mitre Square

Figure 2.8: This tool was adopted from the engineering trades and is now widely favoured on site work for the following reasons: it is robust (the better, more expensive type) and withstands normal site abuse; it can be used for testing or marking narrow rebated edges, as shown, or for testing or marking angles of 90°, 45°, and 135°; the blade can be adjusted from the stock to a set measurement and, with the aid of a pencil, used as a pencil gauge. This facility is useful for marking sawn boards, for example, as opposed to using a marking gauge that may not be clearly visible on a rough sawn surface. The square's stock has an inset spirit vial and can be used for plumbing and levelling – although this is not the tool's best feature. The blade locking-nut

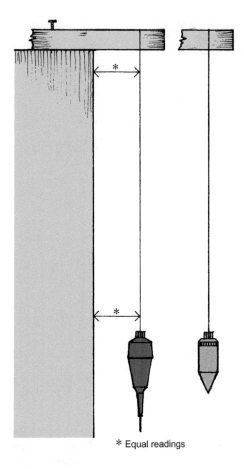

* Equal readings

Figure 2.7 Plumb bobs

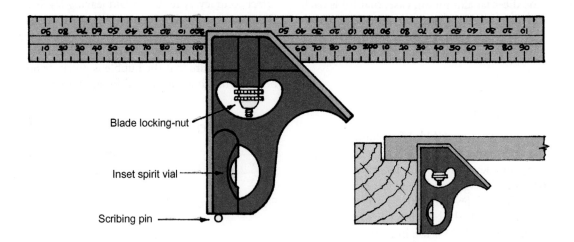

Blade locking-nut

Inset spirit vial

Scribing pin

Figure 2.8 Combination mitre square

should always be tightened after each adjustment, otherwise inaccuracies in the angle between the stock and blade will readily occur, causing errors in marking or testing. Finally, a scribing pin is usually located in the end of the stock. This is a feature carried over from the square's originally intended use as an engineering tool and is used for marking lines on metal.

2.2.9 Sliding Bevel

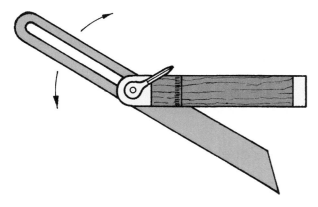

Figure 2.9 Sliding bevel

Figure 2.9: This is basically a slotted blued-and-hardened steel blade, sliding and rotating from a hardwood (rosewood) or plastic stock. The plastic is impact-resistant. The blade is tightened by a screw or a half wing nut. The latter is best for ease and speed, being manually operated. This is an essential tool for angular work, especially roofing if using the *Roofing Ready Reckoner* method. For protection against damage, always return the blade to the stock-housing after use.

2.2.10 Steel Roofing Square/Metric Rafter Square

Figure 2.10: This tool, originally called a *steel square* or *steel roofing square*, is now metricated and referred to as a *metric rafter square*. Its size is 610 × 450 mm. The long side is called the *blade*, the short side the *tongue*. This traditional tool, primarily for developing roofing bevels and lengths (covered in Chapter 13), has a good subsidiary use as a try square, for marking and testing certain right angles with greater opposite sides than the combination square or normal try square can deal with effectively.

2.2.11 Roofmaster Square

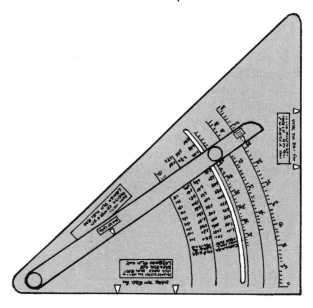

Figure 2.11 The Roofmaster (artistic representation only)

Figure 2.11: This revolutionary roofing square, as mentioned in Chapter 13, is well worth considering as an alternative to a traditional type roofing square. It is a compact, precision instrument, measuring 335 mm on each right-angled side and is of anodized

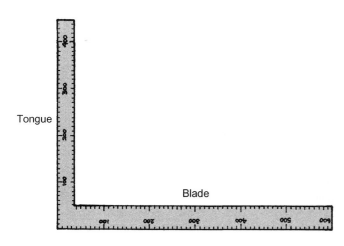

Tongue

Blade

Figure 2.10 Steel roofing square

aluminium construction with easy-to-read laser-etched markings. It gives angle cuts for all roof members *and* the lengths of rafters without the need for separate tables. It is designed for easy use, whereby only the roof pitch angle is required to obtain all other angles and lengths. (Readers wishing to obtain a Roofmaster should contact Kingsview Optical Ltd., Harbour Road, Rye, East Sussex, TN31 7TE, UK, Tel: +44(0) 1797 226202, Fax: +44(0) 1797 226301, Email: Sales@ kingsviewoptical.com for further information.)

2.3 HANDSAWS

2.3.1 Introduction

Traditional handsaws, although still available at a relatively high cost, have been superseded by modern hardpoint, throwaway saws. This is undoubtedly because they are cheap to buy, have a higher degree of sharpness, retain their sharpness for a much longer period of time and, when blunt, can affordably be replaced without the inconvenience – assuming a person has the skill – of re-sharpening. However, I have included the following illustrations and references to traditional saws, for those diehard traditionalists who would still use them – and because the conventions

established with these saws (such as recommended sawing-angles, etc) are still relevant.

2.3.2 Crosscut Saw

Figure 2.12: As the name implies, this is for cutting timber across the grain. Blade lengths and points per 25 mm (pp25) or ppi (points per inch) vary, but 660 mm (26 in) length and 7 or 8 pp25 are recommended. All handsaw teeth on traditional type saws contain 60° angular shapes leaning, by varying degrees, towards the toe of the saw. The angle of lean, relative to the front cutting edge of the teeth, is called the *pitch*. When sharpening saws, it helps to know the required pitch. For crosscut saws the pitch should be 80°. When crosscutting, the saw, as illustrated, should be at an approximate angle of 45° to the timber.

2.3.3 Panel Saw

Figure 2.13: This is a saw for fine crosscutting, which is particularly useful for cutting sheet material such as plywood or hardboard. A blade length of 560 mm (22 in), 10 pp25 and 75° pitch is recommended. When cutting thin manufactured boards (plywood, hardboard, etc.), the saw should be used at a low angle of about 15–25°.

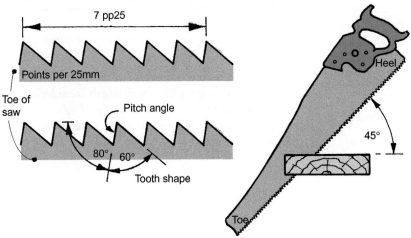

Figure 2.12 Crosscut saw

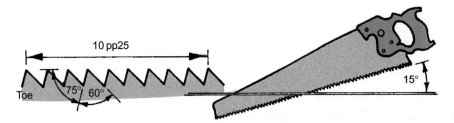

Figure 2.13 Panel saw

2.3.4 Tenon Saw

Figure 2.14 Tenon saw

Figure 2.14: This saw, because of its brass or steel back, is sometimes referred to as a *back saw*. Technically thought of as a general purpose bench saw for fine cutting, it is however widely used on site for certain second-fixing operations involving fine crosscutting of small sections. The brass-back type, as well as keeping the thin blade rigid, adds additional weight to the saw for easy use. The two most popular blade lengths, professionally, are 300 mm (12 in) and 350 mm (14 in). The 250 mm (10 in) saw is less efficient because of its short stroke. On different makes of saw, the teeth size varies between 13 and 15 pp25. For re-sharpening purposes, although dependent upon your skill and eyesight, 13 pp25 is recommended, with a pitch of 75°.

2.3.5 Rip Saw

Figure 2.15: This saw is used for cutting along or with the grain and is not in great demand nowadays because of the common use of machinery and portable powered circular saws on site. However, it is not obsolete and can be very useful in the absence of power. A blade length of 660 mm (26 in), 5 or 6 pp25 and a pitch of 87° is recommended. When ripping (cutting along or with the grain), the saw should be used at a steep angle of about 60–70° to the timber. Because of

the square-edged teeth and pitch angle, this saw cannot be used for crosscutting.

Traditional saws should be kept dry if possible and lightly oiled, but if rusting does occur, soak liberally with oil and rub well with fine emery cloth.

2.3.6 Hardpoint Handsaws and Tenon Saws

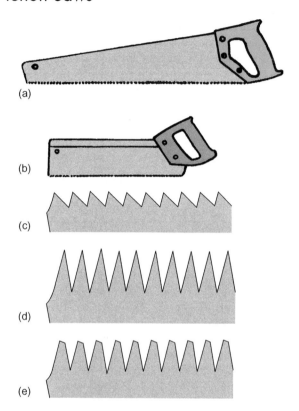

Figure 2.16 **(a)** Hardpoint handsaw; **(b)** hardpoint tenon saw; **(c)** universal tooth-shape; **(d)** fleam tooth-shape; **(e)** triple-ground tooth-shape

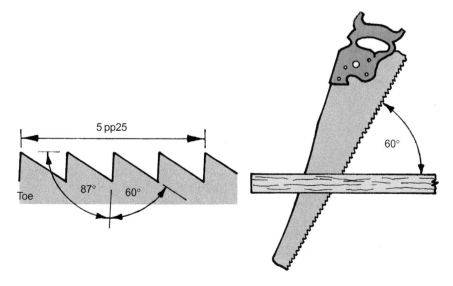

Figure 2.15 Rip saw

Figures 2.16: These modern throwaway saws have high-frequency hardened tooth-points which stay sharper for at least five times longer than conventional saw teeth. Three shapes of tooth exist; the first, referred to as *universal*, conforms to the conventional 60° tooth-shape and 75° pitch; the second, known as the *fleam* tooth, resembles a *flame* in shape (hence its name), with a conventional front-pitch of 75°, an unconventional back-pitch of 80°, giving the fleam-tooth shape of 25°; the third, referred to as *triple ground*, has razor-sharp, circular-saw-type tooth geometry, enabling a cutting action on both the push and the pull strokes. Most of the handsaws are claimed to give a superior cutting performance across and along the grain. Some saws in the range have a Teflon-like, friction-reducing coating on the blade, to eliminate binding and produce a faster cut with less effort.

Range of Sizes

The saws usually have plastic handles – some with improved grip – with a 45° and 90° facility for marking mitres or right angles. Handsaw sizes available are 610 mm (24 in) × 8 pp25, 560 mm (22 in) × 8 pp25, 508 mm (20 in) × 8 pp25, 560 mm (22 in) × 10 pp25, 508 mm (20 in) × 10 pp25, 480 mm (19 in) × 10 pp25, 455 mm (18 in) × 10 pp25 and 405 mm (16 in) × 10 pp25. Tenon saw sizes available are 300 mm (12 in) × 13 pp25, 250 mm (10 in) × 13 pp25, 300 mm (12 in) × 15 pp25 and finally 250 mm (10 in) × 15 pp25. Three recommended saws from this range would be the 610 mm (24 in) × 8 pp25 and the 560 mm (22 in) × 10 pp25 black-coated handsaws and a 300 mm (12 in) × 13 pp25 tenon saw.

2.3.7 Pullsaws

These lightweight, unconventional saws of oriental origin, cut on the pull-stroke, which eliminates buckling. They can be used for ripping or crosscutting. The unconventional precision-cut teeth, with three cutting edges, are claimed to cut up to five times faster, leaving a smooth finish without breakout or splintering. The sprung steel blade is ultra-hardened to give up to ten times longer life and can easily be replaced at the push of a button. Replacement blades cost about two-thirds the cost of the complete saw, but a complete saw is relatively inexpensive.

General Saw and Fine Saw

Figure 2.17: Only two saws from the range are illustrated here. The first is called a *general carpentry saw* and has a 455 mm (18 in) blade × 8 pp25. This model comes in two other sizes, 380 mm (15 in) × 10 pp25 and 300 mm (12 in) × 14 pp25. The latter is

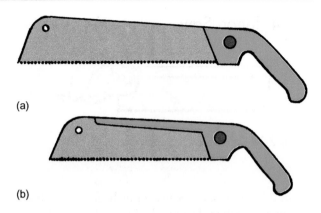

(a)

(b)

Figure 2.17 **(a)** General carpentry saw; **(b)** fine cut saw

recommended for cutting worktops and laminates without chipping. The second model is called a *fine cut saw* and has a half-length back or full-length back support and is said to surpass conventional tenon saws. This model comes in two variations, one with a fine-cut blade of 270 mm × 15 pp25, the other with an ultra-fine blade of 270 mm × 17 pp25.

2.3.8 Coping Saw

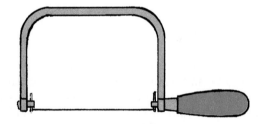

Figure 2.18 Coping saw

Figure 2.18: This traditional tool has not changed or lost its popularity and usefulness over many years. In second-fixing carpentry, it is mainly used for scribing (cutting the profile shape) of moulded skirting boards where they meet in the corners of a room (covered in another chapter), but occasionally comes in useful for other curved cuts in wood or plastic. The saw blades are very narrow with projecting pinned-ends and teeth set at 14 pp25. The blades, although easily broken with rough or unskilled sawing, have been heat-treated to the required degree of hardness and toughness and are obtainable in packs of 10. Although the narrowness of the blade demands that it be set in the frame with the front pitch of the teeth set to face the handle and working on the pull stroke, it can, if preferred, be set to cut on the forward action, providing that a degree of skill has been developed. The blade can be swivelled to cut at any angle to the frame, after unscrewing the handle slightly; the handle should be fully tightened after each adjustment of the blade.

2.3.9 Mitre Saw

Figure 2.19 Mitre saw

Figure 2.19: Nowadays, portable electric mitre saws, often referred to as 'chop saws' (because the rotating saw is brought down onto the timber), have virtually superseded wooden mitre boxes and mitre blocks and offer a variety of other uses. These saws give speedy and effortless precision-cutting and are widely used on site and in small workshops. Although the main uses cover crosscutting, bevel and mitre cutting, as well as compound-mitre cutting, some models also have variable speed control (for low speed work on materials such as fibreglass, etc.) and a grooving stop for grooving and rebating work. There are a variety of makes and different models available. The model drawn here represents a DeWalt DEW 701 with ports for dust extraction.

2.4 HAMMERS

2.4.1 Claw Hammer

Figure 2.20 Metal-shafted claw hammer

Figure 2.20: Although this tool is basically for nailing and extracting nails, it has also been widely used over the years by using the side of the head as an alternative to the wooden mallet. This is an acceptable practice on impact-resistant plastic chisel handles – especially as this type of handle is really too hard for the wooden mallet – but it is bad practice to use a hammer on wooden chisel handles, as they quickly deteriorate under such treatment. However, in certain awkward site situations, the mallet is too bulky and only the side of the claw hammer is effective.

Other Uses

The claw is also used for a limited amount of leverage work, such as separating nailed boards, etc. To preserve the surface shape of the head, the hammer should not be used to chip or break concrete, brick or mortar. When hammering normally, hold the lower end of the shaft and develop a swinging wrist action – avoid *throttling* the hammer (holding the neck of the shaft, just below the head). Choice of weights is between 450 g (1 lb), 565 g ($1\frac{1}{4}$ lb) and 675 g ($1\frac{1}{2}$ lb); choice of type is between steel shaft with nylon cushion grip, steel shaft with leather binding, fibreglass shaft in moulded polycarbonate jacket and the conventional wooden shaft. The latter has a limited life-span on site work. The choice is yours, but the steel-shafted type with nylon cushion grip, 675 g in weight, is recommended.

2.4.2 Mallet

Figure 2.21: The conventional wedge-shaped pattern, made of beech, is rather bulky and not generally favoured for site work, even though the tapered shaft – retaining the head from flying off – can be

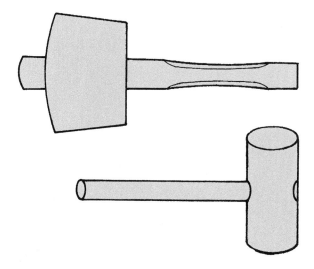

Figure 2.21 Wedge-shape and round-head mallet

removed for easier carriage. A recommended alternative is a round-headed mallet, such as a Tinman's mallet – used traditionally by sheet-metal workers – which has a boxwood or lignum-vitae head of about 70 mm diameter. Finally, wooden mallets should only strike on their end grain, not on their sides.

2.5   SCREWDRIVERS

Figure 2.22: Although power screwdrivers, especially the cordless, battery-operated type, are almost entirely used nowadays, hand screwdrivers are still useful occasionally for certain jobs. To my certain knowledge, the following selection, at least, are still used.

2.5.1   Ratchet Screwdrivers

Ratchet screwdrivers are available with flared parallel-tips (slotted) in four blade-lengths of 75 mm,

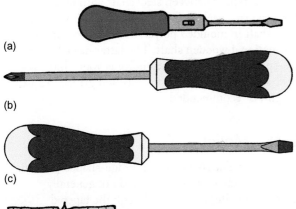

(a)

(b)

(c)

100 mm, 150 mm and 200 mm. They are also available with a No 2 Supadriv/Pozidriv tip and a No 2 Phillips' tip in blade-lengths of 100 mm only.

2.5.2   Pozi-tip and Parallel-tip (Slotted) Screwdrivers

Pozi-tip and parallel-tip (slotted) screwdrivers are available with round-ended soft-grip handles, which provide good torque and comfort – often lacking on plastic-handled screwdrivers. They are also available with magnetic tips for easy pick-up and speedier screw-insertion. Lengths of these screwdrivers (related to the chromium-plated steel bars (stems) that project from their handles) are 75 mm, 100 mm and 200 mm.

2.6   MARKING GAUGES

Figure 2.23: These tools may not have a use on first-fixing carpentry, but will be needed on

Figure 2.22 (a) Ratchet screwdriver; **(b)** Pozidriv screwdriver; **(c)** Parallel (straight slotted) screwdriver

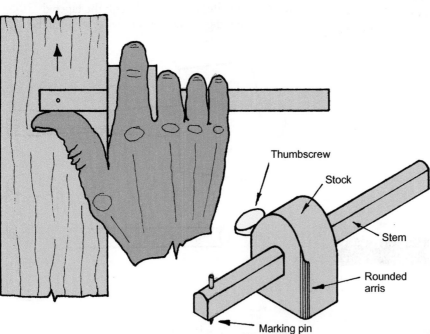

Thumbscrew
Stock
Stem
Rounded arris
Marking pin

Figure 2.23 Marking gauge

second-fixing operations. Although still predominantly made of beech, the thumbscrews are made of clear yellow plastic and – although quite tough – if overtightened, may fracture. To protect the sharp marking-pin and for safety's sake, the pin should always be returned close to the stock after use. To use the gauge, it should be held as shown, with the thumb behind the pin, the forefinger resting on the rounded surface of the stock and the remaining fingers at the back of the stock, giving side pressure against the timber being marked. Always mark lightly at first to overcome grain deviations. The gauge is easier to hold if the face-edge arris – that rubs the inside of the outstretched thumb – is rounded off as shown.

Also, to reduce wear and surface friction, plastic laminate can be shaped and bonded to the face of the stock.

2.7 CHISELS

2.7.1 Firmer and Bevelled-edge Chisels

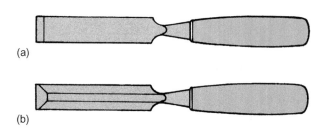

(a)

(b)

Figure 2.24 (a) Firmer chisel; (b) Bevelled-edge chisel

Figure 2.24: Firmer chisels are generally for heavy work, chopping and cutting timber in a variety of operations where a certain amount of mallet/hammer work and levering might be necessary to remove the chopped surface. Bevelled-edge chisels are generally for more accurate finishing tasks – such as paring to

a gauged line – where mallet/hammer work, if any, is limited and levering should be avoided.

2.7.2 Types of Chisels

Although chisels with conventional wooden handles of boxwood or ash are still available, they are more suitable for bench joinery work, where they are less likely to receive rough treatment. Because most site carpenters will hit a chisel with the side of a claw hammer, the modern range of chisels includes: split-proof handled chisels with impact-resistant plastic handles, designed for use with a hammer and guaranteed for life; black non-slip polypropylene impact-resistant handled chisels; and heavy-duty shatterproof, cellulose acetate butyl handled chisels. Blade widths range from 3 mm to 50 mm. Recommended sizes for a basic kit are 6, 10, 12 and 25 mm in firmer chisels (Figure 2.24(a)) and 18 and 32 mm in bevelled-edge chisels (Figure 2.24(b)).

2.7.3 Grinding and Sharpening Angles

Figure 2.25: The cutting edge of chisels should contain a *grinding angle* of 25°, produced on a grindstone or grinding machine, and a *sharpening angle* of 30°, produced on an oilstone or a diamond whetstone. The hollow-ground angle should not lessen the angle of 25° in the concave of the hollow. For extra strength, firmer and mortice chisels should be flat-ground.

2.8 OILSTONES AND DIAMOND WHETSTONES

Artificially manufactured stones, made from furnace-produced materials, as opposed to natural stone, are widely used because of their constant quality and relative cheapness. Coarse, medium and fine grades are available. A *combination stone*, measuring 200 × 50 × 25 mm, is recommended for site work. This stone is coarse for half its thickness, and fine on the alternate side for the remaining half thickness. As these stones are very brittle, they should be housed in purpose-made (or shop-purchased) wooden boxes for protection.

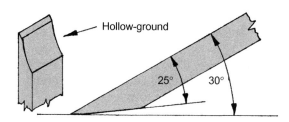

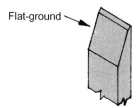

Figure 2.25 Grinding angle 25°; sharpening angle 30°

2.8.1 Oiling the Stone

When sharpening, use a thin grade of oil, animal or mineral, but not vegetable oil, which tends to solidify on drying, so clogging the cut of the stone. Lubricating oil is very good. Should the stone ever become clogged, giving a glazed appearance and a slippery surface, soak it in petrol or paraffin for several hours, then clean it with a stiff brush and/or sacking material and allow it to dry before reusing.

2.8.2 Sharpening

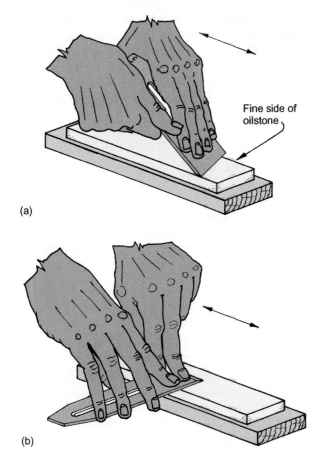

(a)

(b)

Figure 2.26 (a) Recommended hand-hold for sharpening a plane iron; (b) Removing metal burr and polishing underside of cutting edge

Figure 2.26: When sharpening chisels or plane irons, first apply enough oil to the stone to cover its surface and help float off the tiny discarded particles of metal, then hold the tool comfortably with both hands, assume the correct angle to the stone (30°), then move back and forth in an even, unaltering movement until a small sharpened (or honed) edge is obtained. This action produces a metal burr which is turned back by reversing the cutter to lay flat on the stone, under finger-pressure, and by rubbing up and down a few times. Any remaining burr can be removed by drawing the cutter across the arris edge of a piece of wood.

2.8.3 Use of Oilstone

The stone should always be used to its maximum length, the cutter lifted occasionally to bring the oil back into circulation. Narrow cutters, such as small chisels, whilst traversing the length of the stone, should also be worked across the stone laterally to reduce the risk of dishing (hollowing) the stone in its width.

2.8.4 Oilstone Box

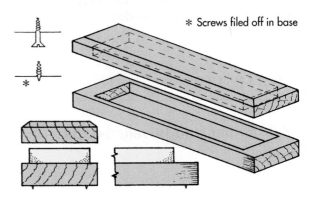

* Screws filed off in base

Figure 2.27 Oilstone box

Figure 2.27: Although now available in tool shops, this was traditionally a hand-made item, usually of hardwood, required to protect the stone from damage and the user from contact with the soiled oil. It is easily made from two pieces of wood, each measuring a minimum of 240 × 62 × 18 mm, to form the two halves of the box. With the aid of a brace and bit and chisel (or a router, if available), recesses are cut to accommodate the stone snugly in the base and loosely in the part which is to be the lid. To stop the box from sliding while sharpening, two 12 mm-long screws can be partly screwed into the underside of the base and filed off to leave dulled (filed off) points of about 2 mm projection.

2.8.5 Diamond Whetstones

Figure 2.28: These modern sharpening surfaces are now popular with many carpenters as an alternative to traditional oilstones. This is because diamond whetstones allow fast and clean removal of the chisel or plane iron's blunt or damaged cutting edges, saving time and reducing or eliminating the need to regrind. They also give an unimpaired, lasting performance under normal working conditions. To help float off the

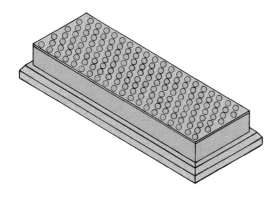

Figure 2.28 Diamond Whetstone

micron-sized, discarded particles of metal, Diamond Abrasive Lapping Fluid should be used instead of oil. Although water can be used instead, this can cause tool-rust. Hardwood boxes or plastic cases are available for the range listed below. Non-slip mats or bench holders can also be used as an alternative.

Colour coding

The DMT® range of 'stones' have colour-coded bases, denoting the grit or micron size of the diamonds. **Green** = extra fine, 1200 grit, 9 micron; **Red** = fine, 600 grit, 25 micron (said to be the most popular for all woodworking tools, including router and auger bits); **Blue** = coarse, 325 grit, 45 micron (for general carpentry tools, including masonry drills); **Black** = extra coarse, 220 grit, 60 micron (for damaged tools, including cold chisels).

Technical details

The silver surface of the 'stones' is a layer of graded mono-crystalline diamonds set two-thirds into nickel, which bonds them to a perforated precision-ground steel base. The base has been injection moulded onto a polycarbonate/glass fibre substrate. This is of steel-like rigidity and strength. Therefore, the whetstones will stay flat and not bend, dish or groove throughout their working life.

2.9 HAND PLANES

The two planes to be recommended as most useful for site work are the No. $4\frac{1}{2}$ smoothing plane with a cutter width of 60 mm and a base length of 260 mm and the No. $5\frac{1}{2}$ jack plane, also with a cutter width of 60 mm, but a base length of 381 mm. Narrower smoothing and jack planes with 50 mm cutter widths and less length and weight, notably the No. 4 and the No. 5, are thought to be more suitable for bench joinery work.

2.9.1 Knowledge of Parts

Figure 2.29 shows a vertical section through a smoothing plane to identify the various parts which need to be named and known for reference to the plane's usage. These named-parts also apply to the jack plane and other planes of this type.

2.9.2 Planing and Setting-up Details

These planes are also available with corrugated (fluted) bases to reduce surface friction, especially when planing resinous or sticky timbers; if not fluted, the sole of the plane can be lightly rubbed with a piece of beeswax or candlewax. When planing long lengths of timber, like the edge of a door, lift the heel slightly at the end of the planing stroke to break the shaving. On new planes, the cutter has been correctly ground to 25°, but not sharpened. To sharpen, remove the cutter from the back iron and carry out the sharpening procedure outlined in the text to Figure 2.26. When reassembling, set the back iron within 1–2 mm of the

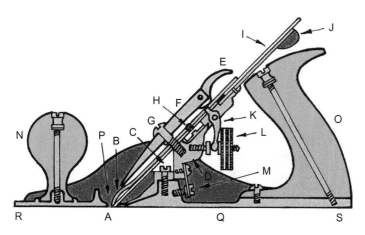

Figure 2.29 Vertical section through smoothing plane. The parts are: A mouth, B back iron, C frog-fixing screws, D frog, E lever, F lever cap, G lever-cap screw, H back-iron screw, I cutting iron (cutter), J lateral-adjustment lever, K cutter-projection adjustment lever, L knurled adjusting-nut, M mouth-adjustment screw, N knob, O handle, P escapement, Q sole (base), R toe and S heel

cutter's edge and, if necessary, adjust the lever-cap screw so that the relaced lever cap is neither too tight nor too loose. Note that when you adjust the cutter to *reduce* its mouth-projection (by turning the round, knurled adjusting-nut seen in Figure 2.29 anti-clockwise), finalize the operation by turning the knurled adjusting-nut clockwise until the adjustment lever has lost its feeling of looseness. I find that this helps to stop the cutter being pushed back via the free-play area in the cutter's slot.

2.9.3 How to Check Cutter-projection

Always check the cutter projection before use, by turning the plane over at eye-level and sighting along the sole from toe to heel. The projecting cutter will appear as an even or uneven black line. While sighting, make any necessary adjustments by moving the knurled adjusting-nut and/or the lateral-adjustment lever. For safety and edge-protection, always wind the cutter back after final use of the plane. Bear in mind that the body is made of cast iron, and if dropped, is likely to fracture – usually across the mouth. Keep planes dry and rub occasionally with an oily rag.

2.10 RATCHET BRACE

Figure 2.30: This item should be carefully chosen for its basic qualities and any saving in cost could prove to be foolish economy. Essentially, the revolving parts – the head and the handle – should be free-running on ball bearings, the ratchet must be reliably operational for both directions and the jaws must hold the

tapered-tang twist bits, and the dual-purpose combination auger bits with parallel shanks, firmly and concentrically. The recommended *sweep* (diameter of the handle's orbit) is 250 mm. Braces with a smaller or larger sweep are available. The advantages of the ratchet are gained when drilling in situations where a full sweep cannot be achieved, such as against a wall or in a corner – or when using the screwdriver bit under intense pressure and sustaining the intensity by using short, restricted ratchet-sweeps.

2.11 BITS AND DRILLS

2.11.1 Twist Bits and Flat Bits

Figure 2.31: Twist bits are also referred to as *auger bits* and traditional types are spiral-fluted, round shanked with tapered tangs. Their disadvantage nowadays is that they will only fit the hand brace and not the electric or cordless drill. However, a set of modern bits, without this disadvantage, is now an option. These bits are also spiral-fluted and round shanked, but are minus the tapered tang and will fit either the electric/cordless drill or the ratchet brace. They are known as combination auger bits. All of these bits are for drilling shallow or deep (maximum 150 mm) holes of 6–32 mm diameter. *Jennings pattern* twist bits have a double spiral and are used for fine work; *Irwin solid-centre* twist bits have a single spiral and are more suitable for general work; *Sandvik combination* auger bits have a wide single spiral with sharp edges and give a clean-cut hole suitable for fine or general work. Seven combination bits are recommended for the basic kit, these being sizes 6, 10, 13, 16, 19, 25 and 32 mm.

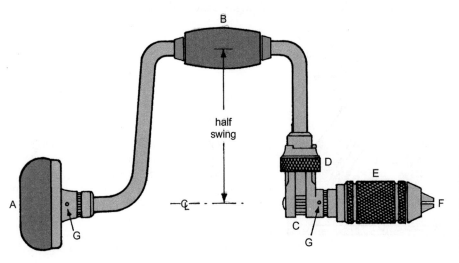

Figure 2.30 Ratchet brace. Parts of the brace are named as follows: **A** head, **B** handle, **C** ratchet, **D** ratchet-conversion ring, **E** jaws-adjustment shell, **F** universal jaws and **G** oil hole

Jennings pattern twist bit

Irwin solid-centre twist bit

Sandvik combination auger bit

Bahco flat bit

Figure 2.31 Twist bits/auger bits and flat bits

Flat bits are ideal for use with electric or cordless drills and are now produced by some manufacturers with non-slip hexagonal shanks (not available on 6 mm diameter), suitable for SDS (Special Direct System) chuck systems, allowing quick release. Another improvement in recent years is the winged shoulders, enabling the bit to score the perimeter of the hole before cutting the material. Because of their flat, simple design, which reduces side friction, they cut a lot faster and with less effort than twist or auger bits; but this simple design feature can be a disadvantage when drilling holes that need to be more precise, at right angles to the surface of the material – and where the bore hole does not wander within the material. However, this tendency (or capability) of wandering when below the surface, can be put to advantage when drilling certain holes in awkward locations; for example, when drilling through the sides of *in situ* floor/ceiling joists in a rewiring job. Sizes of flat bits range from 6 mm to 40 mm.

2.11.2 Drilling Procedure

If the appearance of a hole has to be considered, then care must be taken not to break through on the other side of the timber being drilled. This is usually

achieved by changing to the opposite side immediately the point of the bit appears. Alternatively, drill through into a piece of waste timber clamped onto or seated under the blind side.

2.11.3 Sharpening Twist Bits

Always avoid sharpening twist bits for as long as possible, but when you do, sharpen the inside edges only, with a small flat file; never file the outer surface of the spur cutters. Always take care, when drilling reclaimed or fixed timbers, not to clash with concealed nails or screws, as this kind of damage usually ruins the twist bit.

2.11.4 Countersink Bits

Figure 2.32: These are for screw-head recessing in soft metal and timber. The *rosehead pattern* type is for metals, such as brass or aluminium, although it is also used for softwood (and can be used for hardwood). The *snailhorn pattern* type is used just for hardwood (but can be used for softwood). As illustrated, these are available with a round shank and traditional tapered tang for use with the ratchet brace, or with short or long, round shank only, for use with the hand drill or the electric or cordless drill.

2.11.5 Combined Countersink and Counterbore Bits

Figure 2.33: These two modern drill-bits are useful on certain jobs and although they can be used in traditional hand drills, they will of course be more efficient in electric or cordless drills. The combined countersink bit is available in seven different sizes and is for drilling a pilot hole, shank hole and countersink for woodscrews in one operation. The combined counterbore bit is available in 12 different sizes and is for drilling a pilot hole, shank hole and a counterbored hole (the latter receives a glued wooden pellet after screwing), also in one operation.

2.11.6 Screwdriver Bits

Figure 2.34: There is a wide range of modern bits suitable for electric or cordless drills, in different

(a)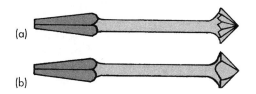

(b)

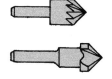

Figure 2.32 (a) Rosehead and **(b)** snailhorn countersink bits

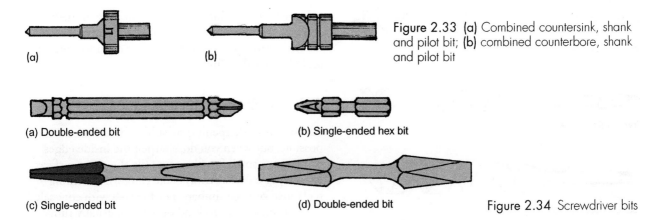

(a) (b)

Figure 2.33 (a) Combined countersink, shank and pilot bit; (b) combined counterbore, shank and pilot bit

(a) Double-ended bit

(b) Single-ended hex bit

(c) Single-ended bit

(d) Double-ended bit

Figure 2.34 Screwdriver bits

lengths (Figures 2.34(a) and (b)) to fit different gauges of screws and different types, such as screws with Supadriv/Pozidriv inserts, Phillips' inserts and slotted inserts. The two traditional screwdriver bits shown in Figures 2.34(c) and (d) have tapered tangs for use with the ratchet brace. These bits are still very useful, mainly for the extra pressure and leverage obtained by the brace and occasionally required in withdrawing or inserting obstinate or long screws. The double-ended bit (d) has a different size slotted tip at each end, in the shape of – and to act as – a tang.

2.11.7 Twist Drills and Masonry Drills

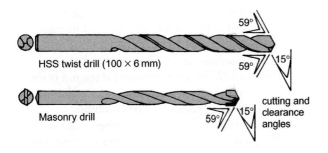

HSS twist drill (100 × 6 mm)

Masonry drill

cutting and clearance angles

Figure 2.35 Twist drills and masonry drills

Figure 2.35: The twist drill is another of those tools adopted from the engineering trades and put to good use in drilling holes in timber. Their round shanks will fit the chuck of the electric, cordless, or hand drill. A set of these twist drills, of high-speed steel (HSS), varying in diameter by 0.5 mm and ranging from 1 mm to 6 mm, is essential for drilling pilot holes and/or shank holes for screws in timber. They may also be used of course for drilling holes in metals such as brass, aluminium, mild steel, etc. (after marking the metal with a *centre punch*). When dull, these drills can

be sharpened on a grinding wheel, but care must be taken in retaining the cutting and clearance angles at approximately 60° and 15°, respectively.

Masonry drill-bits of various diameter for drilling plug holes in brick, block and medium-density concrete, or similar materials, have improved in recent years and are now available with high quality, heavy duty carbide tips.

2.11.8 SDS Drills

Figure 2.36 SDS-Plus drill bit

Figure 2.36: These high-performance drill-bits, manufactured from high grade alloy tool steel, mainly for drilling into dense concrete, or similar material, are for use with the powerful SDS-Plus and SDS-Max range of hammer drills and pneumatic hammer drills with SDS (Special Direct System) chucks. The drill-bits have standardized shanks that slot into the special chuck arrangements with automatic locking devices and quick release.

2.12 INDIVIDUAL HANDTOOLS

2.12.1 Bradawls

Figure 2.37: These are used mainly for making small pilot or shank holes when starting screw fixings. Two different sizes are available and different types. One type of awl has a flat brad-head point which should always be pushed into the timber at right angles to the grain before turning; the other type of awl has a square-sectioned, tapered point which acts as a reamer when turned – ratchet fashion – into the timber.

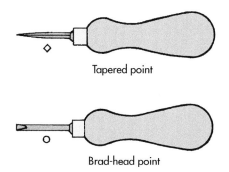

Figure 2.37 Bradawls

Although it seems to be less common, the square-tapered awl (called a 'birdcage awl') is very much recommended.

2.12.2 Pincers

Figure 2.38 Pincers

Figure 2.38: Pincers are used for withdrawing small nails and pins, not fully driven in. Although these are usually extracted by the claw hammer, occasionally a pair of pincers will do the job more successfully. When levering on finished surfaces, a small piece of wood or thin, flat metal placed under the fulcrum point will reduce the risk of bruising the surface. Different sizes are available, but the 175 mm length is recommended.

2.12.3 Wrecking Bar

Figure 2.39 Wrecking bar

Figure 2.39: This tool is also referred to as a *crowbar*, *nail bar* or *pinch bar*. It is not essential, but is very useful in construction work for extracting large nails and for general leverage work. There is a choice of five sizes: 300, 450, 600, 750 and 900 mm. If necessary, leverage can be improved by placing various-size

blocks under the fulcrum point. The 600 mm length is recommended.

2.12.4 Hacksaw

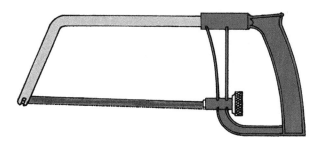

Figure 2.40 Junior hacksaw

Figure 2.40: This is another useful addition occasionally required. Carpenters do not normally need a full-size hacksaw for cutting large amounts of metal objects, so the *junior hacksaw*, with a blade length of 150 mm, is recommended for the limited amount of use involved.

2.12.5 Nail Punches

Figure 2.41 Nail punches

Figure 2.41: These are essential tools, especially in second-fixing carpentry, when used to sink the heads of nails or pins below the surface of the timber, by about 2 mm, to improve the finish when the hole is *stopped* (filled) prior to painting. They are also of use occasionally, by assisting in driving a nail into an awkward position, or in skew-nailing into finished timbers, by switching to a punch to avoid bruising the surface with the hammer. At least three nail punches are recommended for the basic kit, say 1.5, 2.5 and 5 mm across the points. Some points have a concave shape to reduce the tendency to slip.

2.12.6 Cold Chisel and Bolster Chisel

Figure 2.42: A 19 × 300 mm cold chisel and a bolster chisel with a 50 mm and/or 75 mm wide blade are recommended additions to the tool kit, for use on odd

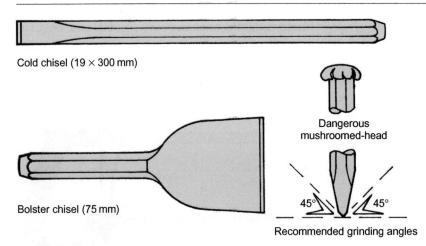

Cold chisel (19 × 300 mm)

Bolster chisel (75 mm)

Dangerous
mushroomed-head

45°   45°
Recommended grinding angles

Figure 2.42 Cold chisel and bolster chisel

occasions when brick or plasterwork requires cutting. Carbon steel chisels are commonly used, although nickel alloy chisels are more suitable. Keep the cutting edges sharp, by file for nickel alloy and by grinding wheel for carbon steel.

Safety Note

Figures 2.41 and 2.42: For safety's sake, grind the sides of the heads before they become too mushroom-shaped from prolonged usage. Also, develop the technique of holding the tool as you would the barrel of a rifle, so that the palm of the hand, not the knuckles, faces the hammer blows. Alternatively, plastic grip hand guards are available separately to fit onto chisels.

2.12.7 Electronic Detectors

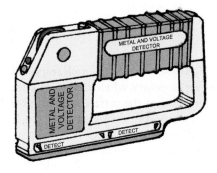

Figure 2.43 Electronic Detector

Figure 2.43: These electronic instruments are very useful for detecting the location of any surface-concealed electric cables, gas and/or water pipes. This preliminary detection is essential on plastered or dry-lined ceilings and walls and screeded or boarded floors, where blind fixings or drilling for fixings may cause a costly and/or messy disaster. There are a variety of these instruments available, although some only detect

power and pipes. Others detect stud or joist positions, with an LED display and buzzer indicator. A few detect all three elements. The model drawn here represents a West WST165 Metal and AC Voltage Detector. This particular model also indicates if cables, power points, junction boxes, etc, are live.

2.13 PORTABLE POWERED AND CORDLESS CIRCULAR SAWS

Figure 2.44(a): These saws are widely used nowadays to save time and energy spent on handsawing operations. Although basically for ripping and cross-cutting, they can also be used for bevel cuts, sawn grooves and rebates. Models with saw blades of about 240 mm or more diameter are recommended for site work.
Figure 2.44(b): Cordless circular saws are now available for site and general use, but because of the limited power supply from rechargeable batteries, may not be constantly powerful enough for heavy ripping work. However, they can be very useful on sites without any electricity supply. The saw recommended and illustrated here is the DeWalt 007K, 24 volt heavy duty model with a 165 mm diameter blade, which gives a 55 mm depth of cut.

2.13.1 Adjusting the Base Plate

Before use, the saw should be adjusted so that when cutting normally, the blade will only just break through the underside of the timber. This is easily achieved by releasing a locking device which controls the movement of the base plate in relation to the amount of blade exposed. For bevel cutting, the base tilts laterally through 45° on a lockable quadrant arm. The telescopic saw guard, covering the exposed blade, is under tension so that after being pushed back by the end of the timber being cut, it will automatically

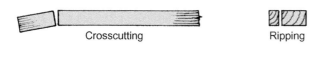

Crosscutting　　　　　　　　Ripping

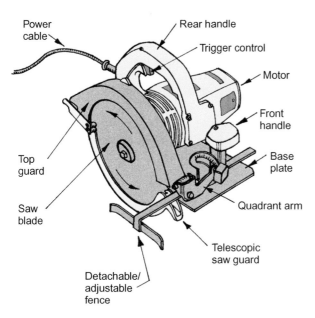

Power cable
Rear handle
Trigger control
Motor
Front handle
Top guard
Base plate
Saw blade
Quadrant arm
Telescopic saw guard
Detachable/ adjustable fence

Bevel cut　　　　Sawn groove　　　　Rebate

(a)

(b)

Figure 2.44 (a) Portable powered circular saw; (b) cordless circular saw

spring back to give cover again when the cut is complete. For ripping, a detachable fence is supplied.

2.13.2 Using the Saw

Before starting to cut, the saw should be allowed to reach maximum speed and should not be stopped or restarted in the cut. Timber being cut should be securely held, clamped or fixed – making certain that any fixings will not coincide with the sawcut and that

there are no metal obstacles beneath the cut. Always use both hands on the handles provided on the saw, so reducing the risk of the free hand making contact with the cutting edge of the blade. At the finish of the cut, keep the saw suspended away from the body until the blade stops revolving.

2.13.3 Safety Note

Additional safety factors include: keeping the power cable clear of the cutting action; not overloading the saw by forcing into the material; drawing the saw back if the saw-cut wanders from the line and carefully re-advancing to regain the line; wearing safety glasses or protective goggles and ear defenders; disconnecting the machine from the supply while making adjustments, or when it is not in use; keeping saw blades sharp and machines checked on a regular basis by a qualified electrical engineer; checking voltage and visual condition of saw, power cable and plug before use; and working in safe, dry conditions.

2.14 POWERED AND CORDLESS DRILLS AND SCREWDRIVERS

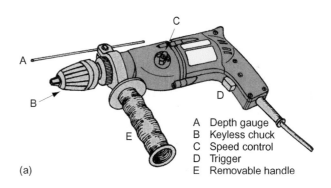

C
A
B
D
E

A Depth gauge
B Keyless chuck
C Speed control
D Trigger
E Removable handle

(a)

(b)

Figure 2.45 (a) Powered drill; (b) cordless drill/screwdriver

Figure 2.45: There is nowadays a wide range of dual- and triple-purpose drills to choose from, starting with the basic *rotary-only drill* and ending with the advanced *electro pneumatic hammer drill*. In the electric-powered range, of either 110 or 240 volts, the following combinations are available: the *drill/screwdriver, drill/impact (percussion) drill/screwdriver, drill/ rotary hammer drill/screwdriver*, and combinations of battery-powered models such as the *cordless screwdriver, drill/screwdriver, drill/impact drill/screwdriver*, and the *drill/rotary hammer drill/screwdriver*. Careful consideration needs to be given in choosing a particular model in relation to the type of work to be done and its location regarding whether power is readily available and if so, whether it is 110 or 240 volts.

2.14.1 Technical Features of Drills and Drivers

- *Drills with rotary impact or percussion action* create a form of hammer action generated by a ridged washer-type friction bearing and should only be used for occasional drilling into masonry or concrete.
- *Drills with rotary hammer action* are designed with an impressive impact mechanism involving a reciprocating piston and connecting rod, which strikes the revolving drill bit at between 0 and 4000 blows per minute on certain models. This makes drilling of masonry or concrete much easier and faster and is recommended when drilling into these materials frequently.
- *Drills with electro pneumatic rotary hammer action* are designed with an impact mechanism which converts the power into a reciprocating pneumatic force to drive a piston and striker at between 0 and 4900 blows per minute on certain models. It develops eight times the power of a normal hammer drill and is recommended for ease, speed and efficiency when drilling frequently or constantly into masonry or concrete.
- *Safety clutches* are torque-limiting mechanisms fitted to impact and hammer drills to protect the user if the drill bit becomes jammed in the material being drilled. They also protect the gears of the machine and the motor from short-term overloading.
- *Chucks* either have three jaws and require a separate chuck key, or are hand-operated and known as *keyless chucks*. Chuck sizes are usually 10, 13 and 16 mm. Some combination drills have an SDS chuck system, using special quick-release keyless chucks that accept only SDS drill bits.
- *The Fixtec system* on some SDS drills enables an SDS chuck to be replaced quickly by a standard three-jaw chuck without the use of tools.

- *Hex in spindle* means that a drill with this facility has a 6.35 mm ($\frac{1}{4}$ in) hexagon recess in the drive spindle (accessible after removing the chuck) to take hexagon shanked screwdriver bits.
- *Drills with variable electronic speed* have an accelerator function for gentle start-up during drilling and screwdriving and, on certain jobs, can be used for driving in screws without first drilling a pilot hole.
- *Drills with adjustable torque control* can be preset and gradually adjusted to achieve precise screwdriving tightness.
- *Reversing rotation* is available with some drills and drivers, which is necessary for removing screws – and desirable when bits get jammed in a drilling operation.

2.14.2 Safety Note

Small pieces of loose timber being drilled should be firmly held or clamped; reliable step ladders, trestles, orthodox platforms or scaffolds should be used when drilling at any height above ground level; safety glasses or protective goggles should be worn; before making adjustments, or when not in use, a machine should be disconnected from the power supply; bits should be re-sharpened or renewed periodically; machines should be maintained on a regular basis by a qualified electrical engineer or by the manufacturer; always check visual condition of machine, power cable and plug or socket before use; always keep a proper hold on the drill until it stops revolving; always work in safe, dry conditions. On certain jobs, wear hard hats/helmets.

2.15 POWERED AND CORDLESS PLANERS

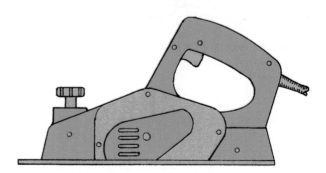

Figure 2.46 Powered planer

Figure 2.46: Powered planers are often used nowadays in conjunction with traditional planes such as the jack and the smoothing plane. They are sometimes preferred on such jobs as door-hanging, to lessen the strenuous task of 'shooting-in' the door by planing its edges. However, on this particular job, it is good practice to finish off with a hand plane, to remove the unsightly rotary-cutter marks – which can be very pronounced if the planer is pushed along at too great a feed-speed.

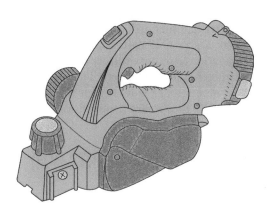

Figure 2.47 Cordless planer

Figure 2.47: Cordless planers are now available for site and general use, but because of the limited power supply from rechargeable batteries (although two are supplied with a charger) and the limited cutting depth of 0.5 mm, they do not have as much industrial appeal as a mains-powered planer. However, with both batteries fully charged, they can be useful for a period of time on sites without any electricity supply. The cordless planer illustrated here is the Makita 1051DWDE 14.4 volt model. It has a no-load-speed of 9000 rpm, maximum planing/cutting depth of 0.5 mm, rebating depth of 15 mm, a planing width of 50 mm and weighs 2.1 kg.

2.15.1 Making the Choice

When choosing a planer for frequent or constant use, avoid models classified in a DIY range; choose one with a professional/industrial classification. The model illustrated in Figure 2.46 is one in a range of three by this particular manufacturer. It has a TCT (tungsten carbide tip) cutter with a width of 82 mm, planing depth of 0–3 mm, rebating depth of 0–22 mm, vee grooves in the base to facilitate chamfering, a safety switch catch to prevent unintended starting and an automatic pivoting guard enclosing the cutter block for increased safety. This latter feature allows the planer to be put down safely before the revolving cutter block has come to a stop. One of the other models of this industrial threesome has an increased cutter width of 102 mm, but a reduced planing depth of

0–2.5 mm. Both models mentioned run at a no-load-speed of 13000 rpm.

2.16 POWERED AND CORDLESS JIGSAWS

Figure 2.48: Whether mains-powered or battery-operated (cordless), the essential purpose for a jigsaw is to enable *pierced work* to be carried out. This self-explanatory term means to enter the body of a material, without leading in from an outside edge. By so doing, small, irregular shapes can be cut. Nowadays, with different blades, this can be done in wood, certain metals and ceramic tiles. Although the initial entry into wood can be done by the saw itself, by resting the saw's front base and tip of the saw on the surface and pivoting the moving saw blade gradually through the material's thickness, the usual way is to drill a small hole to receive the narrow blade. A typical carpentry use for a jigsaw is in fitting a letter plate, described and illustrated here in the chapter on fitting locks, latches and door furniture. The secondary purpose for jigsaws is in cutting curved and irregular

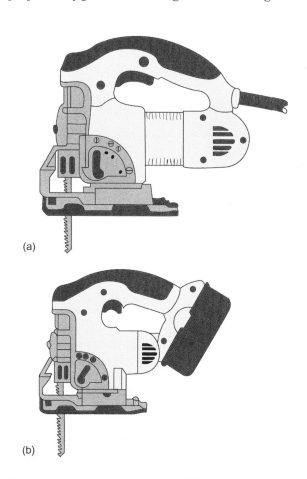

Figure 2.48 (a) Powered jigsaw; (b) Cordless jigsaw

shapes by leading in from an outer edge. The powered jigsaw represented here is a DeWalt model DEW 331K, with a 701 watt power input, 130 mm depth of cut in wood, variable speed, keyless blade change, adjustable orbital action and dust extraction adaptor. The cordless jigsaw represented here is a DeWalt model DEW DC330KA 18 volt with two rechargeable batteries, an impressive 130 mm depth of cut in wood, variable speed, three pendulum cutting actions, a blower and dust extraction hub.

2.17 POWERED AND CORDLESS SDS ROTARY HAMMER DRILLS

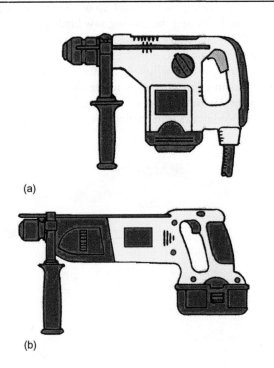

(a)

(b)

Figure 2.49 (a) Powered SDS rotary hammer drill; (b) cordless SDS rotary hammer drill

Figure 2.49: Powered and cordless SDS (Special Direct System) electro pneumatic rotary hammer drills, using special SDS masonry drills for high performance and durability, are essential nowadays for drilling into dense concrete, stone and masonry. Some models come with a rotation-stop facility to enable SDS tools to be used for chiselling operations. The electro pneumatic rotary hammer action has an impact mechanism which converts the power into a reciprocating pneumatic (compressed air) force to drive a piston and striker up to 4900 blows per minute on certain models. It is claimed to develop eight

times the power of a normal hammer drill. There are two types of SDS chuck available for SDS bits. One is the SDS-Plus chuck and the other is the SDS-Max chuck. The SDS-Max is for heavy drilling and core cutting in concrete, and chiselling operations. Bit-shank profiles differ, so an SDS-Plus bit must only be used in an SDS-Plus chuck and, likewise, an SDS-Max bit must only be used in an SDS-Max chuck – although separate adaptors are available. The powered rotary hammer drill represented here is a 900 watt DeWalt DEW D25404K heavy-duty SDS-Plus combination hammer drill, capable of drilling 32 mm diameter holes in concrete. The cordless model represented here is an 18 volt DeWalt DEW 999K2 heavy-duty SDS-Plus rotary hammer drill.

2.18 POWERED (PORTABLE) ROUTERS

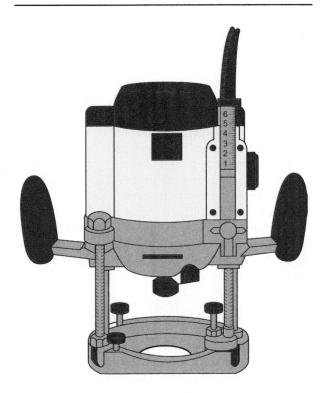

Figure 2.50 Powered router

Figure 2.50: Nowadays, portable powered-routers are often used on site for simplifying and speeding up certain second-fixing jobs. These include cutting recessed housings for door hinges; apertures for letter plates (often referred to as letter boxes); mortises for door locks and latches; joints in laminate kitchen-worktops with postformed edges; end-shaping of laminate worktops; cutting 'dog-bone' or 'T-shaped' recesses for inserting panel bolt connectors (these act like traditional handrail bolts in pulling up and holding worktop joints together); and cutting segmental slots in the joint-edges of worktops for the insertion of so-called 'biscuits'. These elliptical-shaped biscuits are manufactured in different sizes (size 20 is required to suit the biscuit-jointing cutter used in routers) and are pressed and cut to shape from 'feather-grained' beech (a traditional term meaning that the grain is at 45° across the face of the timber). This is done to eliminate the risk of the thin biscuits splitting along the grain when positioned within the joint. Although the insertion of three or four biscuits per worktop joint improves the joint's stability, their main function is to ensure surface flushness of the two joined worktops.

To perform the heavier tasks mentioned above, you will need a plunge router with a power input of at least 1300 watts. The router illustrated here represents a DeWalt DEW629 plunge router with a $\frac{1}{2}''$ (12.7 mm) collet, essential for receiving the $\frac{1}{2}''$ shank router cutters required for the heavier cutting. It has a 1300 watt power input, a no-load-speed of 22000 revs per minute and a plunge-movement depth of 62 mm.

2.19 JIGS

Figure 2.51(a)(b)(c)(d): All of the above operations, except biscuit jointing, are performed with the aid of special patented jigs which are clamped or attached to the work surface. The *hinge jig* represented at (a) is the H/JIG/75 by Trend, for routing 75 mm hinge recesses in internal doors and linings. Made of hard wearing, 12 mm laminate in two lengths for easy handling, it is joined together by a dovetail joint. No marking out of door or lining is needed. It has traditional settings for three hinges, 150 mm down, 225 mm up and middle – middle being optional, of course. There is a swivel end plate to give the 3 mm joint (gap) at the door-head. The jig is positioned by inserting three 10 mm-diameter plastic pins (supplied) along its length, to locate against the edges of the door and lining. Then the jig is secured to the work surface for routing by inserting two bradawls (also supplied) through small holes in its face.

The letter plate template (jig) illustrated at (b) is the Trend TEMP/LB/A, made from 8 mm clear plastic, for cutting three sizes of aperture to suit letter plates 220 × 50 mm, 210 × 60 mm and 200 × 60 mm on doors up to 50 mm thick. The template/jig is secured in its cutting position by small countersunk screws, strategically placed in the two centres required for letter-plate bolts. A $\frac{1}{2}''$ (12.7 mm) collet, plunge router is required.

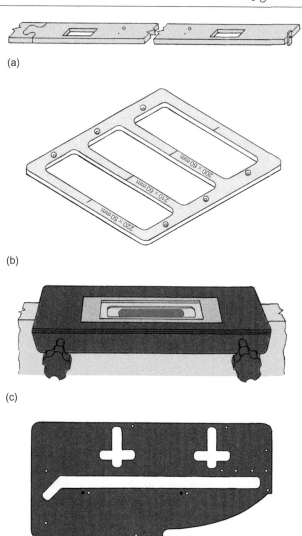

(a)

(b)

(c)

(d)

Figure 2.51 **(a)** Two-part hinge jig; **(b)** letter plate jig; **(c)** lock jig; **(d)** kitchen-worktop jig

The lock jig illustrated at (c) is the Trend LOCK/JIG, which uses a set of separate, interchangeable templates to suit the mortise and face-plate recesses of popular sized door locks. The templates are held in position by two powerful magnets. The jig is fixed to the door, as indicated, by its own clamping devices.

Finally, the kitchen-worktop jig represented at (d) is a Trend COMBI 651 jig, made from 16 mm thick, hard-wearing solid laminate. The holes that receive location-bushes for the different functions are colour-coded for quick and easy use and three alloy bushes are supplied with a width-setting stop. Worktops from 420 mm to 650 mm width can be jointed. There are two template-apertures for cutting T-shaped recesses for the panel bolt connectors and a radius edge for forming peninsular end-cuts.

2.20 NAILING GUNS

2.20.1 Nailing Guns for Wood to Wood Fixings

Figure 2.52(a)(b)(c): Although carpenters' claw hammers are still basically essential tools, nail guns are being used additionally on sites nowadays – especially on new-build projects. The reason for this is that these tools eliminate the effort involved in repetitive nailing and speed up the job. The three main types of nail gun to choose from are:

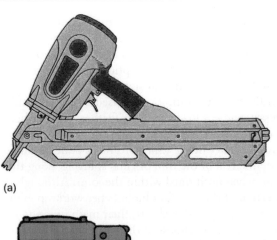

(a)

- (a) Pneumatic nail guns, involving air hoses and a portable compressor. The advantages with these tools include a capability of nail-size up to 100 mm (4″) and lower maintenance and running costs. The obvious disadvantage is the handicap of working with and around air hoses and a compressor – although portable, 10-bar compressors are available in quite small units, weighing from 16 to 25 kg, and hoses up to 30 m long.
- (b) Cordless nail guns using gas fuel cells, a spark plug and a rechargeable battery. A triggered spark ignites fuel in a combustion chamber, forcing a piston to drive the loaded-nail to a pre-set depth. The main advantage with these tools is the fact that they are without restriction by virtue of being cordless. The disadvantages are that their capability of nail-size is limited to 90 mm ($3\frac{1}{2}''$) and they require higher maintenance and running costs.
- (c) Cordless nail guns operated by rechargeable batteries **only**. The weight of this tool is about 1.5 kg heavier than its equivalent in group (b) and about 2.4 kg heavier than its equivalent in group (a), otherwise it has the same advantage as group (b) of being cordless and unrestricted. It does not use gas fuel cells, so running costs will be lower and there should be less maintenance involved. However, nail-size capability is limited to 63 mm ($2\frac{1}{2}''$).

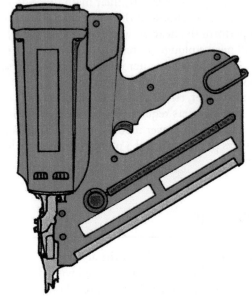

(b)

The (a) and the (b) types mentioned above, both provide a so-called 'Finish Nailer' for second-fixing operations and a 'Framing Nailer' (that holds larger nails) for first-fixing operations. The third type (c) is limited to only a second-fixing Finish Nailer that fires 16-gauge nails of 32 mm to 63 mm length.

The general features of these nail guns include a rubber pad at the base of the nose of the Finish Nailer, to eliminate bruising to the timber; small spikes at the base of the nose of the Framing Nailer, to locate its position and ensure no slipping when firing at an angle; angled nail-magazines – holding angled glue- or paper-collated strip nails – to enable skew-nailing

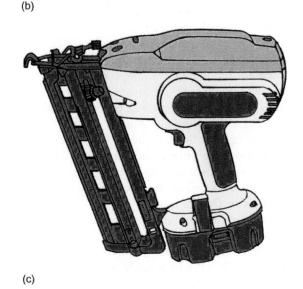

(c)

Figure 2.52 (a) Pneumatic Framing Nailer; **(b)** cordless gas/battery Framing Nailer; **(c)** cordless battery-only Finish Nailer

and fixing in confined areas; depth-of-drive adjustment; a capability of firing up to 4000 fixings per battery-charge and between 850 and 1000 fixings per fuel cell for type (b), about 800 per battery charge for type (c) and, of course, no limitation on the amount of fixings from the constant compressed air supply for type (a); a capability of fixing at a speed of two to three nails per second, depending on the fixing situation and the skill developed in the use of the tool; the nails available for use in these guns are smooth or ring-shank, in bright, galvanized or stainless steel.

The nail guns suggested for consideration in type (a) are 1) Ace & K's Angle Finish Nailer, model TYI 261650AB, that fires 25 mm to 50 mm × 16 gauge nails, for use in second-fixing operations and 2) Ace & K's Angle Framing Nailer, model TYI 10034FN, that fires 50 mm to 100 mm heavy-gauge nails, for use in first-fixing operations.

The popular models in type (b) seem to be 1) the ITW (Illinois Tool Works) Paslode Impulse, cordless Angle Finish Nailer, model IM65A (recently changed in model-number-only from IM250A), that fires 32 mm to 63 mm × 16-gauge nails for use in second-fixing operations and 2) the ITW Paslode Impulse, cordless Angle Framing Nailer, model IM350/90 CT, that fires 50 mm to 90 mm heavy-gauge nails, for use in first-fixing operations.

The model for consideration in type (c) is the DeWalt 18 volt, cordless Angle Finish Nailer, model DEW DC618KA, that fires 32 mm to 63 mm × 16-gauge nails, for use in second-fixing operations.

2.20.2 Nailing Guns for Wood to Hard-Surface Fixings

In the (a) and (b) types, there are also nailing guns available which are capable of fixing relatively thin timber (up to 20 mm thick) to hard surfaces such as concrete, steel, masonry and blockwork, etc. One to consider from each type is as follows:

- (a) In the pneumatic nail gun range, there is the Ace & K Multi T Nailer, model TYI 2.2–2.5/64T, that fires 25 mm to 64 mm masonry-type nails into sand-and-cement screeds, brick and blockwork. This tool also uses 14 gauge nails, known as Maxi Brads, for wood to wood fixings.
- (b) In the cordless, gas/rechargeable-battery type nail-gun range, there is the ITW Spit Pulsa 700P nail gun, that fires special nails into concrete, steel, hollow- or solid-brick. This tool has automatic power-adjustment, a 20-nail magazine (with a 40-nail magazine accessory) and is capable of 650 fixings per fuel cell and 1000 fixings per battery

charge. The manufacturers claim that it does not split concrete, has no recoil and makes less noise.

2.20.3 Cartridge (Powder-Actuated) Guns for Wood to Hard-Surface Fixings

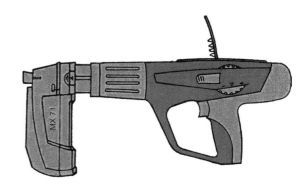

Figure 2.53 Powder-actuated tool

Figure 2.53: These tools are of great value on certain construction sites where they might be needed for fixing larger sections of timber, steel sections and anchors to *in situ* concrete, mild steel and brickwork. They are actuated by an exploded charge of gunpowder when the triggered firing pin strikes a loaded cartridge. The driven fixings are either special steel nails or threaded studs. Originally, all fixings from these tools were fired by the high-velocity principle, whereby the nail travelled along the barrel at a velocity of about 500 metres per second. The nail, travelling at such a speed, was potentially lethal; hazards such as *through-penetration, free-flighting* and *ricochets*, could easily occur through fixing into thin or weak materials, or striking hard aggregates or reinforcing rods.

Figure 2.54: To overcome these dangers, manufacturers developed the low-velocity principle in their tools. This reduces the muzzle velocity from 500 metres to between about 60 to 90 metres per second without any loss of power. This has been achieved by introducing a piston in the barrel between the cartridge at the rear and the nail in front. The nail, supported and guided by the cartridge-actuated piston, safely ceases its journey forward at the same time that the captive piston comes to rest.

The powder-actuated tool represented here is the Hilti DX 460 with an MX72 magazine attached. This tool has now replaced the long-established DX 450 model and is now fully automatic with nails in magazine strips of 10 (in nail-lengths of 14 to 72 mm) and cartridges also in strips of 10. It has a universal piston and less recoil for operator safety and comfort.

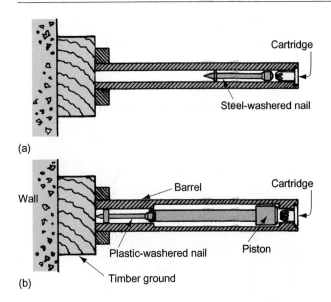

Figure 2.54 (a) High-velocity principle; (b) low-velocity principle

2.20.4 Cartridge Tool Fixings

Figure 2.55: There should be few problems if fixing into mild steel, *in situ* concrete or medium-density bricks such as Flettons, but difficulties will be experienced if attempts are made to fix into high yield steel, cast iron, cast steel, glazed tiles, high-density precast concrete, engineering bricks, soft bricks such as London Stocks and most natural stones – all of which may have to be drilled and anchored with some form of bolt or screw fixing.

If cartridge-tool users are ever in doubt, it is usually possible to arrange for a manufacturer's representative to visit a site to give advice and/or test materials for a required fixing system.

The recommended allowances for penetration by nail or stud are: steel, 12 mm; concrete, 20 to 25 mm; brickwork, 25 to 37 mm. Note that if the steel is only 6 mm thick, 12 mm should still be allowed for penetration, even though the point will protrude through the steel. When fixing to the lower flange of a steel 'I' beam, as illustrated, avoid central fixings which might coincide with the web above. If concrete is weakly compacted (less dense), use a longer fastener.

To calculate the length of nail required for fixing a timber batten or *ground* to concrete, deduct 5 mm for countersinking of nail into 25 mm timber, add 20 mm for minimum penetration, then adjust, if necessary, up to the nearest available nail size.

Cartridges are colour-coded to denote their strength, indicated by a touch of coloured lacquer on the crimped end. For the Hilti DX 460 tool, the cartridges available are: Green = light strength, Yellow = medium, Red = heavy, Black = extra heavy.

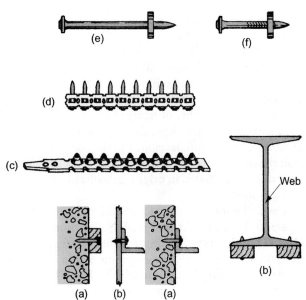

Figure 2.55 (a) Fixings to concrete; (b) fixings to steel; (c) cartridge strip; (d) magazine nail-strip; (e) nail for concrete and masonry; (f) nail for steel

Fixing too near the edges of base materials should be avoided, or spalling may occur; minimum distances should be: concrete, 75 mm; brick, 63 mm; steel, 12 mm. The minimum distance between fixings should be: concrete, 50 mm; brick, 63 mm; steel, 20 mm. Avoid fastening into concrete when its thickness is less than 3 × length of fastener-penetration; and avoid fastening into steel which is thinner than the shank-diameter of the fastener.

2.20.5 Safety Notes

The same general safety precautions used on mains powered tools, including the protection of eyes and ears, should be adhered to, as well as the following:

- Carefully read all the instructions provided in the manufacturer's literature before using the tool.
- Obtain thorough training in the safe use of the tool, either from the manufacturer's representative or from a competent person who has received training himself from the manufacturer.
- Ensure that the tool is well maintained and regularly serviced, in accordance with the manufacturer's recommendations.
- Wear heavy-duty rigger gloves and good-quality safety goggles (these are usually supplied by the tool manufacturer).
- When a battery pack reaches the end of its life cycle and needs replacing, it should be fully discharged and disposed of safely or recycled via your dealer. NEVER dispose of it in normal rubbish-disposal bins or on domestic waste-disposal sites.

3

Carpentry Fixing-Devices

3.1 INTRODUCTION

Knowing what fixings to use to fix timber together or to other materials and surfaces such as walls, ceilings and floors, etc, comes with few rules other than trade experience and is not normally specified by the architect or structural engineer. However, there are some exceptions to this, usually when structural stability is dependent upon certain fixings. For example, the maximum spacing of galvanized restraint straps at roof level is not determined by the carpenter who fixes them, but by British Standards' Codes of Practice and the Building Regulations. The majority of fixings, therefore, are determined by the carpenter and details of these will be stated as we progress through each section of this book. Fixing devices such as framing anchors, joist hangers, restraint straps, metal banding and panel adhesive, will also be detailed and dealt with in other chapters.

3.1.1 Modern Improvements

Fixing devices have improved with modern technology in the last few decades and although nails (with some changes in design) are still being driven in traditionally by hammer, purposely-designed T-headed nails are also being fired in by Strip or Coil Nailers (nail guns) using glue- or paper-collated strips of nails (see Chapter 2).

Screws have changed considerably with twin thread, deep-cut thread and steeper thread angles extending along the entire shank to provide a better grip and speedier driving performance. They also now have at least seven variations in slotted head-design, ranging from the traditional straight-slotted head to the Torx- or T Star-slotted head. With the revolutionary change from hand screwdrivers to electric and cordless drill/drivers, the development in slotted-head types was necessary to create a more secure, non-slip driving attachment. Improved slotted heads such as Uni-Screw and Torx/T-Star promote non-wavering driving alignment, virtually eliminating so-called *camout* situations. Like nails, screws also come in collated strips for use in mains powered, automatic-feed screwdrivers. The screws for these drivers have a bugle-shaped head, which is an innovative countersunk shape allowing them to sit at the correct depth in plasterboard without tearing the paper. These auto-feed drivers have been marketed particularly for drywall operations – fixing plasterboard to lightweight steel or timber stud-partitions – but can be used on other repetitive screwing operations such as fixing sheet material to wood floors prior to tiling, or sheet material to flat roofs prior to felt or GRP roofing. For these operations, collated floorboard-screws, with ordinary countersunk heads, are available.

3.2 NAILS AND PINS

- *Figure 3.1(a)*: **Round-head wire nails** are available in galvanized, sherardized and bright steel; sizes range from 25 mm × 1.80 mm Ø to 200 mm × 8.00 mm Ø. The nail-sizes most commonly used in first-fixing jobs such as roofing and stud partitioning are 75 mm × 3.75 mm Ø and 100 mm × 4.5 mm Ø. In certain fixing situations, even when using the softest of softwoods, splitting may occur with such large diameter nails and it will be wise to rest the head of

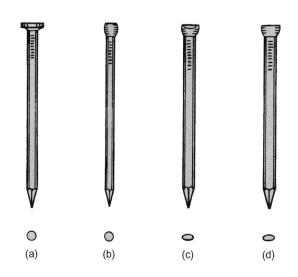

Figure 3.1 **(a)** Round-head wire nail; **(b)** lost-head wire nail; **(c)** brad-head oval nail; **(d)** lost-head oval nail

the nail on a hard surface and blunt the point of the nail with a hammer. This helps to push the fibres forward, instead of sideways in a splitting action.

- *Figure 3.1(b)*: **Lost-head wire nails** are available in galvanized, sherardized and bright steel; sizes range from 40 mm × 2.36 mm Ø to 75 mm × 3.75 mm Ø. The nail-sizes most commonly used in first-fixing jobs are 50 mm × 3 mm Ø and 65 mm × 3.35 mm Ø for floor-laying with T&G boards. The rare rule used here is that the nail length should be at least two-and-a-half times the thickness of the floorboard.

- *Figure 3.1(c)*: **Brad-head oval nails** are available in galvanized, sherardized and bright steel; sizes range from 25 mm to 150 mm. The nail-sizes most commonly used in second-fixing operations, such as fixing architraves and door stops, are 40 mm and 50 mm. The oval shape reduces the risk of splitting, but the sharp point of the nail may still need to be blunted in certain situations (for example, when fixing the head of an architrave, near the mitre), depending on the density of the timber.

- *Figure 3.1(d)*: **Lost-head oval nails** are available in bright steel; sizes range from 40 mm to 75 mm. As at (c) above, the nail-sizes most commonly used are 40 mm and 50 mm for fixing architraves, etc. The advantage that these nails have over the brad-head oval nails is that they are much easier to punch in below the surface, as the nail punch is seated on a flat head, rather than a thin-ridged head.

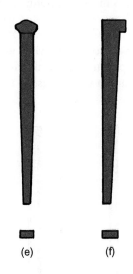

(e)   (f)

Figure 3.2 (e) Cut clasp nail; (f) cut floor brad

- *Figure 3.2(e)*: **Cut clasp nails** in black iron; sizes range from 40 mm to 100 mm. In second-fixing operations, these nails were originally used for fixing skirting boards to wood-plugged mortar joints and for making direct (unplugged) fixings into mortar joints, soft Stock bricks and breeze

blocks. When inner-skin walls changed to light-weight foamed building blocks, such as the original Thermalite blocks, direct fixings were still successful; but with the introduction in recent years of ultra-lightweight and high-thermal foamed blocks – which either do not hold or will not accept unplugged, direct cut-clasp-nail fixings – these nails lost their original advantage and were superseded by other fixings. However, because of their good holding-power and direct-fixing ability, they are still very useful in fixing temporary work (i.e. datum battens and cut stair-strings, etc) to the joints of brickwork. Research indicates that they are also being used for their original fixing uses in the large number of old properties undergoing modernization and conversion. The nail-sizes most commonly used seem to be 50 mm, 75 mm and 100 mm.

- *Figure 3.2(f)*: **Cut floor brads** in black iron; sizes range from 50 mm to 100 mm. As the name implies, these traditional nails are for fixing floorboards and, like cut clasp nails, they still seem to be used in refurbishment work on old properties. (A local ironmongery supplier confirmed that they still sell large quantities of these nails.) On new-build projects, however, cut floor brads have been superseded by lost-head wire nails for fixing T&G floorboarding and annular-ring shank wire nails for fixing chipboard flooring panels – but only the latter type are equal in holding-power to cut floor brads. For this reason, boards fixed with lost-head wire nails should be nailed dovetail-fashion, as illustrated in the chapter on flooring. The nail-sizes most commonly used seem to be 50 mm and 65 mm.

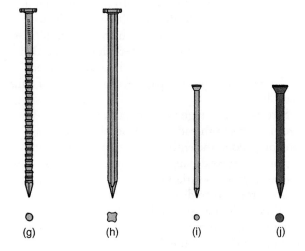

(g)   (h)   (i)   (j)

Figure 3.3 (g) Annular-ring shank nail; (h) grooved-shank nail; (i) panel pin; (j) masonry nail

- *Figure 3.3(g)*: **Annular-ring shank nails** are available in sherardized and bright steel; sizes range from 20 mm × 2 mm Ø to 100 mm × 4.5 mm Ø. The nail-sizes most commonly used seem to be the 65 mm × 3.35 mm Ø sherardized for first-fixing operations such as the fixing of 22 mm thick wind-bracing to trussed rafters and the 50 mm × 2.65 mm Ø bright steel nails for sub-flooring sheet materials.

- *Figure 3.3(h)*: **Grooved-shank nails** are available in galvanized steel; sizes are 40 mm, 50 mm, 60 mm, 75 mm and 100 mm. These newly marketed nails have four fluted grooves around the shank, running along their entire length. They are claimed to provide excellent grip, lighter weight – and therefore produce more nails per kilo when compared with round-head wire nails. Made from high-grade steel, BBA tested, they are also claimed to be bend-resistant.
- *Figure 3.3(i)*: **Panel pins** are available in sherardized and bright steel; sizes range from 15 mm to 50 mm. The sizes occasionally used in second-fixing carpentry are 20 mm and 25 mm × 1.4 mm Ø, 30 mm and 40 mm × 1.6 mm Ø, for fixing small beads and thin panels, etc.
- *Figure 3.3(j)*: **Masonry nails** in zinc plated and phosphate finish hardened steel; available sizes are 25 mm × 2.5 mm Ø, 40 mm, 50 mm, 65 mm and 75 mm × 3 mm Ø and 100 mm × 3.5 mm Ø. These nails are useful if fixing into sand-and-cement rendering and screeds, mortar joints and high-thermal foamed blocks – but will not penetrate concrete or hard bricks like Flettons.

3.3 SCREWS AND PLUGS

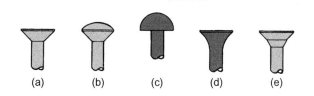

Figure 3.4 **(a)** Flat countersunk head; **(b)** raised countersunk head; **(c)** round head; **(d)** bugle countersunk head; **(e)** double-pitched countersunk head

Figure 3.4(a)(b)(c)(d)(e): Screws are sold in cartons or boxes of 100 and 200 (and smaller amounts in packages for DIY jobs) and are referred to by the amount required, the length, gauge, head-type, metal-type/ finish and driving-slot type, usually in that order – by carpenters – although screw-suppliers put the gauge before the length. As illustrated, the three most commonly used head-types in carpentry are (a) *flat, countersunk (CSK) head*, (b) *raised (RSD), countersunk head* and (c) *round (RND) head*. Two new heads recently added to the conventional *flat, countersunk-head type* are (d) *bugle countersunk head* and (e) the *double-pitched*

countersunk head. The raised or round part of the head is not included in the stated length. The lengths of screws generally range from 9 mm ($\frac{3}{8}''$) to 150 mm (6''), but only relatively short screws are available in RSD- and RND-head types. CSK screws predominate for their structural use; RSD- and RND-head screws are mostly used for fixing ironmongery. The different types of metal and treated-finish used (either for strength, cost, visual or anti-corrosion reasons) include: steel (S), galvanized steel (G), stainless steel (SS), bright zinc-plated (BZP), zinc and yellow passivated (YZP), brass (B), sherardized (SH), aluminium alloy (A), electro-brassed (EB), bronze (BR), black phosphate (BP) and black japanned (J). Most of the abbreviations given here are used in suppliers' catalogues and sometimes on the screw-box labels.

3.3.1 Understanding the Gauge-size of a Screw

A screw-manufacturer's given-gauge of a screw refers to the diameter of its shank in millimetres, which usually dictates the size (diameter) of its head and its length, i.e., the longer the screw, the larger the gauge must be structurally; and screw-heads, which are usually formed at an angle of 45° to the shank, end up being approximately twice the diameter of the shank. And carpenters and woodworkers often need to know this head-size, to ensure that screws will fit correctly in countersunk holes in ironmongery such as butt hinges, where the flat-headed screw should sit flush or slightly below the surface.

So, if we need to know the diameter of a screw's head for a certain job, we must either measure the actual screw, or double-up on the manufacturer's given shank-gauge. So, a screw-box label describing 4.0 × 40mm CSK screws is calculated as 4.0 × 2 = 8mm Ø (approx.) screw-head size.

3.3.2 Types of Driving Slot or Recess

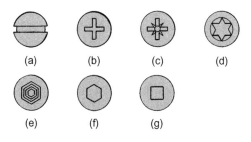

Figure 3.5 **(a)** Straight slotted; **(b)** Phillips recess; **(c)** Pozidriv recess; **(d)** Torx or T-Star recess; **(e)** Uni-Screw recess; **(f)** Allen (Hex) recess; **(g)** square recess

Figure 3.5(a)(b)(c)(d)(e)(f)(g): Carpenters nowadays need to carry a good-quality Screwdriver-Bits Set, containing a quick-release bit holder and a set of driver bits to cater for the majority of the following variations in slotted/recessed head design: (a) Straight slotted, (b) Phillips recess, (c) Pozidriv recess, compatible with Supadriv and Prodrive, (d) Torx or T-Star recess, (e) Uni-Screw recess, (f) Allen (Hex) recess, (g) Square recess.

3.3.3 Conventional, Twinfast and Superfast Screws

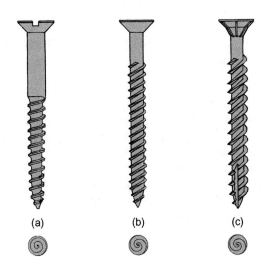

Figure 3.6 **(a)** Conventional, single-spiral thread; **(b)** twinfast, double-spiral thread; **(c)** superfast, double-spiral thread with extra features

Figure 3.6(a)(b)(c): In engineering terms, woodscrews have spaced threads and are grouped into three types: (a) Conventional, (b) Twinfast and (c) Superfast.

(a)   Conventional, single-thread (single-start) screws are comprised of a single spiral thread that takes up two-thirds the length of the screw. The remaining unthreaded shank has a diameter equal to, or slightly larger than, the major diameter of the thread. Because of this, these screws often require a shank hole and a smaller pilot hole to be drilled into timber to timber fixings.

(b)   Twinfast, double-thread (double-start) screws are comprised of a twin, parallel, spiral thread that either takes up the entire length of the shank – for increased holding-power in man-made materials such as chipboard – or – above a certain length of screw – has a portion of unthreaded shank left under the head. The diameter of the

remaining shank is smaller than the thread, so there is a reduced risk of splitting the timber near edges, etc. Also, this feature makes separate shank-and-pilot-holes redundant, as only one shank/pilot hole is required when there is a need to drill.

(c)   Superfast screws have additional innovative features other than the steeper, twin thread that pulls the screw in faster with less effort. These include a sharper, deeper-tread thread for increased holding-power; a specially designed self-drilling, sharp-cutting point for easier location and to enable screwing without (in some cases) the need to drill a pilot hole; specially hardened and heat-treated steel with screw-heads as strong as the tip of the driver bit. Other features available on this type of screw are: specially designed, rifled shanks with an indented spiral that adds to the drill-like cutting and screwing action; double-pitched countersunk heads, giving more torque and strength; self-countersinking ribs that protrude on the underside of the counter-sunk head.

3.3.4 Screwing

Figure 3.7(a)(b): When screwing items of ironmongery to timber, in most cases only a small pilot hole is required and can easily be made with a square-tapered shank, *birdcage awl*. When pushed into the timber and turned ratchet-fashion, it will ream out the fibres (unlike a brad-headed awl, which will push them aside). Drilled pilot holes, if preferred, should be about one-third smaller than the average diameter of the threaded shank. If using conventional screws when fixing timber to timber, a countersunk shank-hole (also called a clearance hole) and a smaller pilot (thread) hole, as illustrated, may be required. Depending on the density of the timber, often the small pilot/thread hole can be omitted. When using Twinfast or Superfast screws with screw-threads protruding from the shank, as opposed to being cut into them, only a single, combined shank/pilot hole may be necessary, as illustrated.

When using powered or cordless drivers, always start with a low torque setting and adjust it gradually to bring the screw flush to the surface. Further screw fixings of the same type, in the same area should be possible without altering the final torque setting. If using brass screws (for example, with brass butt hinges), it is a wise practice to use steel screws first, then replace them with the brass screws. This greatly reduces the risk of these soft-metal screws snapping under pressure. Alternatively – or additionally – rub beeswax, candle wax or tallow on the thread to reduce the tension on the screw.

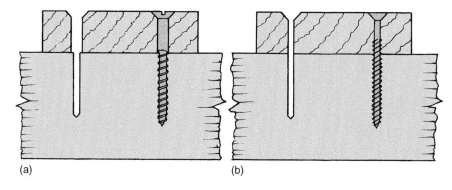

Figure 3.7 **(a)** Preparation for conventional screws; **(b)** preparation (where required) for Twinfast-threaded screws

3.3.5 Wall Plugs

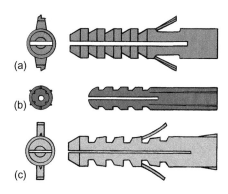

Figure 3.8 **(a)** 40 mm × 8 mm Ø brown-coded polythene plug; **(b)** 35 mm × 6 mm Ø red-coded polythene plug; **(c)** 40 mm × 8 mm Ø grey-coloured nylon plug

Figure 3.8(a)(b)(c): Plastic wall plugs have been the foremost screw-fixing device for many years now (at least three decades) and are extremely reliable if used properly when fixing timber or joinery items, etc., to solid walls. They are either made of high-density polythene or nylon. The nylon types are usually grey-coloured. The polythene types are colour-coded in relation to their screw-gauge size. The size of masonry drill required is given with each group of plugs – although I would be wary of these recommendations and have often found that, on certain 'soft' walls, a smaller drill was better suited to a particular plug. The coding is: Yellow = gauge No. 4 and 6; Red = gauge No. 6, 8 and 10; Brown = gauge No. 10, 12 and 14; Blue = gauge No. 14, 16 and 18. In my experience, the most common colour-coded plugs used in second-fixing carpentry are brown, which are 40 mm long × 8 mm Ø, and red – of less use – which are 35 mm long × 6 mm Ø. If plugs of a certain gauge are required to be longer, to suit the characteristics of the wall and/or the length of the screws required, long strips of polythene plug-material can be purchased and cut to size by the user. These plastic strips are

colour-coded as follows: White = screw gauge 4 to 6 mm, drill size = 8 mm Ø; Red = screw gauge 7 to 8 mm, drill size = 10 mm Ø; Green = screw gauge 9 to 10 mm, drill size = 12 mm Ø; Blue = screw gauge 11 to 14 mm, drill size = 16 mm Ø.

Finally, plastic-strip plugs tend to rotate more easily than purpose-made plugs – and for this reason they must fit snugly in the plugged wall; some wall plugs have a small, flexible barb projecting from each side, like the head of an arrow, to create an initial anchorage in the plugged wall. This allows the screw to be turned without the negative effect of rotating the plug.

3.3.6 Plugging the Wall

Figure 3.9(a)(b): When fixing to traditionally-plastered walls of so-called float-render-and-set finish plaster (usually just referred to as *solid plaster*), it is important that the fixing has at least 50% of its wall-depth anchored in the underlying brickwork or blockwork. This is especially so in older-type properties where the plasterwork is relatively soft. If you have marked the wall through shank holes drilled in a piece of timber or in a joinery item prior to plugging, mark these points further with a pencilled cross, so that they can be used as cross-sights when drilling. Be prepared

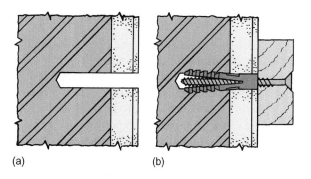

Figure 3.9 **(a)** Plastered wall drilled ready for plugging; **(b)** exaggerated effect of plug-expansion to highlight good anchorage beyond the plastered surface

to control the drill to keep on target. It can be done – about 90% of the time! The procedure is as follows:

- Select the required size of tungsten-carbide tipped masonry drill and secure it in your rotary hammer drill (see drill-types described in Chapter 2).
- Work out the amount of screw that should be in the wall as per the following example: If fixing a 19 mm thick timber shelf-bearer to a solid wall, plastered with 15 mm thick, relatively soft Carlite plaster, the length of screw in this situation should equal at least three times the bearer thickness, making it 19 × 3 = 57. The nearest screw is 60 mm and if we use a No. 5 shank-gauge CSK screw (measuring 10 mm across its head) and a brown-coded plug, which is 40 mm long, we have 60, minus 40, leaving 20 mm. This is 1 mm more than the bearer thickness, so we must allow the screw to go through the plug by this amount. In practical terms, we need to increase this by, say, another 4 mm (to accommodate any trapped debris in the plug hole and to allow for any under-surface countersinking). So we have 60 minus 19 = 41 + 4 mm tolerance = 45 mm drilling depth. If you drill slightly more than this depth, it is not critical – but if you drill excessively more, you run the risk of the plug being pushed in too deeply by the screw.
- Wrap a piece of masking tape around the drill bit to determine the depth (or set the depth-gauge attachment on the hammer drill); wear your safety goggles, hold the drill visually square to the wall – vertically and horizontally – and start drilling.
- When deep enough, relax the pressure and with the drill rotating negatively, quickly work it in and out of the hole for a few seconds to clear the debris.
- Drill all the required holes in a similar way and, when finished, insert the plugs flush to the wall. You are now ready for screwing.

3.3.7 Nylon-Sleeved Screws

Figure 3.10(a)(b)(c): Nylon-sleeved (grey coloured) **Frame-fix**, **Hammer-fix** and **Window-fix** screws are popular nowadays. They provide the advantage of enabling a door frame or lining, a window frame, a kitchen unit or whatever else, to be fixed whilst the item is in the required fixing-position. There is no need to mark and plug the wall separately. This is achieved by drilling through the pre-positioned joinery-item and the wall in one operation. Then the sleeved Frame-fix, Hammer-fix or Window-fix screw is inserted as a one-piece, combined fixing. Once located in the two aligned holes, the protruding

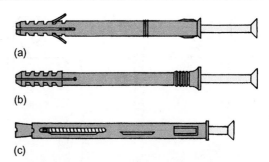

Figure 3.10 Sleeved screws: **(a)** Frame-fix screw; **(b)** hammer-fix screw; **(c)** window-fix screw

screw-head is either screwed in or driven in by hammer, according to the type of fixing used. If necessary, all three types can be unscrewed once fixed.

The Frame-fix and Hammer-fix types work on the principle of the screw being driven into a decreasing, tapered hole in the nylon sleeve, which causes the end portion to expand out into an alligator shape, creating excellent anchorage. The Window-fix screw works on a similar principle, except that the alligator effect is achieved by a tapered, cork-shaped nylon-end being pulled up into the sleeve when the screw is tightened. It has to be said that the Frame-fix and the Window-fix screws seem to achieve more holding power than the Hammer-fix type.

Although the through-drilling is described as one operation, in practice it should be two. First, the joinery item is drilled through with a twist drill (carefully stopping short of piercing the wall), then the job is completed by switching to a masonry drill which is equal in diameter to the twist drill.

3.3.8 Frame Screws

Figure 3.11: So-called Frame screws are now available and are described as a masonry screw for fixing into 'most substrates'. No plugs are needed, but a 6 mm Ø pilot hole must be drilled through the frame into the concrete, brickwork, etc. The gauge of these screws is 7.5 mm, which means that 0.75 mm of thread cuts into the 6 mm hole in the substrate to create its own mating thread and fixing. These screws are zinc plated and yellow passivated, with T30 Star-drive recesses. Their lengths range from 42 mm to 182 mm in increments of 20 mm up to 122 mm, then 30 mm up to

Figure 3.11 Non-sleeved, self-tapping Frame screw

182 mm. White or brown T30 plastic caps are available for these screws, if required.

3.4 CAVITY FIXINGS

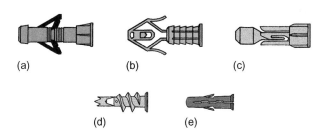

(a)   (b)   (c)

(d)   (e)

Figure 3.12 (a) Plasplugs Super Toggle Cavity Anchor; (b) Plasplugs Plasterboard Heavy-Duty Fixing; (c) Fischer Plasterboard Plug; (d) Nylon or Metal (zinc-plated) Easi-Driver; (e) Rawlplug 'Uno' Plug

Figure 3.12(a)(b)(c)(d)(e): Cavity Fixings are used to create reliable fixings into hollow walls and ceilings. Therefore, the so-called substrate is usually plasterboard, of which there are two thicknesses: 9.5 mm and 12.5 mm. However, they may be used for fixings into any other relatively thin lining material. With the introduction of dry-lined walls more than three decades ago – which have now mostly superseded solid-plastered walls – there has been more demand for reliable fixings and most manufacturers have risen to this with a wide variety of different devices now on the market. Because of this large selection, only five types that seem to be popular, will be covered here:

- (a) **Plasplugs Super Toggle Cavity Anchors** are made of nylon and are self-adjusting to lining thicknesses of between 7 and 14 mm. The nylon anchor-arms are drawn up to the back of the plasterboard as the screw is tightened. A 10 mm Ø hole is required to accommodate the plug, which takes a No. 4 gauge screw.
- (b) **Plasplugs Plasterboard Heavy-Duty Fixings** provide an extra strong expanding grip with their multiple barbs and screw-tensioning calliper anchors. These are squeezed in when inserting the

plug through a 10 mm Ø hole and – like the above plug – are drawn up to the back of the plasterboard as the No. 4 gauge screw is tightened.
- (c) **Fischer Plasterboard Plugs** are made of nylon to provide a short expansion zone to take high pull-out loads. There are two sizes; one requires an 8 mm Ø hole and takes a No. 4 gauge, single-thread screw, the other requires a 10 mm Ø hole and takes a No. 5 gauge, single-thread screw.
- (d) **Nylon and Metal (zinc-plated) Easi-Drivers** provide rapid and effective fixings into plasterboard, without the need for drilling separate location holes. The screw-fixing itself acts as a pilot drill, cutting-thread and securing-thread. They are driven easily into the plasterboard by using a cross-head screwdriver or a powered Pozidriv bit. The metal Drivers are more suitable for heavier loads than the nylon Drivers. They are supplied with screws, but their limitation, in carpentry terms, is that the maximum thickness of material to be fixed is recommended by the manufacturers to be 10 mm for nylon Drivers and 15 mm for metal Drivers.
- (e) **Rawlplug 'Uno' Plugs** have a universal function. Their patented zigzag expansion pattern allows them to form a secure fixing into hollow walls, or any other type of wall, ceiling or floor. They are suitable for all light to medium weight applications and are colour-coded in Yellow, Red and Brown for screw gauges ranging from 3 mm to 6 mm, with plug lengths of 24, 28, 30 and 32 mm. Corresponding drill-holes are 5, 6, 7 and 8 mm diameter.

Two final points on this subject are: (1) it should be remembered that when fixing fixtures to a plaster-boarded, stud-partition wall, or a plaster-boarded ceiling, there are concealed studs or joists which would provide excellent fixings – if the fixture happened to coincide with them, or if the fixture's position was not critical and could be altered to suit them; (2) Nylon-sleeved Frame-fix and Window-fix screws make a very good alternative heavy-duty fixing for use on dry-lined block walls. This is because they can bridge the small cavity behind the plasterboard and take their anchorage from being drilled and fixed into the blockwork. This feature is graphically illustrated in Figure 4.5(b) in Chapter 4 of this book.

4

Making and Fixing Shelving Arrangements

4.1 INTRODUCTION

Although designing shelving arrangements is not usually a carpenter's concern, a love of the craft and an acquired understanding of wood and wood-derived products can inspire many of us to create our own designs. However, even if our designs are not revolutionary, we do need to consider shelf thicknesses in relation to the weight of the books or other items to be placed upon them; and in so doing, we become involved with the principles of structural mechanics. This is because the shelves, usually required aesthetically to be kept thin in depth in relation to their span, are likely to bend (deflect) in their middle area under their load. This is rarely of concern regarding the actual shelves reaching a point of collapse, but it must be realized that bending is the most severe form of stress in a shelf. And to avoid the unacceptable appearance of excessive deflection, we can work out the required mechanical thickness of a shelf by applying simple or complex mathematical formulae – in a similar way to working out the depth of timber beams or floor joists. Not being a structural engineer myself, but having designed and built many successful shelf arrangements over the years, the following notes and illustrations are based on practical intelligence and common-sense mechanics related to *simple* formulas.

4.2 SHELF MATERIAL

Although this chapter will refer to engineered-wood products such as plywood and blockboard, and wood-derived products such as chipboard and MDF, etc., which could be used for shelves, the calculations given here for the critical shelf-thickness (depth) are mainly based on the use of good quality, solid redwood (pine) or a hardwood species of at least equal density and strength.

4.3 BEAMS OR JOISTS LIKENED TO SHELVES

Because the structural stability of suspended wooden beams or joists is mostly dependant on their cross-sectional depth, suspended shelves – by comparison – are like ignorantly positioned, suspended wooden beams of rectangular cross-section, lying wrongly on their widest sides – and yet they have to support so-called *live loads* (be it only of books, etc.) similar to floor joists that carry a *live load* of people and furniture, etc. As beams or joists, therefore, shelves are structurally weak and – to avoid excessive deflection under load – must be limited in span in relation to their thickness and their intended load. All relatively thin, loaded wooden shelves will suffer a downwards deflection to some extent, but, with calculated design, this deflection should not exceed (in my opinion) 1 in 300 (i.e. a shelf with a 900 mm span should not have more than a 3 mm mid-area deflection under load).

4.4 UNDERSTANDING BASIC MECHANICS

Figures 4.1(a)(b)(c): The fibres of loaded beams or shelves, as illustrated below at 1(a) and (b), are subject to a crushing effect referred to as *compressive stress (compression)* across their horizontal width below the topside – and to being torn apart (stretched) by *tensile stress (tension)* across their horizontal width above the underside. Being at maximum stress on the top and bottom outer-surfaces, where these two

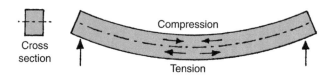

Figure 4.1 (a) Exaggerated deflection of a simply supported beam, with arrowed indication of the compressive and tensile stresses; the broken line through the centre represents the neutral layer. The extreme vertical arrows depict the bearing points

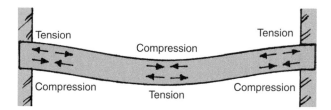

Figure 4.1 (b) Exaggerated serpentine-shape deflection highlighting the changes in the compressive and tensile stresses on a beam with built-in end-bearings

Figure 4.1 (c) A simply supported, overloaded beam displaying exaggerated deflection and a theoretical impression of the effects of *horizontal shear*

theoretical forces meet in the horizontal middle area of a beam or shelf, they neutralize and have no stress. This area is referred to as the *neutral layer*, or *neutral axis*. Because the compressive stress in the layer of wood-fibres above the neutral axis causes a limiting- or balancing-effect on the tensile stress in the layer of wood-fibres below the neutral axis, it follows that the greater the thickness of the wood-fibres above and below the neutral layer, the less bending (deflection) will occur – even though the weight of the beam or shelf will increase.

Another way to appreciate the validity of the limiting- or balancing-effect of the opposing compression and tension forces in a rectangular beam is to study the purposely exaggerated concaved beam-shape at 4.1(a) and question its ability to actually (physically) compress and *shorten* its length on top and stretch and *expand* its length on the bottom (as, geometrically, any such concentric curvature demands) without fracturing (rupturing) the beam on the underside. In theory, of course, an excessive dead load on a *simply-supported beam* (the ends of the beam resting

on supports, but not built-in) can also cause *horizontal shear*, as illustrated at 4.1(c). This causes a separation of the beam's horizontal fibres (to a greater or lesser extent), which tend to slide over each other as the top of the beam is shortened and the bottom lengthened by bending.

Understanding these simplified explanations of the theory of basic beam-mechanics should help in visualizing its modifying effect on the different shelf-supports described below. The supports are also referred to as *bearings* and – depending upon which of the three usual kinds are used – it has to be realized that each of them affects the degree of *stiffness* or *elasticity* of the shelf and its ability to resist deflection under load – as mentioned above.

4.5 SHELF SUPPORTS

4.5.1 Built-in End-bearings

Figure 4.2 (a) Part of a side cheek showing a stopped-housing and part of a shelf showing the recess-shouldered front-end positioned for assembly

Figure 4.2(a): There are traditionally three different ways to support wooden shelves – apart from modern innovations by manufacturers and individual designers. The first, illustrated at 4.2(a), is where the ends of shelves are housed (usually stop-housed) into the cheeks (sides) of a shelf unit by one-third the cheek's thickness – the housing-stop distance equals the thickness of the shelf away from the edge. Note that if this is done neatly – i.e. the shelves fit tightly in the housings – you will achieve more mechanical support by having altered the tensile and compressive stresses, as indicated in Figure 4.1(b) above. Another advantage is that the restraint created by the housings will restrict warping or cupping of the shelves.

4.5.2 Unfixed End-bearings

Figures 4.2(b)(c)(d): The second traditional support, illustrated at 4.2(b), is where the shelves rest on – or can be fixed to – wooden end-bearers. However, it must be realized that shelves *simply supported* in this way (whether fixed or not) are entirely or mostly subject to full-span stresses of compression and tension, as described above and indicated in Figure 4.1(a).

Another type of bearing, illustrated below at (c) and (d), for simply supported shelves is available in the form of vertical metal-strips that are pierced with closely spaced, small slots. The slots support small, flat-lying metal lugs (referred to as *clips*) that act as interchangeable bracket-bearers, giving infinitely variable shelf-spacing options. The vertical strips – two normally required at each end of the shelves – are either of a shallow, inverted u-shaped surface-fixing type, or are composed of flat strips of metal that require housing into double-grooved, vertical channels to bring them flush with the surface of the bookcase cheeks. Note that the stepped inner groove accommodates the bracket-ends of the shelf-supporting clips. The flush-finishing type is more professional-looking – especially when used with hardwood.

An online search for these so-called *Tonk strips* at www.ironmongerydirect.com revealed that these traditional items of ironmongery are still available in both types and are described as *Tonk Flat Bookcase Strip* in lengths of 1829 × 19 mm and *Tonk Raised Bookcase Strip* in lengths of 1829 × 24 mm. They are available in a variety of finishes, namely zinc-plated, brown, electro brass, satin-nickel plated, polished solid brass and polished chrome (on brass). Note that the last two types are not available in the Raised Strip range. Four clips are required per shelf – and a specially designed router cutter is available for forming the stepped, vertical grooves in the cheeks.

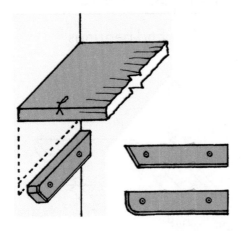

Figure 4.2 (b) Part of a simply supported shelf shown in a raised position to allow a view of the 45° splay-chamfered bearer fixed to the wall; two alternative end-shapes to edge-chamfered bearers are shown alongside

4.5.3 Indented and Mid-area Bearings

Figure 4.2(e): The third traditional form of support – used predominantly for storage shelves of short or long lengths – is right-angled brackets. Although these can be made of wood, they need a diagonal brace and – unless the brace is ornate – they tend to look bulky and ugly. Also, making small, wooden brackets is tedious and time-consuming. Therefore the cantilever-type metal brackets – especially the so-called *London-pattern* type, illustrated below – are quite commonly used. They can be screwed directly onto the wall, via their three fixing holes (two at the top, one at the bottom), but it is more practical for them to be fixed onto pre-positioned, plugged-and-screwed vertical wooden cleats. The cleats, usually made from

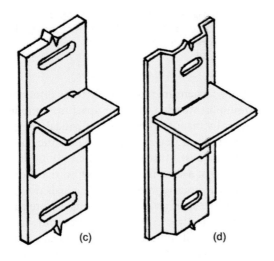

Figure 4.2 (c) Part isometric-view of a Tonk Flat Bookcase Strip and **(d)** a part view of a Tonk Raised Bookcase Strip; the clips shown are a heavy-duty type

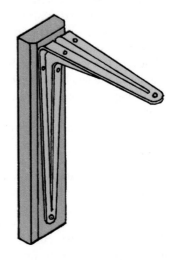

Figure 4.2 (e) London pattern, pressed-steel bracket fixed to vertical wooden cleat

ex. 50 × 25 mm softwood, should be neatly chamfered on the two face-side edges and bottom edge.

4.5.4 Spacing of Indented End-cleats and Brackets

Regardless of whether the ends of these shelves butt up against walls or not, the end-cleats and brackets are normally set in (indented) from the shelves' ends by a certain amount, thereby creating cantilevered ends. And if the amount of indent is calculated, rather than guessed at, the strength of the remaining shelf-lengths between the brackets will be significantly improved – as will the degree of deflection. The simple formula for a *uniformly distributed load* is:

Indent = 0.207 × Overall Length of shelf.

For example, a shelf with an overall length of 1500 mm would be 0.207 × 1500 = 310.5 (say 310 mm) indented bracket each end. Note, though, that the formula is for a theoretical *uniformly* distributed load – and yet in practical terms, the shelves are more likely to be *unevenly* loaded. I would recommend, therefore, that the formula for *unevenly distributed loads* should produce an *indent* reduced by about 25%. This would produce a formula of:

Indent = 0.155 × Overall Length of shelf.

Therefore, the same length shelf of 1500 mm would be 0.155 × 1500 = 232.5 (say 232 mm) indented bracket each end – instead of 310 mm for shelves with a uniformly distributed load. Note again though that the remaining 1036 mm between the brackets should be checked out against the shelf-span formula below, as certain lengths of shelves will obviously require intermediate brackets.

4.5.5 Back-edge Bearers

Figure 4.2(f): Note that shelves being supported by vertical cleats and brackets can be additionally

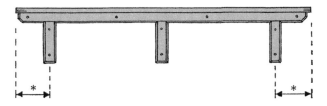

Figure 4.2 (f) Front elevation of a mid-area shelf fixed to a back-edge bearer seated on vertical cleats. The three brackets are omitted. (*) The asterisks highlight the calculated indents for the brackets at each end. Note that – because of the chamfer – the cleats should be scribed or housed-in to the underside of the bearer.

strengthened by pre-fixing horizontal shelf-length bearers to the wall, directly below the shelves and seated on the cleat-tops, as illustrated below. The back-edge bearers would be of the same section and chamfered detail as the cleats and the shelves should be nailed or screwed to them – especially if using relatively weak shelving material, such as chipboard, Contiboard, OSB (oriented strand board) or MDF (medium-density fibreboard).

4.5.6 Splayed Bearings to Shelf-ends and Sides

Figures 4.2(g)(h): The traditional method for joining solid- or slatted-shelves at right-angles to each other – as in a small storage cupboard or (in the case of slatted shelves) in an airing cupboard – is to splay the ends to an angle of about 60° to 70° and form splayed housings in the shelf-sides to accommodate them. These splayed joints (or bearings) are usually dry-fixed and panel-pinned, but could be glued as well.

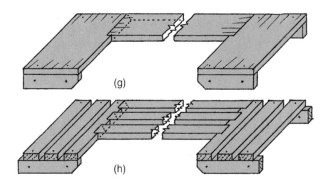

Figure 4.2 (g) Three-sided shelf arrangement in a cupboard, with a splay-ended shelf between the outer shelves; note that the back-edge bearer (only partly shown) should be taken right across the wall, under the three shelves; (h) Slatted shelves

4.6 LOADS ON SHELVES LIKENED TO LOADS ON FLOOR JOISTS

The so-called *live-loads* on floor joists in residential properties refer to the *aggregate* weight of furniture, fittings, etc., and the number of people who are likely to use the dwelling – and the formulae to determine the sectional size of the joists required over certain spans incorporate tested margins of safety and do not have to take into account the *actual* weight of the furniture or fittings, the *actual* number of occupants and visitors likely to load the floors, or their *actual* en masse bodyweight related to anorexia or obesity. Neither, therefore, does the formula given below for *residential (non-commercial or industrial)* shelf-thicknesses related to span have to take into account the *actual* size and weight of the books, magazines and other items to be placed upon them.

First, though, having likened the shelves to floor joists and because the formula for the shelves evolved from the joist-formula, I give below an example of the simple rule-of-thumb formula that exists for depth of *floor-joists* related to span. The formula was traditionally expressed in feet and inches, but was metricated by technical authors in the 1970s.

4.6.1 Depth of Floor-joists Related to Span

The formula for this is easily remembered as: *span over two, plus two*, or:

$$\frac{\text{Span}}{2} + 2$$

This is expressed literally as: Depth of joists in centimetres equals span in decimetres divided by two, plus two. Or, as an equation as:

$$\frac{\text{Depth of joists}}{\text{in centimetres}} = \frac{\text{span in decimeters}}{2} + 2$$

For example, with a joist-span of 4 metres (40 decimetres), the equation would be:

$$\frac{\text{Depth of joists}}{\text{in centimetres}} = \frac{40}{2} + 2 = 22 \text{ cm (220 mm)}$$

Note that The Building Regulations' Approved Document A1/2 Floor Joist Tables for a small house recommends *195 mm × 50 mm C16* structurally graded joists for a 4 m clear span, which highlights a 25 mm difference between the more precise Joist-Tables and the rule-of-thumb formula. This is because the latter has to slightly overcompensate to make up for its mathematical elementariness.

4.6.2 Thickness of Shelves Related to Span

The formula for this is easily remembered as: *span over fifty, plus two*, or:

$$\frac{\text{Span}}{50} + 2$$

This is expressed literally as: Thickness of shelves in millimetres equals span in millimetres divided by fifty, plus two. Or, as an equation as:

$$\frac{\text{Thickness of shelf}}{\text{in millimetres}} = \frac{\text{span in millimetres}}{50} + 2$$

For example, with a shelf-span of 900 mm, the equation would be:

$$\text{Thickness of shelf} = \frac{900}{50} + 2 = 20 \text{ mm}$$

Two further examples, with shelf-spans of 1200 mm and 450 mm would be:

$$\text{Thickness of shelf} = \frac{1200}{50} + 2 = 26 \text{ mm}$$

$$\text{Thickness of shelf} = \frac{450}{50} + 2 = 11 \text{ mm}$$

Note that calculated shelf-thicknesses usually have to be slightly increased or decreased by a millimetre or two to suit commercial timber sizes and manufactured-board thicknesses; but if a heavy load of large hardback books or glossy magazines and the like are anticipated for the shelves, decreases should be avoided.

4.7 BLOCKBOARD OR PLYWOOD AS SHELVES

(i)   (j)

Figure 4.2 (i) 9 mm-thick, lipped face-edge of 18 mm-thick blockboard shelf; and (j) Similar lipping bonded to face-edge of 18 mm-thick plywood shelf. Note the sharp arrises removed from the lipped edges

Figures 4.2(i)(j): Although I devised the shelf formula for good quality redwood (pine) and hardwood shelves, I find that it also works well on such material as blockboard (when the core strips run parallel to the span) and plywood (when the face grain of the outer plies runs parallel to the span). Also, these materials have the advantage of being more stable and will not suffer from warping or cupping. Note, though, that as illustrated at (i) and (j), shelves made from these materials need to be *lipped* on their face-edges to improve their finished appearance. Ideally, the lipping should be glued on without pinning, but can be glued and pinned with, say, 25 mm panel pins, punched in to allow for fillings.

4.8 MID-AREA SHELF-SUPPORTS

If a shelf or an arrangement of shelves has multiple bearing points or brackets, i.e. one or more inner supports in addition to a support at each end, as in Figure 4.2(f) above and Figure 4.3(b) below, the shelving will take slightly more load. This is because of the cantilever effect acting on each side of the mid-area bearing points, related to compressive and tensile stresses explained at the start of the chapter. Alternatively, in these situations, the shelving could be of slightly less thickness than the calculated thickness determined by the actual span between each of the multiple supports. These strength-considerations also apply to bookcase shelves when their rear edges are held by pinning or screwing through built-in back panels.

4.9 LADDER-FRAME SHELF UNIT

Design was mentioned at the beginning of this chapter and is now the subject at the end of it. The illustration below shows part of a simply constructed ladder-frame shelf unit that I first designed and made over three decades ago. Similar units made in recent years have been only slightly modified.

4.9.1 Original Concept

Figure 4.3(a): Wall-to-wall shelving was needed in the recesses on each side of my family's living room chimney-breast, above newly built-in storage cupboards of about 900 mm height. Aesthetically, shelves supported by wooden end-bearers or indented brackets were not an option – and the other extreme of using solid bookcase-type cheeks was not quite what was wanted either. So, with simplicity and economy in mind, I thought of a ladder-frame design that could be fixed to each side-wall so that the *rungs* could support the shelf-ends.

4.9.2 Frame Material

The ladder-frame uprights are made from ex. 25 × 25 mm (usually reduced by timber merchants to a 20 × 20 mm par finish) redwood or hardwood,

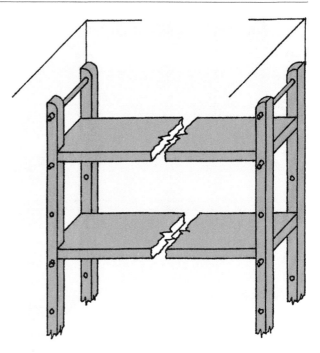

Figure 4.3 (a) Original concept of adjustable wall-to-wall shelving. Over the years this design – still using the same sectional-sized timber – has evolved into fixed, floor-standing units of varying heights (mostly floor-to-near-ceiling or coving), usually with one or two mid-area ladder-frames, depending on the overall unit width required in relation to the span calculations

which must be fairly straight-grained, straight and knot-free. If either one or both of the end ladder-frames are not up against a return wall, short lengths of the frame material are also required for modifications to the shelf-ends and rear-edges, as detailed below. These modifications create more connectivity between the fixed frames and the unfixed shelves. The horizontal rungs are made from 9 or 10 mm Ø (diameter) dowel rod – which is commercially available in redwood or light-coloured hardwood in standard lengths of 2.4 m.

4.9.3 Construction of the Ladder-frames

Figures 4.3(a)(b)(c)(d)(e)(f)(g): As illustrated, each pair of uprights are drilled right through their faces with 9 or 10 mm Ø dowel-holes, equally spaced at 75 mm centres throughout the entire height from bottom- to top-shelf dowel-bearers. Prior to drilling, though, the bottom parts of the uprights against the wall have to be scribed or modified to accommodate the projecting skirting member; and the top parts can be prepared for alternative finishes at the same time. Both of these items are detailed separately in Figures 4.3(c) to (g).

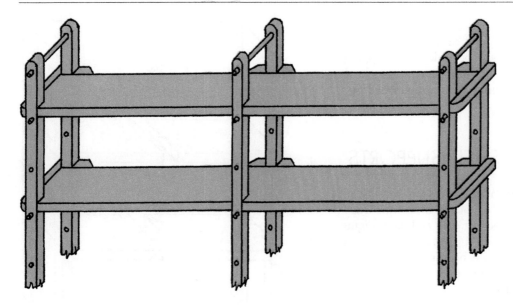

Figure 4.3 (b) The upper part of a floor-standing, open-ended shelf unit with one mid-area ladder-frame. (No additional dowel-rungs shown)

Each pair of uprights becomes a complete frame by inserting three end-prepared (bullnose-rounded) projecting dowel-rungs in position and fixing them through the frame-sides with 18 mm sherardized panel pins. The three positions for the fixed dowels are the extreme top, the extreme bottom and a mid- or near mid-area location.

4.9.4 Dowel-rung Details

The length of all dowel-rungs equals the overall width of a frame (dependent upon the shelf-width decided upon), plus a small front dowel-projection of 6 mm, plus a 2 mm tolerance-allowance. After cutting the dowels to length with a fine saw, *all* ends must be rounded or chamfered to a good visual finish. This can be done with a rasp and glasspaper or – if available – by rotating each dowel-end safely on the bed of a fixed disc-sander machine. The essential number of rungs required is one for every shelf bearing – the idea being that if one or more of the shelves ever require repositioning, the rungs can be slid out and replaced in a lower or higher position. And the maximum optional number of rungs is one for every pair of frame-holes – thereby using them as a visual feature. I usually allow at least one additional rung per frame, per shelf, as – with floor-standing, open-ended frames – these retain and support the end books.

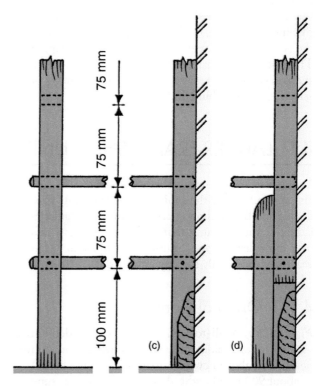

Figure 4.3 (c)(d) Ladder-frame upright scribed (shaped) to override the skirting; and **(d)** the alternative treatment of modifying the upright to step over the skirting. Note the pinning of the bottom dowel-rung

4.9.5 Skirting Allowances

Figures 4.3(c)(d): As illustrated below, the ex. 25 × 25 mm frame member against the wall will usually require modifying to avoid clashing with a projecting skirting board. If the skirting is of a modern, simple design – usually with a 16 × 70 mm finish – the base of the upright can be scribed to override it, as at (c), or it can be stepped over with a short, additional piece of framing glued to the upright, as at (d).

4.9.6 Alternative Finish to Frame-tops

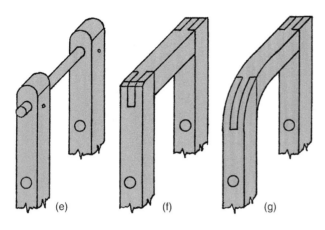

Figures 4.3 (e)(f)(g) Alternative top-members fixed to ladder-frame uprights

Figures 4.3(e)(f)(g): One of the simplest ways to finish the frame-tops is with a semi-circular shape on the end of each upright, as at (e), with the fixed dowel-rung centred at 20 mm below. Alternatively, a piece of framing can be used as a top member, bridle-jointed and glued to both uprights, as at (f). With this finish, instead of three fixed dowel-rungs, only two are required. Finally – although other ideas could be used – a good visual effect can be achieved by bridle-jointing a quadrant-shaped diagonal to each upright, as at (g). Again, only two fixed dowel-rungs per frame are required.

4.9.7 Shelf Details

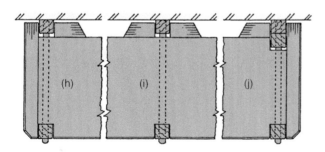

Figures 4.3 (h)(i)(j) Horizontal section through a ladder-framed unit showing: (h) Common shelf-end detail; (i) Common mid-shelf detail; (j) Extra recessed-notch detail to bottom shelf wherever the frame is stepped over the skirting board

Figures 4.3(h)(i)(j)(k)(l)(m): As previously mentioned, if the end-frames are not fixed to return walls, more connectivity has to be created between the frames and the shelves. As illustrated above at (h) to (k), this is achieved by gluing and pinning (or screwing)

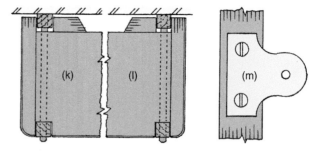

Figure 4.3 (k) and (l) Horizontal section showing bullnose-rounded side fillets on both ends of a framed shelf, as an alternative to the 45° bevel-ended fillets shown at (h) and (j) above; (m) Part rear-elevation of a ladder-frame upright and a brass picture/mirror plate (facing into the shelves) for wall-fixings; three plates per upright are recommended

Figure 4.4 (a) Part of a newly built bookcase

shelf-thickness fillets to the end-and rear-edges of the shelves. Note that the bottom shelf requires extra notched recesses, as at (j), if the stepped framing at 4.3(d) is used. All notched recesses should be 0.5 to 1 mm wider than the 20 mm wide frame uprights to allow easy positioning and – if required – repositioning of shelves.

Figure 4.4 (b) Close-up of the stepped frame over the skirting and the additional notched recess in the rear edge of the bottom shelf

Figure 4.4 (d) An end-view of a book-laden shelf arrangement comprising three ladder-frames

Figure 4.4 (c) One of the picture/mirror-plate fixings facing into the shelves

Figure 4.4 (e) A wall-to-wall softwood-stained and sealed book-laden shelf unit (designed and made in 1994), displaying quadrant-shaped frame-tops, as detailed at 4.3(g) above

4.9.8 Dry-lined Wall Fixings for Independent Shelving Arrangements

Figures 4.5(a)(b): Fixing horizontal timber shelf-bearers and vertical bracket-cleats, etc. to traditional solid-plastered walls is simply achieved by drilling

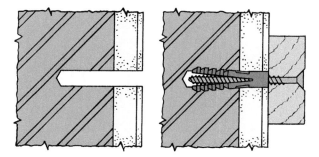

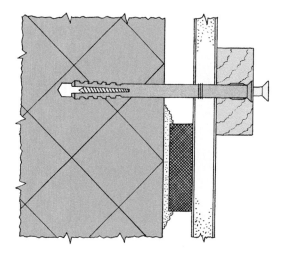

Figure 4.5 (a) A drilled-and-plugged solid-plastered wall, showing the exaggerated effect of the plug's expansion to highlight good anchorage beyond the plastered surface

and plugging the plastered brick- or block-work, as covered in Chapter 3 and shown here in Figure 4.5(a). But, where the unplastered, solid brick- or block-walls have been *dry-lined* – covered with sheets of plasterboard, which are fixed by a so-called *dot–and-dab* technique that leaves a cavity between the dry-lining and the solid wall – problems of fixing shelves, etc. to the walls are created. Because of this, only *patent cavity fixings* seem appropriate, but these are not strong enough to support shelves or their loads. Therefore, the technique that I recommend, as illustrated in Figure 4.5(b), is to override the cavity by using nylon-sleeved Frame-fix screws, which are available in various lengths.

Note that the 'dot-and-dab' technique of fixing the plasterboards involves a number of small, square pieces of bitumen-impregnated fibreboard (referred to

Figure 4.5 (b) A drilled-and-plugged dry-lined wall, showing the exaggerated effect of the nylon-sleeved Frame-fix screw's expansion to highlight good anchorage beyond the cavity and the plasterboard; note that the fibreboard 'dot', shown here as being plaster-bedded in the cavity just below the Frame-fix screw, is for illustration purposes – the unseen dots do not necessarily need to be near the fixing-points

as *dots*) being bedded strategically onto the solid wall with *dabs* of plaster (as illustrated in Figure 4.5(b)), to 'true-up' the vertical- and horizontal-alignment of the structural wall-surface before larger dabs of plaster are applied to the wall (between the dabs) to receive and adhere the plasterboard-linings, tight up against the cavity-creating dots.

5

Making Site-Equipment Items

5.1 SAW STOOL

5.1.1 Introduction

Apart from metal and plastic stools now available, a wooden saw stool, sometimes called a saw horse or trestle, is still preferred – to my certain knowledge – by a number of traditional tradesmen. And, because the stool's simple geometrical setting-out is a good introduction to roofing-bevels and rafter-lengths, the making of them is still taught in some vocational colleges. There are a few variations in design, but the one shown in Figure 5.1 is most common. The length and height can vary, although the height should not be less than that shown, otherwise on any hand rip-sawing operations (still done occasionally on short lengths of timber) – when the saw should be at a steep angle of about 60–70° – the end (toe) of the saw may hit the floor.

5.1.2 Material Required

The material used is usually softwood and can be a sawn finish or planed all round (par). The latter reduces the risk of picking up splinters when handling the stool. Material sizes also vary, often according to what may or may not be available on site, or, for design reasons, in consideration of the weight factor of the stool.

Typical sizes for a sturdy stool are given in the illustrations, showing the top as ex. 100 × 50 mm, legs ex. 75 × 50 mm, end-cleats 6, 9 or 12 mm plywood, MDF or OSB (oriented strand board).

5.1.3 Angles of Legs

The angles of the legs are not critical to a degree and are usually based on the safe angle-of-lean used on ladders, which is to a ratio of 4 in 1 (4:1). This refers to a slope or gradient measured by a vertical rise of four units over a horizontal distance of one unit. (This ratio works out to give approximately 76°.)

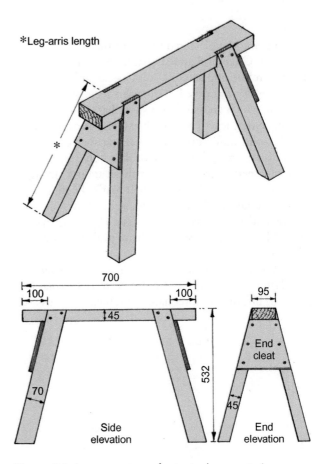

Figure 5.1 Isometric view of a typical carpenter's saw stool

5.1.4 The Need For Geometry

By following very basic criteria and guesswork, saw stools are often roughly made without any regard for the simple geometry involved. This is partly to save time, of course, but also reflects a lack of knowledge of the subject. Making a saw stool should be a quick and simple operation, joinery and geometry-wise (it can be done in about one hour).

5.1.5 Length of the Legs

Before metrication of measurement in industry, the height of saw stools ranged between 21 and 24 in. The approximate metric equivalent is 532 and 610 mm. The lesser height is used here for preference. Once the vertical height has been decided (and because the legs are at an angle, thereby reducing that chosen height), the length of the legs has to be worked out. Basically, without adding waste-allowances, this is a measurement along the outside corner of the leg (indicated in Figure 5.1 with an asterisk) known as the *leg-arris length (arris* is a French/Latin-derived word used widely in the trade to define sharp, external angles).

5.1.6 Determining the Leg-arris Length

The leg-arris length can be worked out in three different ways. These different methods are (A) by practical, on-the-job geometry, (B) by drawing-board geometry and (C) by a method of calculation. Only method (A) is used here.

Practical On-the-job Geometry

Figure 5.2: Finding the required leg-arris length by this method can initially be *thought of* as leaning a piece of timber (like a saw-stool leg of, say, 600 mm length) against a wall at a 1 in 4 ladder-angle, to realize that you have formed a right-angled triangle with a measurable base-measurement from the wall, a vertical height against the wall, and a hypotenuse opposite this right angle, in the form of the piece of leg-timber. Understanding this is the key to the simple geometry, but it needs to be developed. It would not give us the required height of the stool (because we leant the visualized 600 mm leg away from the wall, to a 1 in 4 angle, thereby reducing its height). So, to clarify our understanding so far, Figure 5.2 illustrates the triangles we need to deal with (be it just one of them) and shows that by dividing the required height by four (vh ÷ 4), the required base measurement is found to be 133 mm. But, as indicated in Figure 5.3, it is not the base-measurement that we *finally* need. We need the *diagonal measurement* of the square formed by them. And once this is known, the length of the timber on the hypotenuse (the *leg-arris length*) can be worked out.

Using Base Measurements

Figure 5.3: Now that the base measurement of the side and end elevation triangles is known to be 133 mm, this information can be used to find the unknown base measurement of the leg-arris triangle. Because the leg is leaning at the same angle in both elevations, the

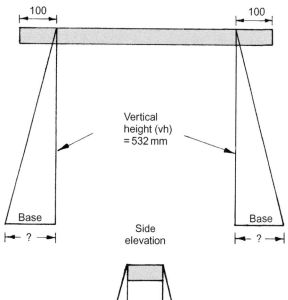

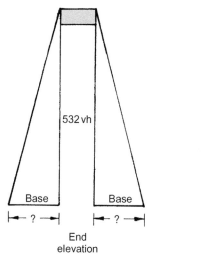

Figure 5.2 With 4 in 1 leg-angle, the base equals vh ÷ 4 = 532 ÷ 4 = 133 (base measurement = 133 mm)

leg-arris base must be a 45° diagonal within a square of 133 × 133 mm.

Finding the Leg-arris Length by Practical Geometry

Figure 5.4: All that needs to be done now, therefore, is to form a true square with these measurements, draw a diagonal line from corner to corner and measure its length. The product is the base measurement of the leg-arris triangle. This measurement is then applied to the base of another right-angled setting out, the stool's vertical height is added, and the resultant diagonal on the hypotenuse measured to produce the true leg-arris length.

Two practical ways of doing this are to use a steel roofing square (now called a metric rafter square by manufacturers), or to use the right-angled corner of a piece of hardboard or plywood.

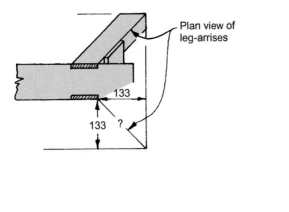

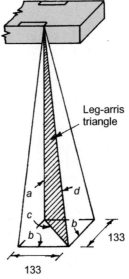

Figure 5.3 Base measurements, 133 × 133 mm, forming a square, give required diagonal base-measurement of leg-arris triangle *c*

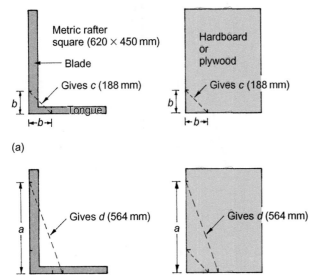

Figure 5.4 (a) Set *b* on each side and measure the diagonal *c*; (b) Set *c* and *a* on each side to find length of *d*

As illustrated, the base measurement *b* of 133 mm is set on each side of the angle to enable the diagonal *c*, the base of the leg-arris triangle, to be measured. This measurement of 188 mm forms the base of another setting out on the roofing square or hardboard, in relation to the vertical height of the stool *a*, of 532 mm, being placed on the opposite side of the angle. The diagonal *d* is then measured to produce the true leg-arris length of (to the nearest millimetre) 564 mm.

5.1.7 Adding Cutting Tolerances

Now that the true leg-arris length has been determined, at least another, say 35 mm at the top, plus say 11 mm

at the bottom, cutting tolerance has to be added to the leg as an allowance for the angled setting out involved. The length of legs, therefore, equals the true leg-arris length, 564 mm, plus 46 mm tolerance, equals 610 mm.

5.1.8 Setting Up the Bevel(s)

Figure 5.5: The legs and stool top can now be cut to length, ready for marking out. The marking out can either be done with a carpenter's bevel (or bevels), or with a purpose-made template. As illustrated, the 4 in 1/1 in 4 angles are set out against the square corner of a piece of hardboard, plywood or MDF board, to a selected size. 300 mm × 75 mm is advisable as a minimum ratio if the setting out is to be cut off and used as a template in itself. If carpenters' bevels are to be used, set up one at 4 in 1 and the other at 1 in 4, as illustrated. If only one bevel is available, set and use at the 4 in 1 (76°) angle before resetting for use at the 1 in 4 (14°) angle.

5.1.9 Alternative Setting Out

Figure 5.6: This shows an alternative way of setting out the two required angles, to enable a carpenter's bevel to be set up. First, mark a square line from point a on the face of a straightedge or on the underside-face of the stool-top material. Then, from point a, mark point b at 4 cm (40 mm) and point c at 1 cm (10 mm). Draw line c through b to establish the first required angle of 4 in 1 (76°). From point a, mark point d at 1 cm and point e from either side of a, at 4 cm. Draw line e through d to establish the second angle of 1 in 4 (14°). The advantage of setting out on the underside of the stool's top, is that the setting out will remain visible for years, ready at any time for making the next stool.

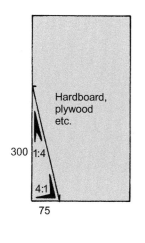
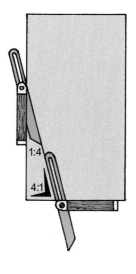
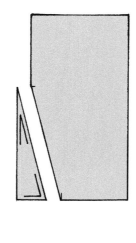

Figure 5.5 Mark 4 in 1/1 in 4 angle, set up bevels, or cut out as template

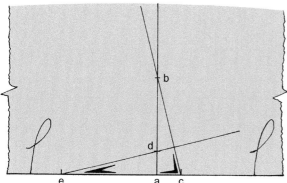

Figure 5.6 Alternative setting out

5.1.10 Marking Out the Legs

Figure 5.7(a): Before attempting to mark out the birdsmouth cuts, it is essential to identify and mark the starting point on each leg to avoid confusion. As illustrated, this point is on each inside-leg edge, about 6 mm down, and relates to the uppermost point of the compound splay cut in relation to the top surface of the stool. This mark is referred to as zero point.

5.1.11 Two-slopes-down, Two-slopes-up Formula

Figure 5.7(b): From zero point, the 4 in 1 angle always slopes down on edge A, turns the corner and, likewise, slopes down on face-side B, turns again and this time slopes up on edge C, and up on back-face D, to meet the first point (zero) on face-edge A. So, on the four sides of each leg, the marking sequence leading from one line to the other, is two slopes down, two slopes up, back to zero point on leg-edge A.

5.1.12 Waste Material

Note that in Figure 5.7(b) the shaded areas represent the timber on the waste side of the cut. Graphic

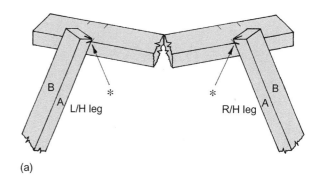

Figure 5.7 (a) Marking out of legs starts from the points marked*; **(b)** plan view of the marking sequence

demarcation such as this – but roughly marked – is often used by carpenters and joiners to reduce the risk of removing the wrong area of material.

5.1.13 Marking Zero-point Arrises

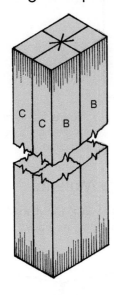

Figure 5.8 Marking zero-point arrises

The zero-point arris can be quickly determined on each leg by placing the four legs together, with the inside-leg edges and back-faces (A and D) touching, and by marking the top corner of each leg on the middle intersection, as illustrated in Figure 5.8.

5.1.14 Use of Template – First Stage

Figure 5.9: By using the template positioned on its 4 in 1 (76°) angle, surfaces A, B, C and D of the leg are first marked as illustrated here and as described in Figure 5.7(b).

5.1.15 Use of Template – Second Stage

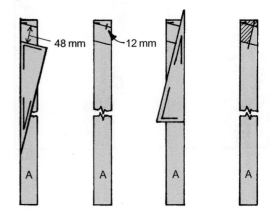

Figure 5.10 Marking out of the legs – second stage

Figure 5.10: Returning to surface A, the template is again positioned and marked on the 4 in 1 angle, set at 48 mm down, measured at right-angles to the first line. This represents the stool-top thickness, 45 mm, plus 3 mm to offset the fact that the stool-top thickness registers geometrically at a greater depth on the 4 in 1 angled legs. Next, measure 12 mm in on the top line to represent the beak of the birdsmouth-cut and use the 1 in 4 (14°) angle of the template to mark the line as shown.

5.1.16 Use of Template – Third Stage

Figure 5.11: To complete the marking out of the first leg, the lower line on surface A, representing the stool-top thickness, is picked up on surface B by the template and marked parallel to the line above. It is picked up on surface C and again marked parallel to the line above, then the 1 in 4 (14°) angle is marked in at 12 mm to complete the beak of the birdsmouth

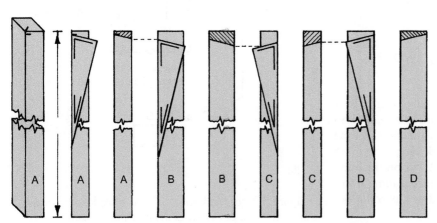

Figure 5.9 Marking out of the legs – first stage

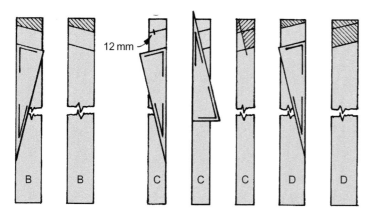

Figure 5.11 Marking out of the legs – third stage

(this birdsmouth-marking is on the opposite side to the one marked on surface A). Finally, the lip of the birdsmouth is picked up from C and marked on surface D, yet again parallel to the line above.

5.1.17 Complete the Marking Out

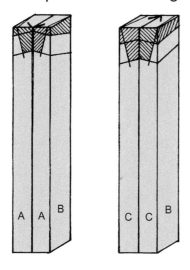

Figure 5.12 The four legs marked out

Figure 5.12: As illustrated, complete the marking out of the four legs, being careful to follow the marking sequence already explained, i.e. slope down from zero point on surface A of each leg to produce two opposite pairs of legs.

5.1.18 Cutting the Birdsmouths

Figure 5.13: By using a suitable handsaw (a sharp panel saw is recommended, as some hardpoint saws are ineffective for ripping), the four birdsmouths should be carefully cut to shape by first ripping down from the beak and then by cross-cutting the shoulder line on the lower lip. This can be done on a stool,

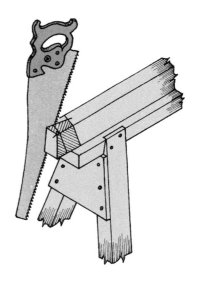

Figure 5.13 Cutting birdsmouths to shape

if one is available, or on any convenient cutting platform – failing access to such equipment as a a Workmate vice or a joinery-shop vice, etc. The cross-cutting of the shoulders can also – more accurately – be cut with a tenon saw.

5.1.19 Checking the Matched Pairs

Figure 5.14: After cutting, check that the legs match in pairs and hold each birdsmouth against the side of the stool-top material at any point to check the cut for squareness. If necessary, adjust by chisel-paring; inaccuracies, if any, are usually found on the inner surface of the end grain of the lower lip.

5.1.20 Marking the Leg-housings

Figure 5.15: Next, as illustrated, the birdsmouthed legs are held in position on the side of the stool top, 100 mm in from the ends, and the sides marked to indicate the housings.

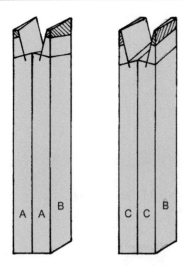

Figure 5.14 The completed, matched pairs

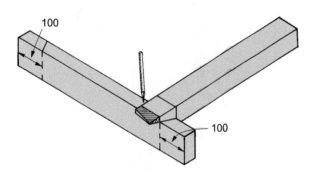

Figure 5.15 Mark leg-housings

5.1.21 Cutting the Leg-Housings

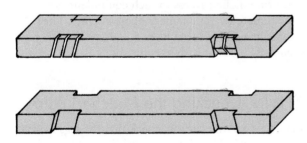

Figure 5.16 Gauge housings to depth, cut and pare out with wide bevel-edged chisel

Figure 5.16: When marked and squared onto the face-sides and underside, the housings are then gauged to 12 mm depth on both sides of the stool top, ready for housing. The angled sides of the housings can be cut with a tenon saw or panel saw and pared out with a wide bevel-edged chisel. The recommended chiselling technique, for safety and efficiency, is to pare down (known as vertical paring) – either by hand pressure

or with the aid of a mallet – onto a solid bench or boarded surface from alternate faces of the stool top.

5.1.22 Alternative Leg-fixings

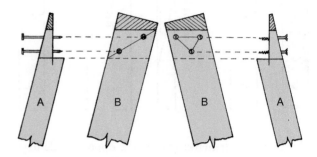

Figure 5.17 Alternative fixings

Figure 5.17: The stool top can now be set aside and the legs prepared for fixings – if screws are to be used. As illustrated, two or three fixings may be used and these are usually judged rather than marked for position. However, for strength, the fixings should neither be too near the edges nor too close together. Shank- or pilot-holes should be drilled and countersunk for, say, 5.0 gauge × 50 mm screws. If nails are used instead of screws, 63–75 mm round-head wire nails are preferred.

Take care when screwing or driving-in fixings, to ensure that they are parallel to the stool-top's surface, i.e. at a 4 in 1 angle to the surface of the legs.

5.1.23 End Cleats

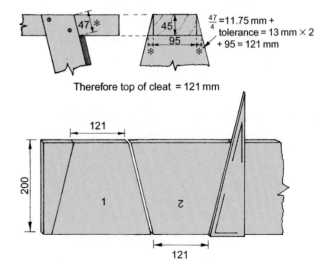

Figure 5.18 End cleats

Figure 5.18: To enable the assembly of the stool to flow without interruption, it will now be necessary to prepare the end cleats. By using the template as illustrated, these can be set out economically on board material or plywood and cut to shape with the panel or hardpoint saw. The size and shape of the cleats is critical, as the assembled legs are unlikely to assume the correct leg-spread on their own and will rely on the end cleats to correct and stabilize their posture.

A Common Error

It is a common error in stool-making to fix the legs and mark the shape of the cleats unscientifically from the shape of the distorted leg assembly. This is done by marking the outer shape of the stool's end onto the cleat material laid against it.

5.1.24 Working Out the Cleat Size

The setting out of the cleats, therefore, should be marked from the template (or bevel) in relation to a measurement at the top of the cleat. This measurement is made up by the width of the stool top and the visible base of the birdsmouth beak on each side. This can either be measured in position or determined by dividing the 4 in 1 diagonal thickness of the stool top by four. That is, as illustrated in Figure 5.18, $47 \div 4 = 11.75\,\text{mm} \times 2 = 23.5\,\text{mm}$, plus width of stool top (95 mm) giving 118.5 mm. Add 2.5 mm as a working tolerance and for final cleaning up, therefore, top of cleat equals 121 mm.

5.1.25 Fixing the Legs

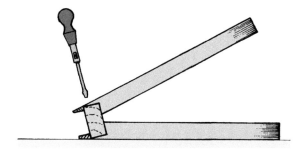

Figure 5.19 Legs screwed into housings

Figure 5.19: Now ready for assembly, the legs are fitted and screwed or nailed into the housings, making sure that the lower lip of the birdsmouth cut fits tightly to the underside of the stool top.

5.1.26 Fixing the Cleats

Figure 5.20: Next, the top edges of the cleats are planed off to a 4 in 1 angle, to fit the underside of

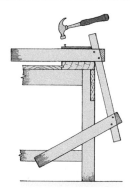

Figure 5.20 Fix plywood cleats with 50 mm round wire nails or 4.0 gauge × 40 mm CSK screws

the stool top, then the cleats are nailed or screwed into position, using 50 mm round-head wire nails or 4.0 gauge × 40 mm screws. During the fixing operation, to avoid weakening the leg connections, the stool should be supported on a bench, as illustrated, or any other suspended platform.

5.1.27 Removing the Ears and Marking the Leg-waste

Figure 5.21: Next, the projecting ears are cut off from the legs, with a slight allowance, say 1 mm, left for planing to an even finish with the stool top. After cleaning-up the top to a flat finish, measure down each leg and mark the leg-arris length (previously determined) to establish the stool's height. Join these marks with a straightedge and mark the line of feet on the ends and sides of the stool – or, if preferred, use the template or bevel instead of the straightedge.

5.1.28 Removing the Leg-waste

Figure 5.22: Finally, lay the stool on its alternate sides, and alternately end for end, up against a wall, bench, etc., and carefully cut off the waste from the legs with a panel or hardpoint saw. Clean off waste material from the sides of cleats and clean off any splintering arrises to finish.

Figure 5.23: If it is discovered that the stool is uneven and wobbly, turn it upside down and, as illustrated, place *winding sticks* (true and parallel miniature straightedges for checking twisted material) on the feet at each end and sight across for alignment. If in line, then the floor is uneven; if out of true, plane fractions off the two high legs and re-sight and re-plane, if necessary, until the legs are even.

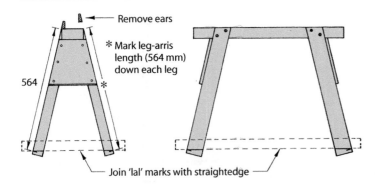

Remove ears

* Mark leg-arris length (564 mm) down each leg

564

Join 'lal' marks with straightedge

Figure 5.21 Join 'lal' marks with straightedge (or use template or bevel) to establish line of feet

Figure 5.22 Lay the stool on its side, end against the wall, and remove leg-waste

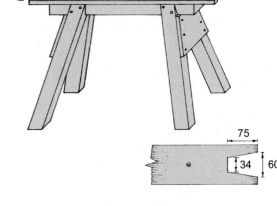

Stool-top length +100 mm each end

95

75

34   60

Figure 5.24 Optional vee-ended top

Uneven legs

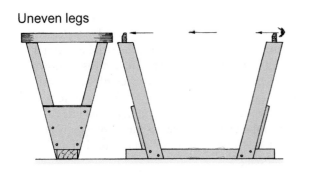

Figure 5.23 Using winding sticks on uneven legs

5.1.29 The Vee-ended Top

Figure 5.24: Sometimes, a vee-shape is cut in the end of a stool to facilitate the holding of doors-on-edge when they are being 'shot-in' (planed to fit a door-opening). To accommodate this, the legs ought to be set in 150 mm from each end of the stool top, instead of 100 mm.

This allows the top of the stool to touch the wall at one end, while the other holds the door in the vee cut without touching the bottom of the cleat. However, this does create a potentially dangerous stool if used as a 'hop-up', and you step along the stool towards an over-extended end.

Alternatively, as illustrated, a separate vee-ended board, of ex. 25 mm material, can easily be made and fitted to the top of the stool – and removed again, when not required.

5.2 HOP-UPS

5.2.1 Introduction

Even though 'wet plastering' seems to have disappeared since *drylining with plasterboard* became economically more popular, hop-ups may still be of use on Maintenance & Repair Works, involving replacement of solid plaster. They are very useful when rendering/floating and setting walls. In these situations, the plasterer uses a floating or skimming trowel, a *handboard* (hawk) with which to repeatedly carry the plaster to the wall, and a *board-and-stand*, from which to feed the material onto the hawk. Because the loaded hawk is in one hand and the trowel in the other, plasterers cannot easily – or safely – climb step-ladders if plastering to a height beyond their reach – although they might be able to use a pair of *plasterer's stilts*. So, I believe that there is still a need for an easily movable (repositioned by foot), easily ascendable (stair-like rise), and non-restrictive piece of equipment such as a hop-up.

5.2.2 Traditional Hop-up No. 1

Figure 5.25: As illustrated, this is simple in design and structure, built up of square-edged or tongued-and-grooved boarding, cleated and clench-nailed (protruding nail-points bent over for a stronger fixing). The 100 × 25 mm sawn boarding, shown here, is arranged to form a two-step hop-up with a step rise of 225 mm.

First, two side-frames are constructed of vertical boards and horizontal cleats clench-nailed together with 56 mm round-head wire nails or cut, clasp nails (clench-nailing, by bending protruding nail-ends over, is usually stronger than screw-fixings). These frames are then joined together by the tread boards being nailed into position and two cross-rails at low level, one at the front, the other at the back. These should be fixed with 63 mm round-head wire nails or oval nails. Finally, a diagonal brace of 50 × 25 or 75 × 25 mm section is also fixed at the back.

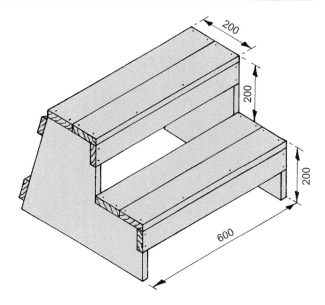

Figure 5.26 Traditional hop-up No. 2

bracing. On a hop-up of this width (600 mm) 18 mm WBP plywood could also be used as tread boards.

5.3 BOARD-AND-STAND

5.3.1 Introduction

A board-and-stand is also a piece of site equipment still required in a wet-plastering operation. It acts as a platform upon which the mixed plastering materials are deposited and are more easily trowel-fed onto the hawk placed under the board's edge. There are two types of stand that can be made, one being simplistic and rigid in construction, the other, folding. Both types have a similar, separate mortar board which lays in position on top without any attachment to the stand.

5.3.2 Rigid Stand

Figure 5.27: The height of this can vary from 675 to 750 mm and the width and depth is usually about 600 × 600 mm. This allows the board to overhang the stand to facilitate the loading of the hawk. The material used can be sawn or prepared softwood. First, 50 × 50 mm legs and 75 × 25 mm rails are cut to length to form two frames. On each frame, one rail is nailed to the top of the legs, the other is nailed about 100 mm clear of the bottom. Each frame is then braced diagonally with 50 × 25 mm or 75 × 25 mm bracing material and the two frames joined with the remaining cross-rails at top and bottom. Finally, the two remaining braces are fixed; 50–63 mm round-head wire nails

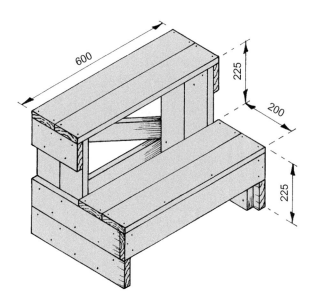

Figure 5.25 Traditional hop-up No. 1

5.2.3 Traditional Hop-up No. 2

Figure 5.26: The hop-up shown here is simplified by using 18 mm WBP plywood or Sterling board as side frames, boarded steps, cross-rails and diagonal

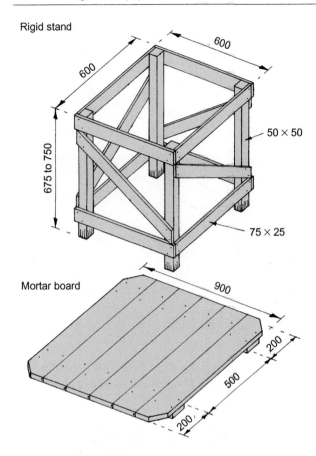

Figure 5.27 Plasterer's stand and board

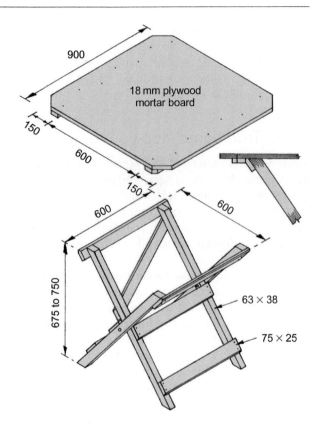

Figure 5.28 Folding stand

were traditionally used throughout, but screws would be better.

5.3.3 Folding Stand

Figure 5.28: When in the open position, the height, width and depth can be similar to those given for the rigid stand. The leg-length can be worked out as already explained for the saw stool, either by practical geometry or by calculation (by using, say, 750 mm as the vertical height and 600 mm as the base measurement and adding at least 70 mm allowance for splay cuts). Alternatively, the side elevation showing the crossed legs could easily and quickly be set out at half-scale or full-size, so that the exact details of legs are available.

Construction Details

The legs, of 63 × 38 mm or similar section, with central holes drilled for 9 mm diameter coach bolts, are fixed together with 75 × 25 mm rails to form two interlocking frames. The inner frame must be minus a tolerance allowance in width to enable it to fit easily into the outer frame. Ideally, the allowance, say

2–3 mm, should, on assembly, be taken up with a washer each side, between the frames.

Once the frames are bolted together, they are partly retained in the open position by the middle rails on each side, but mostly rely on the top rails fitting between the cleats of the mortar board. Accordingly, the cleats on the board must be so positioned as to leave a clear middle area of 600 mm. Before bolting the frames together, diagonal braces should be added, as indicated, for extra strength. Nails or screws can be used for fixing the rails and braces, although with this type of construction screws are advisable. Finally, the ends of the bolts, if protruding more than a reasonable amount, should be close-cut and burred over with a hammer.

5.3.4 Mortar Boards

These can be made from 900 mm to 1 m square, either from tongued-and-grooved boarding, cleated and clench-nailed together or from sheet materials such as resin-bonded plywood, MDF board, Sterling board or similar. As indicated, the sharp corners are usually removed. Whether made from sheet material or T&G boards, cleats, as indicated, will still be required to retain the board's position on the stand. Cleat material is usually 75 or 100 × 25 mm. Mortar boards used by bricklayers – often referred to as *spot boards* – only vary by their size, which is usually between 600 and 760 mm square.

5.4 STRAIGHTEDGES, CONCRETE-LEVELLING BOARDS AND PLUMB RULES

5.4.1 Straightedges

Figure 5.29: These are boards used by various tradespeople for setting out straight lines, checking surfaces for straightness, and levelling and plumbing with the addition of a spirit level. As the name implies, the essential feature of these boards is that the two edges are straight and parallel to each other. This can, of course, be done by hand-planing, but is best achieved on surface planer and thicknessing machines.

Lengths of straightedges vary from 1 to 2 m, with sectional sizes of ex. 100 × 25 mm to ex. 150 × 25 mm. Large straightedges of 3 m or more in

length are usually made from ex. 200 × 32 mm, or 225 × 38 mm boards. Holes of about 38 mm diameter were traditionally drilled through the large straightedges, at about 900 mm centres along their axis. This was done to establish the board visually as a proper straightedge and to discourage anybody on site from claiming it for other uses. Prepared softwood, of a reasonable quality, should be used and periodic checks for straightness are advisable.

5.4.2 Concrete-levelling Boards

Figure 5.30: These are usually about 5–6 m long, made from 225 × 38 mm sawn softwood. As illustrated, a handle arrangement is formed at each end to assist in easier control and movement of the board during the levelling operation. If a bay of concrete was to be laid within a side-shuttered area, or within the confines of a brick upstand, the levelling board would have to be long enough to rest on the shutters or brickwork each side. As the concrete was being placed, a person at each handle would tamp the concrete and pull the board, zig-zagging back and forth across the surface.

5.4.3 Traditional Plumb Rule

Figure 5.31: This traditional piece of equipment, in its original form as illustrated, is here for reference purposes only, as it is obsolete nowadays, even though plumb bobs themselves are still very useful in some carpentry operations. When used with a plumb rule,

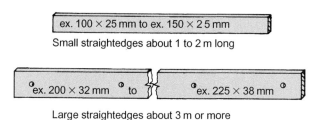

Small straightedges about 1 to 2 m long

Large straightedges about 3 m or more

Figure 5.29 Straightedges

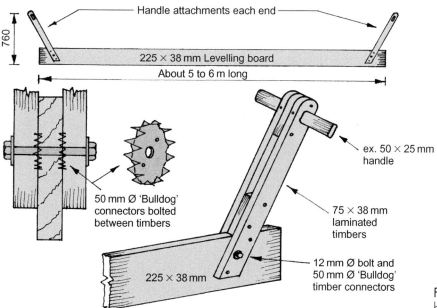

Figure 5.30 Concrete-levelling boards

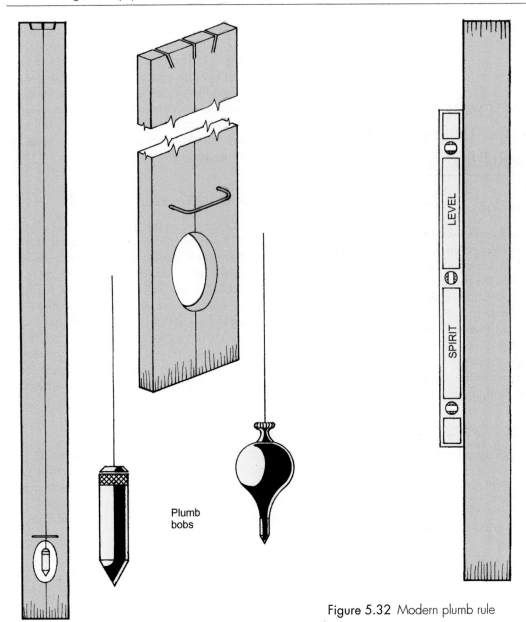

Plumb
bobs

Figure 5.31 Traditional plumb rule and plumb bobs

Figure 5.32 Modern plumb rule

waiting for the plumb bob to settle against the gauge line, to indicate plumbness, was a tedious and slow operation – although very accurate, if carefully done.

5.4.4 Modern Plumb Rule

Figure 5.32: Although extra-long spirit levels are now obtainable to replace plumb rules, the modern plumb rule consists of a straightedge of about 1.675 m length of ex. 125 × 25 mm selected softwood, with a shorter (say 750 mm/31 in) spirit level placed on its edge when in use. When fixing door linings, or striving for accuracy in plumbing on any similar operation, the straightedge and level combined are preferable to the relatively short spirit level on its own. This is because any inaccuracies in the shape of hollows or rounds in the surface of the item being plumbed/fixed, show up easily because of the plumb rule or straightedge's greater length.

6

Fixing Doorframes, Linings and Doorsets

6.1 INTRODUCTION

Wooden frames and linings are fixed within openings to accommodate doors which are to be hung at a later stage in the second-fixing operation. This is necessary when wet trades, such as bricklaying and plastering and early-stage rough building operations, are involved. However, where dry methods of construction are used (or achieved, as described later), *doorsets* are sometimes used. These are made up of linings or frames with pre-hung doors attached, locks or latches and architraves in position. This reduces site work by eliminating the conventional second-fixing door-hanging operation.

6.1.1 Lining Definition

Figure 6.1: A door lining, by definition, should completely cover the reveals (sides) and soffit (underside of the lintel) of an opening, as well as support and house the door. Nowadays, this coverage only usually occurs where the opening is within a block- or stud-partition wall.

6.1.2 Frame Definition

Figure 6.2: A doorframe should be of sturdier construction, strong enough to support and house the door without relying completely on the fixings to the structural opening. Unlike linings, which have loose doorstops, frames are rebated to receive the door and usually have hardwood sills – if used on outer walls.

6.1.3 Internal or External

Generally speaking, frames are used for external entrance doors and linings for internal doors. Apart from this, doorframes are usually set up and built in at the time the opening is being formed, whereas a lining, because of its thinness (usually 21–28 mm) and flexibility, should only be fixed after the opening is formed. Linings also require a greater number of fixings and a more involved fixing technique than doorframes, as described later.

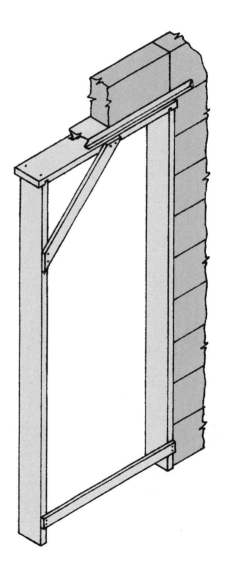

Figure 6.1 Door lining

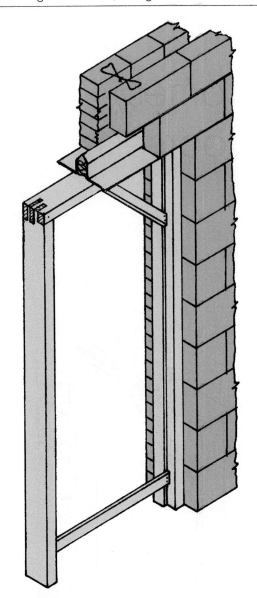

Figure 6.2 Doorframe

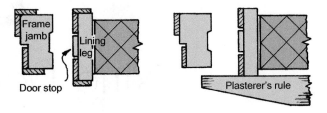

Figure 6.4 Protection strips

risk of decay, built-in frames should be treated with preservative before being fixed, preferably at the manufacturing stage. Usually, at the time of being built in, the abutment-surfaces of the frame have a continuous strip of damp-proof material such as bituminous felt, plastic film (to BS 743) or waterproof building paper fixed to the full width, as in Figure 6.3(a), or fixed to the inner edges, as in (b) and (c). This shows the use of plastic film fixed to frame-jambs and sandwiched between brickwork and blockwork to stop any moisture in the outer wall from bridging the cavity – although, nowadays, patented *cavity closers*, with insulation added, are also used.

6.1.5 Protection Strips

Figure 6.4: The inner edges of the jambs (sides) of frames and the legs (sides) of linings, especially at low level, should be protected from wheelbarrow damage and other careless movement of material and plant, by being covered with temporary wooden strips. These can be of whatever size; the illustration shows 38 × 12 mm strips fixed lightly with 38 mm oval nails to face and edges of the jambs and legs. Of course, the strips fixed to the edges would have to be removed if a wet plastering operation were to take place, as opposed to modern *dry lining*. This, as shown, would allow the plasterer's rule (straightedge) access to a guiding edge.

6.1.6 Setting up the Frame

Figure 6.5: Built-in doorframes are set up immediately before starting to build the walls, or sometimes after the first brick-course has been laid. The position of each door-opening reveal is set out and the frame is stood up in position with one or two scaffold boards supporting it at the head. Two boards (or two nails in one board) are better, especially if the frame is twisted. A 75 mm wire nail driven through the top of the board(s) holds the frame. The nail is driven through before lifting the board into position and the nail-point rests on the frame – traditionally, it was not driven in.

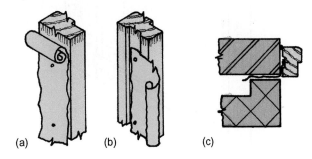

(a) (b) (c)

Figure 6.3 Frame protection material applied (a) full width; (b) and (c) to inner edges and cavity

6.1.4 Protecting External Frames

Figure 6.3: To inhibit initial moisture penetration into the timber from wet trades and to offset any future

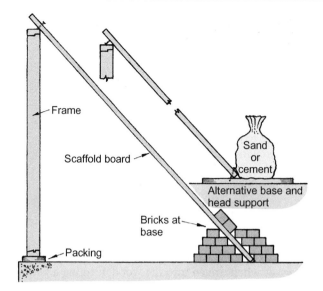

Figure 6.5 Setting up the frame

Safe-nailing Technique

Alternatively, as illustrated, the nail is only driven *into* the board – not through it – and the *head* of the nail is used to hold the frame. This is quite effective and a much safer practice if removal of the nail is eventually neglected.

The jambs are plumbed by spirit level, then bricks or blocks (or any heavy material) are piled at the foot of the scaffold boards to hold them in position. The head of the doorframe must be checked for level and adjusted if necessary.

6.1.7 Vertical Adjustments

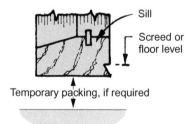

Figure 6.6 Vertical adjustments

Figure 6.6: These adjustments, under the sill, must also take into account the eventual finished floor level (ffl). It is also important to ensure that the head of the frame meets brick courses. If the brickwork is flush or slightly higher than the top of the frame, this will suit the seating of the open-back or classic profile shape cavity lintel. Any slight gap between frame-head and the underside of the lintel will eventually be sealed when the jambs and head are cartridge-gunned around with a flexible frame sealant.

6.2 FIXING DOORFRAMES

6.2.1 First Set of Fixings

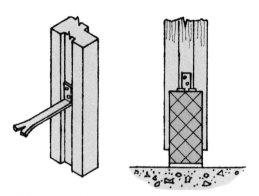

(a) Frame-cramp lug recessed ideally in grooved jambs

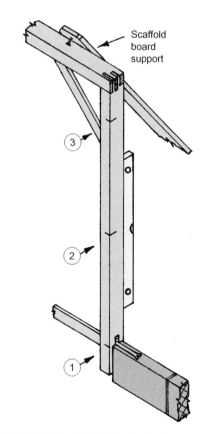

(b) Check frame for plumb with spirit level

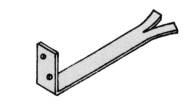

(c) Galvanized steel frame cramp

Figure 6.7 Fixing ties

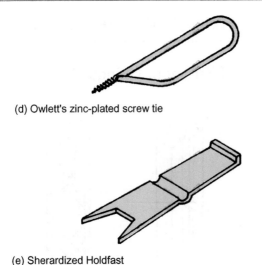

(d) Owlett's zinc-plated screw tie

(e) Sherardized Holdfast

Figure 6.7 (continued) Fixing ties

Figure 6.7 (a)(b)(c): As the brickwork or blockwork proceeds, metal ties are usually fixed to the jambs and built into the bed joints of the mortar. The first set of ties (one to each jamb) must be fixed at low level, on the first course of blockwork, or on the second or third course of brickwork – no higher (Figure 6.7(a)).

6.2.2 Second and Third Set of Fixings

Figure 6.7(b): As the work progresses, the frame must be checked for plumbness from time to time in case the supporting scaffold boards have been accidentally knocked. At least two more sets of ties are fixed, at middle and near-top positions. For storey-height frames (from floor to ceiling, with fixed glass or fan-light above the door opening), a total of four sets of ties is an advisable minimum.

Different types of tie are available, each with points for and points against, as listed below.

Galvanized-Steel Frame Cramps

Figure 6.7(c): These provide good fixings and, because they are screwed to the frame, any cramps (or ties) already fixed and bedded in mortar are not disturbed by hammering. Also, by resting on the last-laid brick, the next brick above the tie is easily bedded. The disadvantages are: handling small screws; requiring a screwdriver and bradawl (especially if no carpenters are on site yet, or they are sub-contract labour, not wanting to be involved); if no groove is in the frame, the upturned end of the cramp inhibits the next brick from touching the frame; doubts as to whether rust-proofed screws were used.

Owlett's Zinc-plated Screw Ties

Figure 6.7(d): These also provide good fixings and are screwed to the frame without vibration from hammering. They do not require screws, bradawl or screwdriver and can be offset or skewed to avoid the cavities in hollow blocks. On the debit side, the brickwork or blockwork has to be stopped one course below the required fixing to allow rotation of the loop when screwing in, then the brick or block beneath the tie is bedded – with some difficulty and loss of normal bedding-adhesion.

Sherardized Holdfast

Figure 6.7(e): Holdfasts are fixed quickly and easily, being driven in by hammer. The spiked ends spread outwards when driven into the wood, forming a fish-tail with good holding power. The main disadvantage is that the hammering disturbs the frame and perma-nently loosens any Holdfasts already positioned in the still-green (unset) mortar.

Marking the Positions

Whatever type of frame cramps or ties are used, their intended positions, relative approximately to bed-joints, are best boldly marked on the jambs with a soft pencil (as seen at points (1), (2) and (3) of Figure 6.7(b)), to act as a reminder to the bricklayer as the work rises.

Braces and Stretcher-ties

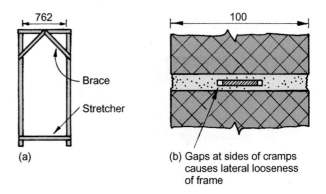

(a)

(b) Gaps at sides of cramps causes lateral looseness of frame

Figure 6.8 (a) Braces and stretcher; (b) effect of early removal

Figure 6.8(a): Usually, one or two strips of sawn tim-ber, of about 50 × 18 mm section, acting as temporary diagonal *corner braces*, are nailed lightly to the frame to keep it square at the head. Another piece, called a *stretcher*, is fixed near the bottom to keep the jambs set apart at the correct width. Ready-made frames deliv-ered to the site, already have braces and a stretcher – or a sill attached.

Disadvantages of Early Removal

Figure 6.8(b): Although there is good reason on site to remove the braces and stretcher at the outset, because they obstruct easy passage through the opening, they should not be knocked or removed until the surrounding brickwork or blockwork is set. The illustration at 6.8(b) shows gaps at the sides of a frame cramp which will cause lateral looseness of the frame. This is due to the frame being accidentally knocked or the braces and/or the stretcher being removed (knocked off) too early, before the mortar was set.

6.2.3 Alternative Method of Fixing

Figure 6.9 Frame-fix screw

Figure 6.9: A popular practice used nowadays for fixing doorframes is to position and build them in without any fixings, care being taken with level and plumbness – especially lateral plumbness, affecting the brick reveals. Eventually, when the brickwork is completed and set, the frame is checked for plumbness on its face edges, minor adjustments made, then drilling and fixing to the brick reveals is carried out with nylon-sleeved Frame-fix or Hammer-fix screws. At this stage, rather than risk the quality of the fixing in the recently set mortar joints, more reliable fixings will be achieved by drilling into the bricks.

6.3 FRAME DETAILS

6.3.1 Weathering the Sill

Figure 6.10 (a): For purposes of weathering and structural transition between exterior and interior levels and finish, external door frames usually have hardwood sills – sometimes referred to as thresholds. Traditionally, water bars, with a sectional size of $25 \times 6\,mm$, ran along a groove in the top of the sill, protruding $12\,mm$ to form a water check/draught excluder. These bars were made of brass or galvanized steel, but when this form of weathering is used nowadays, the bar is available in grey nylon or brown-coloured rigid plastic with a flexible face-side strip which acts as a draught seal against the rebated edge of the door. In this arrangement, a weatherboard must be fitted to the bottom face of the door.

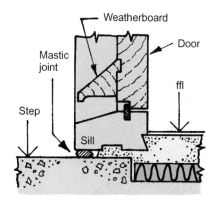

(a)

(b)

Figure 6.10 (a) Traditional sill with water bar; (b) weather seal

6.3.2 Modern Weathering Method

Figure 6.10(b): There are now very effective weather seals available for fixing to the threshold or sill of exterior frames. The wooden sill does not require the groove for a water bar, as before, but if a groove is present, it should be filled with a frame-sealant compound. These weather seals/draught excluders, resembling an open channel, are usually made of extruded aluminium in natural colour or with a brass effect finish.

Fitting and Fixing

When being fitted, the weather-seal channel is simply cut to length, fitted with rubber seals to the manufacturer's instructions and screwed in position to the sill or threshold. The door usually has to be reduced by about $25\,mm$ at the bottom and the metal channel may require to be sealed at each end with silicone or a frame-sealant compound. This prevents moisture seepage from the channel and possible wet rot to frame or sill. Any rainwater that does enter the channel should drain out to the exterior, via weep holes in the inside front-edge of the channel.

6.3.2 Anchorage of Jambs

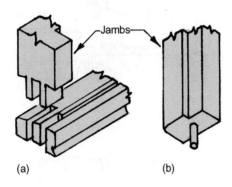

Figure 6.11 Anchorage of jambs: **(a)** double tenon; **(b)** dowel

Figure 6.11 (a)(b): Sills also provide excellent anchorage for the feet of the jambs which are double-tenoned into them with comb joints (Figure 6.11(a)). Although usually through-jointed, the softwood jambs would be better protected if only stub-jointed into the hardwood sill. If the frame is of the type without a sill, 12 or 18 mm diameter metal dowels can be used to secure the jambs (Figure 6.11(b)). A hole is drilled up into the foot of each jamb with an electric or cordless drill and a combination auger bit, or with a brace and a Jennings' twist bit, and the dowels are hammered in to leave a 40–50 mm protrusion. When the frame is being set up, the protruding dowels are either set into dowel holes or they are bedded in concrete or sand-and-cement floor screed, etc. Galvanized steel pipe, of a suitable diameter, can also be used for dowels.

6.3.3 Sunken Rebates or Planted Stops

Figure 6.12 Rebates and stops

Figure 6.12: External frames should be rebated from solid wood to house the door (Figure 6: 12(a)), but internal types may have separate door stops fixed to the jambs and underside of the head to form a rebate (Figure 6.12(b)). The former are referred to as *sunken* or *stuck* rebates and the latter as *loose* or *planted* door stops.

6.3.4 Door-frame Joints

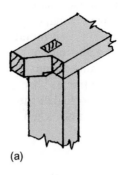 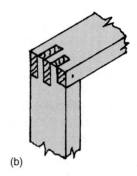

Figure 6.13 Door-frame joints

Figure 6.13 (a)(b): Traditionally, mortice and tenon joints were used to join the jambs to the head and sill, as at (a). Projections of the head and sill, called *horns*, were left on to strengthen the joint and to anchor the frame when built into the brickwork. However, horns do not lend themselves so well to cavity-wall construction and bricklayers tend to cut them off anyway. In line with this practice, the jambs of frames are now comb-jointed to the head and sill and pinned through the face, producing strong corner joints without the need for horns, as at (b).

6.4 FIXING DOOR LININGS

6.4.1 Choice of Fixings

Linings may be fixed to timber stud partitions with 75 mm oval nails, brad-head or lost-head type or, traditionally, with cut clasp nails – which, when punched in, leave a larger hole to fill, but provide a really secure fixing. Alternatively, linings may be screwed (with the heads taken slightly below the surface for filling) or counterbored, screwed and pelleted. Screwing and pelleting is usually restricted to the fixing of hardwood linings, but can be justified on good-quality softwood jobs.

6.4.2 Fixing-Problems

Although cut clasp nails may also be used when fixing to medium-density partition blocks, other aerated building blocks do not hold these fixings very well. Therefore, when fixing linings to walls such as this, it is better to use mid-width positioned *through fixings* (drilled and fixed through timber and wall in one operation), such as nylon-sleeved Frame-fix screws.

6.4.3 Screwing and Pelleting

Figure 6.14(a)–(f) outlines the various steps involved with screwing and pelleting in relation to the following points:

(a) The lining is counterbored at the selected fixing points on each leg with a 12 mm diameter centre or twist bit, about 9 mm deep;

(b) The shankholes are drilled to suit the screw-gauge;

(c) The lining is screwed into position, taking care not to damage the edges of the counterbored holes with the revolving blade of the screwdriver;

(d) The pellets are then glued, entered lightly into the holes, lined up with the grain direction and driven in carefully;

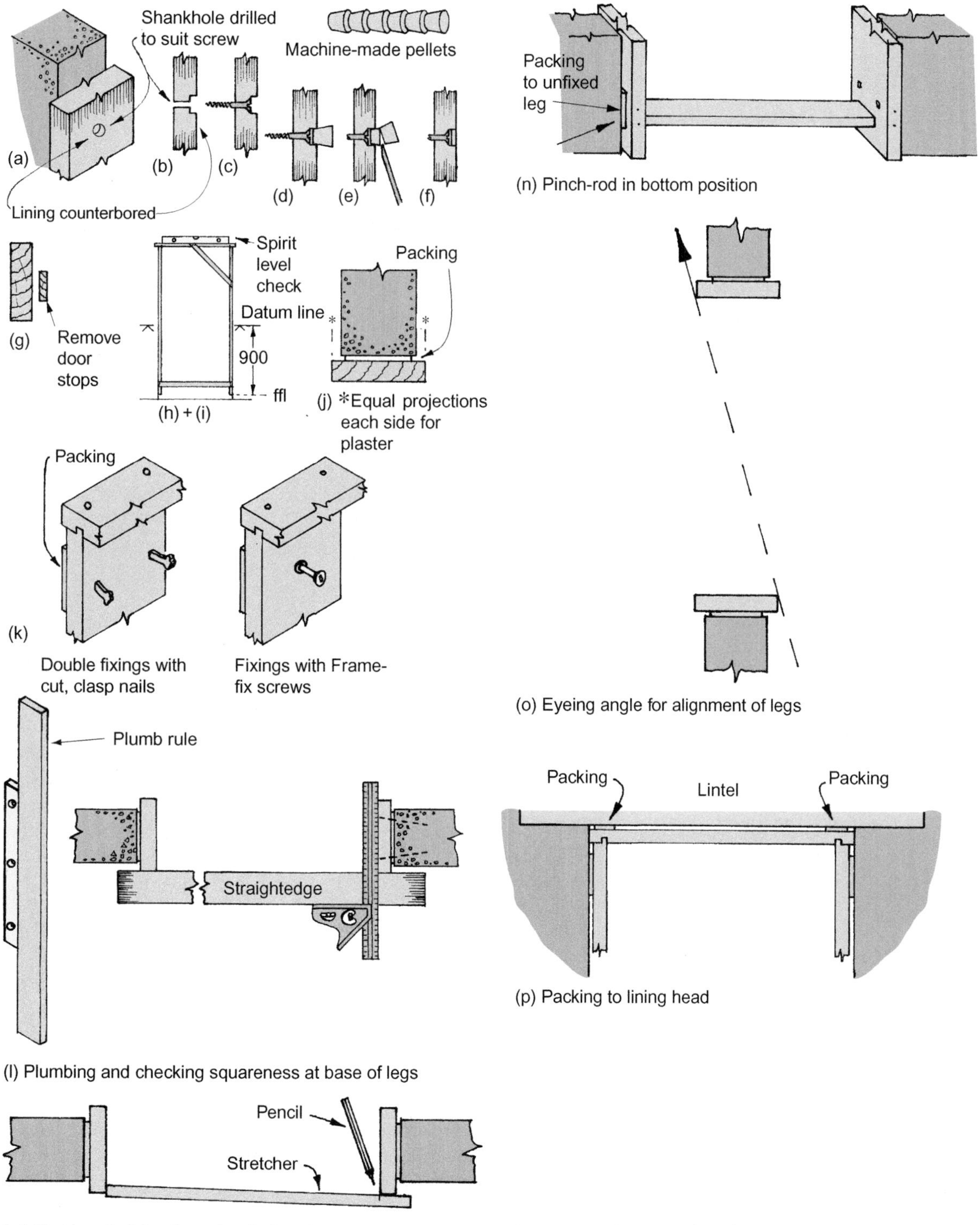

(a) Lining counterbored
Shankhole drilled to suit screw
Machine-made pellets
(b) (c) (d) (e) (f)

(g) Remove door stops

(h) + (i)
Spirit level check
Datum line
900
ffl

(j) *Equal projections each side for plaster
Packing

(k)
Packing
Double fixings with cut, clasp nails
Fixings with Frame-fix screws

(l) Plumbing and checking squareness at base of legs
Plumb rule
Straightedge

(m) Marking stretcher to make pinch-rod
Pencil
Stretcher

(n) Pinch-rod in bottom position
Packing to unfixed leg

(o) Eyeing angle for alignment of legs

(p) Packing to lining head
Packing   Lintel   Packing

Figure 6.14 Fixing door linings

(e) The bulk of the pellet-surplus is removed with a chisel;

(f) The remaining pellet-surplus is cleaned off with a block plane or smoothing plane.

6.4.4 The Fixing-technique for Linings

Owing to a lining's relative thinness and flexibility, the fixing operation can be problematic and unmanageable unless a set procedure is adopted. The following fixing technique, illustrated by Figures 6.14(g)–(p), is therefore recommended.

Initial Stages

(g) First remove the door stops which are usually nailed lightly in an approximate position on the lining legs and set aside.

(h) If working on unfinished concrete floors, check the finished floor level (ffl) in relation to the base of the lining. As illustrated, this is best done by measuring down from a predetermined datum line set at 900 mm or 1 m above ffl. Place packing pieces under the lining's legs, as necessary.

Wedging and Packing

(i) Stabilize the lining in an approximate position by placing small wedges temporarily above each leg, in the gap between the lintel and lining-head, then check the head with a spirit level and adjust at the base, if necessary. If gaps exist between the structural opening and the back of the lining's legs, as is usual when a tolerance has been allowed, pack these out with plastic shims, obtainable in varying colour-coded thicknesses, or with pieces of non-splitting material such as hardboard or plywood, initially on each side of the top fixing positions only.

Adjusting and the Initial Fixing

(j) Adjust the top of the lining to establish equal projections on each side of the opening for eventual plaster-thickness on the wall surfaces, or to form equal abutments for the edges of dry-lining boards.

(k) Now fix the lining near the top, *through* the packings on each leg, either with two nails per fixing or with mid-width screw fixings.

Plumbing, Squaring and Fixing

(l) Now plumb the lining on the face sides and edges with a long (1.8 m) spirit level, or alternatively, a 1.8 m straightedge with a short spirit level placed on it. Pack the bottom position each side as required and fix through one packing only. Check for squareness of the fixed leg at the base with a straightedge and try-square, as illustrated, then pack and complete the intermediate fixings on the same leg, checking before and after each fixing with the long level or straightedge. The amount of fixing points on each side should ideally be five, but not less than four. These should be placed at about 100 mm from the underside of the head, 100 mm from the ffl, and two or three intermediate points on each side.

Converting the Stretcher to a Pinch Rod

(m) Next, remove the stretcher from its position at the base of the lining, denail it, hold it up to a position just below the lining head, mark the exact inside lining-width and cut to make a *pinch rod*.

(n) Fit the pinch rod in the bottom section of the lining-legs, as shown, and pack accordingly behind the lower fixing point on the unfixed leg.

Checking, Aligning and Final Fixing

(o) Check the plumbness of the unfixed leg for correct sideways position and check the alignment by sighting across the face edges as illustrated. Now fix the bottom, then the intermediate points, moving the pinch rod to each fixing area and packing out, if necessary, before fixing. The lining head is not normally fixed to the lintel unless the opening exceeds the normal width.

Packing-out the Lining Head

(p) However, the head does require to be held firmly by replacing the initial temporary wedges mentioned in point (i) with packing or plastic shims driven into the gap between the lintel and head, at the two extreme points only, immediately above each leg. Failure to complete this detail can result in the head becoming partially disjointed from the legs, at the tongued-housing joint, during final nailing of the head door-stop at a later stage.

The Finishing Touches

Remove the corner brace or braces (although this could have been done earlier, after points (g) or (h) to facilitate easier working). If nailed, punch in all nail fixings to about 3 mm below the surface. If screwed with nylon-sleeved Frame-fix screws, check that these are at least flush or slightly below the surface (having been fixed in the mid-width area, they should be covered by the door stops). Or, if counterbored for pellets, complete the pelleting operation as outlined in stages

(d) to (f). Finally, replace the door stops in their temporary position and fix protection strips, if considered necessary, as mentioned earlier in the text relating to Figure 6.4.

6.5 SETTING UP INTERNAL FRAMES PRIOR TO BUILDING BLOCK-PARTITIONS

6.5.1 Introduction

In these situations, the first step is to set out the wall positions on the concrete floor in accordance with the architect's drawing. Providing the structural walls are square to each other, the various internal partitions required need only be measured out at two extreme points from any wall to form parallels and right angles.

6.5.2 Spotting

Figure 6.15 Spotting with a mortar-sliver

Figure 6.15: This setting out is best done with a steel tape rule and a method of marking known as *spotting*. A spot of mortar (about half a trowelful) is placed in position on the floor, then trowelled down to form a thin sliver. The tape rule, preferably being held at one end, is pulled taut over the mortar spot and the trowel-tip is cut through it at the required measurement. Right-angled trowel-tip marks are also needed, as illustrated, to mark doorway openings.

6.5.3 Positioning the Frame

Figure 6.16 (a)(b)(c): Frames should be stood in position to relate to the setting out – preferably when the mortar spots are set – and some means of holding them at the foot and head must be devised. The following methods can be used.

Holding the Frame at Its Base

Figure 6.16(a): The first course of blocks can be laid and when set will act as a means of steadying the feet in one direction, while loose blocks on either side will hold the position in the other direction. Alternatively, notched pieces of wood can be placed against the feet or against the protruding metal dowels on the opening side and fixed to the concrete floor by means of a cartridge tool.

Holding the Frame at Its Head

Figure 6.16(b) and (c): The head of each frame can be held by a leaning scaffold board or boards – but a far better method is to use wall cleats and a system of top braces. The braces, of say 50 × 18 mm sawn timber, must be triangulated or placed in such a way as to create stability. The wall cleats, about 300 mm long, of say 100 × 25 mm timber, can be fixed to the walls with 75 mm cut clasp nails, with heads left protruding for easy removal later. If these fixings are unsuitable for the walls, masonry nails may do the job.

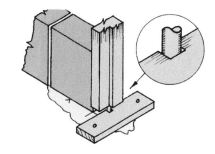

(a) Holding the frame at its base

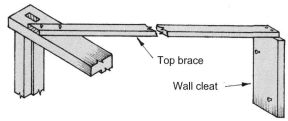

Top brace

Wall cleat ⟶

(b) Holding the frame at its head

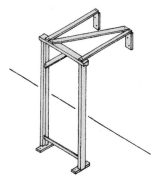

(c) An isometric view of a braced frame

Figure 6.16 Positioning the frame

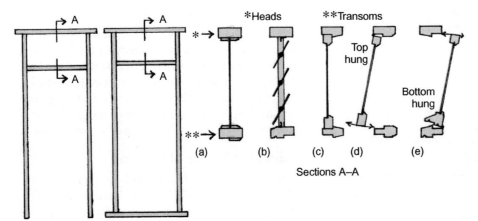

Figure 6.17 Storey frames

6.6 STOREY FRAMES

6.6.1 Internal and External Storey Frames with Fanlights

Figure 6.17: Storey frames may be internal or external and, as the name implies, fully occupy the vertical space between the floor and ceiling. The frame comprises two extended jambs, a head, a transom above the door and, usually, a hardwood sill on external types. The frame-space above the door can (a) be directly glazed, (b) contain louvres, (c) house a fixed sash, or (d) contain an opening sash (fanlight), opening outwards, or (e) opening inwards.

6.6.2 Extended Jambs Without a Fanlight

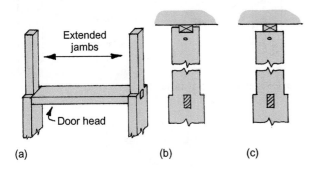

Figure 6.18 (a) Extended jambs without fanlight; (b) jambs notched over a fixing-batten; (c) jambs butted and fixed to the batten

Figure 6.18(a): Another type of storey frame which can be used on block partitions of less than 100 mm thickness, would be minus the fanlight aperture, but have extended jambs to allow for some form of fixing to the ceiling. This would give greater rigidity

to the thinner wall whose strength might otherwise be impaired by the introduction of an opening. The jambs protruding above head level should be reduced by the plaster thickness on each outer edge and, after the wall is built, these edges should be covered with a strip of expanded metal lath – unless dry-lining methods are to be used.

6.6.3 Fixing to Ceiling

Figures 6.18(b) and (c): When the ceiling above the storey frame is timber-joisted, the extended jambs are fixed to the sides of the joists or, more likely, to purpose-placed noggings between the joists. When the construction is of concrete, a timber batten or ground can be 'shot' onto the ceiling with a cartridge fixing tool and the jamb-ends fixed to this by (b) notching (cogging) or (c) butting and skew-nailing.

6.7 SUBFRAMES

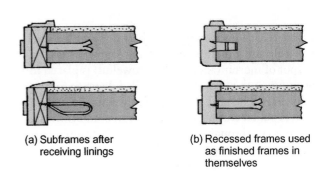

(a) Subframes after receiving linings

(b) Recessed frames used as finished frames in themselves

Figure 6.19 Subframes

Figure 6.19: If subframes are used, they should be recessed to take the thickness of the blocks. The recess helps to stabilize the block wall during construction and the subframe is ideal in providing an eventual means of fixing the lining and architraves – especially if these were in hardwood. Sub-frames, if built in as illustrated, also alleviate the fixing problems experienced with lightweight aerated block-work. When being built in, the frames can receive metal ties or frame cramps as normal and may be used as (a) sub-frames to receive linings or (b) frames in themselves.

6.8 BUILT-UP LININGS

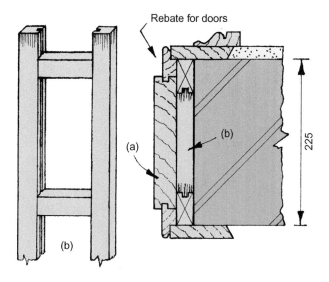

Figure 6.20 50 × 25 mm framed grounds for built-up linings

Figure 6.20: Although built-up linings, covering the full width of the reveals and soffit of door-openings, are not normally used nowadays, the subject is mentioned here in case it should be met on repair or conversion work. These linings, labelled (a) on the figure, are built up on grounds (b). Grounds are foundation battens which, if set up accurately, packed and fixed properly in the first-fixing operation, provide a good and true fixing base for the separate parts of the lining and/or architraves in the second-fixing operation.

6.8.1 Framed Grounds

Figure 6.20: Traditional framed grounds, illustrated at (b), consisting of two verticals and multi-spaced horizontal members morticed and tenoned together, looking like a ladder, were joinery shop-made and fixed on site to suit built-up linings. These linings, used on walls of 225 mm thickness and above, were so constructed to minimize the effects of shrinkage across the face of the wider timber. The fixing technique for the three (two sides and a head) sections of framed grounds, would be similar to that used in fixing linings.

6.8.2 Separate Architrave-grounds

Figure 6.21: Apart from the advantages of providing a wider fixing area for architraves, these traditional grounds also protected lining-edges from becoming swollen by the wet plastering operation. As illustrated, these grounds were bevelled to retain the plaster on the outer edge. The width of the ground was such as to allow a minimum 6 mm overlap of the architrave on to the plaster surface. Cut clasp nails, in sizes of 50, 63 and 75 mm, were commonly used to fix the grounds to the mortar joints of the wall.

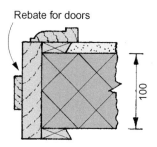

Figure 6.21 Traditional grounds for architraves

6.9 MOISTURE EFFECT FROM WET PLASTERING

Figure 6.22(a): When frames, linings or grounds are set up and wet plastering methods are used, the effect of this should be realized as it is often detrimental to the finished work. In the first instance, excessive moisture from the wet rendering/floating coat, labelled (1) in the figure, against the timber lining, causes the timber to swell (2). While still in this state, the plasterer usually applies the setting/finishing coat of plaster (3), flush to the swollen edges. The timber eventually loses moisture and shrinks back to near normal (4), leaving an awkward ridge between the wall surface and lining (5), which upsets the seating of the architrave and the trueness of the mitres at the head.

6.9.1 Solving the Mitre Problem

Figure 6.22(b)–(d): Geometrically, when architraves are not seated properly, true mitre-cuts will appear to be out of true, touching on the outer (acute) points and open on the inner (obtuse) surfaces (Figure 6.22(b)). If the plaster is only slightly proud of the lining edges, sometimes it can be tapered off with a scraping knife (or the edge of the claw hammer). This will seat the architrave better, but minor adjustments

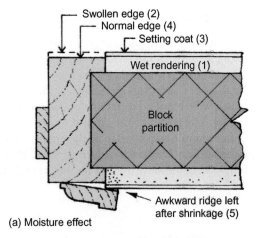

(a) Moisture effect

Swollen edge (2)
Normal edge (4)
Setting coat (3)
Wet rendering (1)
Block partition
Awkward ridge left after shrinkage (5)

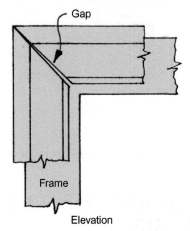

Gap
Frame
Elevation

(b) Appearance of true-mitred architrave seated on raised plaster edge

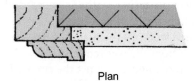

Plan

(c) Trimming plaster edges

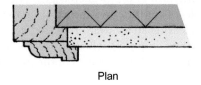

Plan

(d) Rebating architrave edges

Figure 6.22 Moisture effects and their solution

to the mitre-fit may still have to be made. If so, a sharp smoothing plane or block plane is used. Figure 6.22(c) shows how the problem can also be solved by trimming the plaster edges more drastically with a bolster chisel, which is not ideal and involves making good, or by rebating the edges of the architraves (Figure 6.22(d)).

6.10 DOORSETS

6.10.1 Introduction

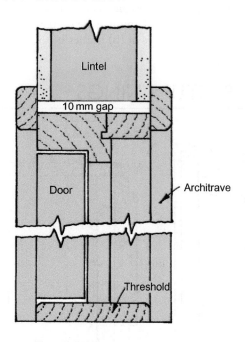

Lintel
10 mm gap
Door
Architrave
Threshold

Figure 6.23 Vertical section through Doorset

Figure 6.23: Doorsets comprise linings or frames with pre-hung doors attached, locks or latches and one or two sets of architraves in position. These units are supplied by specialist firms producing Doorsets as a factory operation. The main advantage of this modern practice is a reduction in time-consuming site work by eliminating conventional door-hanging, architrave and lock-fitting and fixing.

6.10.2 Suitability of Doorsets

Because Doorsets are complete units, they are not immediately suitable where conventional methods of construction, involving wet trades, are to be used. The issues against this practice include the protrusion of the architraves and hinges, which inhibits conventional plastering methods, greater risk of damage to doors, and possible distortion of door-jointing tolerances due to moisture absorption by the lining from wet plaster.

6.10.3 Variations now Available

The possible variations now available are, therefore:

1. fixing coventional linings in situations where 'wet trades' are involved;
2. fixing conventional linings to openings in dry-lined walls;

3. using Doorsets and fixing to openings in dry-lined walls; or
4. modifying the wet trade operation to enable the use of Doorsets to openings in plastered walls.

6.10.4 Fixing Doorsets

Figure 6.24(a): The fixing of Doorsets is somewhat similar to the fixing technique already covered for fixing conventional linings, except that Doorsets are usually fixed after the dry-lining operation has been completed. This means that care must be taken around the opening to ensure that the finished wall thickness meets the exact width of the Doorset lining. One way of doing this is to fix temporary profile boards to the sides of the opening, similar to those shown in Figure 6.24(b), as a guide for the dry-lining fixer to work to. Other variations in technique include using the pre-hung door to check door-jointing tolerances as final confirmation of level and plumbness before completing the fixings.

6.10.5 Fixing Profiles

Figures 6.24(b)–(d): The modification mentioned in Section 6.10.3, variation 4, to enable the use of Doorsets to openings in plastered walls, involves producing plywood profiles of minimum 12 mm thickness, cut to the finished wall thickness and fixed

around the opening like a traditional lining. This can be done either temporarily as a guide for the wet plaster (Figure 6.24(b)), to receive Doorsets after removal (Figure 6.24(c)), or permanently as an initial guide for plaster and subsequent subframe for the Doorset fixing (Figure 6.24(d)).

6.11 FIRE-RESISTING DOORSETS

Figure 6.25: Doorsets with fire-resisting doors and frames are available. If they have been tested to the latest British Standards specification, instead of being referred to as ½-hour and 1-hour firecheck doors and frames, as in previous years, they should be referred to as 'fire-resisting Doorsets' with a quoted stability/integrity rating. This rating is expressed in minutes, such as 30/30 or 30/20, meaning 30 min stability/20 min integrity. *Stability* refers to the point of collapse, when the Doorset becomes ineffective as a barrier to fire spread. *Integrity* refers to holes or gaps concealed in the construction when cold, or to cracks and fissures that develop under test.

6.11.1 Frame and Door Details

Figure 6.25: As illustrated, fire-resisting Doorsets are identifiable by frames with sunken rebates of 25 mm depth and varying thicknesses of door, according to

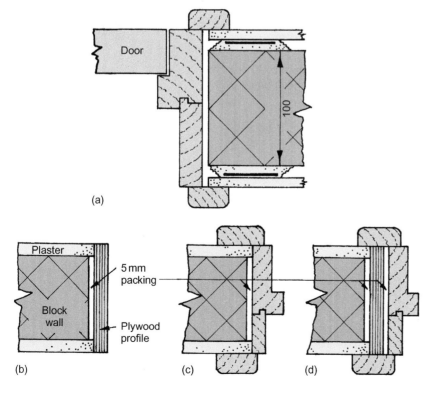

Figure 6.24 (a) Horizontal section through Doorset within dry-lined wall; (b) profiles ensure correct wall thickness; (c) receipt of Doorset after profile removal; (d) act as subframe for Doorset

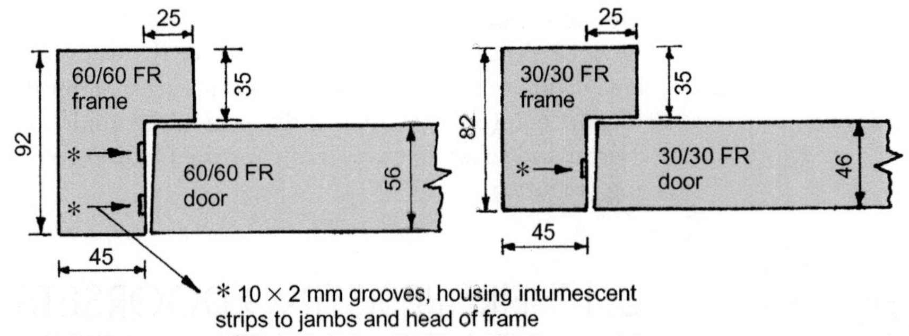

25   60/60 FR frame   35   92   * →   * →   45   60/60 FR door   56

25   30/30 FR frame   35   82   * →   30/30 FR door   46   45

* 10 × 2 mm grooves, housing intumescent strips to jambs and head of frame

Figure 6.25 Fire-resisting Doorsets

the amount of fire resistance required. Fire-resisting doors are usually:

1. made up of solid-core timber construction, clad with thin plywood, looking like thick blockboard;
2. built up of ply-clad framing with mineral infill; or
3. made up of timber frame, plasterboard, fibreboard and bonded plywood facings.

6.11.2 Intumescent Strips

As illustrated, the gaps (joints) between door and frame usually contain intumescent strips which swell up when heated, thereby sealing the top and side edges of the door to increase the fire resistance. Intumescent strips give a fairly good seal to hot smoke, but as they do not become active until temperatures of 200–250°C are reached, they have no resistance to cold smoke.

6.11.3 Final Details

Figure 6.26: When fitting a fire-resisting Doorset, before fixing the second set of architraves, pack the gaps between the frame and wall with mineral wool or similar fire-resisting material.

The latest TRADA (Timber Research and Development Association) Wood Information Sheet on fire-resisting Doorsets recommends that narrow – not broadleaf – steel hinges should be used, to allow continuous intumescent strip to jamb edges. Slim locks, preferably painted with intumescent paint

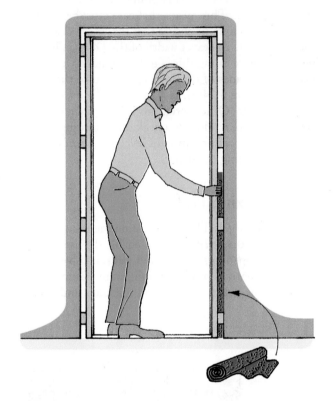

Figure 6.26 Mineral wool packing to gaps of fire-resisting frame

or paste, should be fitted; the thickness and thermal mass of these locks must be minimal. Over-morticing must be avoided, otherwise these hidden gaps will, in effect, reduce the *integrity* rating of the Doorset.

7

Fixing Wooden and uPVC Windows

7.1 INTRODUCTION

An important consideration which determines the method of fixing windows, is whether they are to be built-in as the brickwork proceeds, or fixed afterwards in the openings formed in the brickwork. This decision is related to the type of windows being installed and whether they are robust enough to withstand the ordeal of being used as profiles at the *green* brickwork stage.

7.2 CASEMENT WINDOWS

7.2.1 Wooden Casement Windows

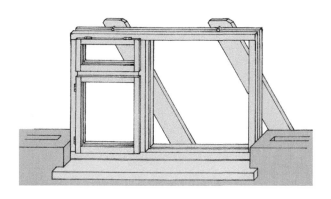

Figure 7.1 Built-in window frame

Figure 7.1: Casement windows made of wood are usually built in as the brickwork proceeds. They are secured with separate fixing devices, traditionally referred to as *frame cramps*, which are covered in detail in Chapter 6. Essentially, according to the type used, they are either screwed or hammered into the wooden side-jambs as the brickwork rises, to be built into the

bed joints. Two or three cramps each side is usual. Like built-in door frames, these windows, after being positioned, are plumbed and supported at the head with one or two weighted scaffold boards pitched up from the oversite sub floor. If the windows have a separate sill of stone or pre-cast concrete, usually these must be bedded first and protected with temporary boards on their outer face sides and edges. Projecting sills, formed with sloping bricks-on-edge, are usually built at a later stage, the windows having been packed up accordingly to allow for this procedure.

7.2.2 uPVC Casement Windows

Windows made of uPVC are usually fixed after the opening is formed, by screw fixings drilled through the box-section jambs into the masonry reveal on each side, or – if these fixings clash with the cavity seal – by screwing into projecting side lugs which have been pre-cut from multi-holed galvanized strap and screwed to the sides of the window before insertion. To allow for expansion and fitting, either the window openings are built with a 6 mm tolerance added in height and width, or the windows are ordered with a 6 mm tolerance *deducted* in height and width. In either case, to ensure the correct size of opening is built, temporary wooden profile frames made of 50 × 50 mm or 75 × 50 mm prepared softwood, with 6 mm WBP plywood corner plates, as illustrated, are constructed and placed in position during the brickwork and blockwork operation.

Profile Frames

Figure 7.2: The temporary profile frames may be made on site or in the workshop and can be removed soon after the brickwork/blockwork has set, or left in position until the windows are to be installed. After careful removal, the frames may be dismantled, stored or reused immediately. When the uPVC windows are eventually

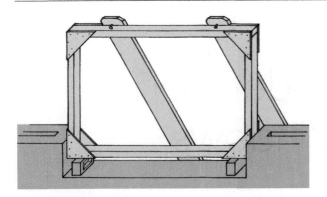

Figure 7.2 Temporary profile frame

fitted, the expansion/fitting tolerance is taken up equally all round – or as equally as possible – with special plastic shims. On the sides of the window, these shims, which are 'U' shaped, are slid around each screw fixing before the screw is fully tightened. Upon completion, or when all other building work is complete, any projecting shim is trimmed off and the gap around the window is cartridge-gunned around with a silicone sealant.

7.3 GLAZING

7.3.1 DIY Glazing

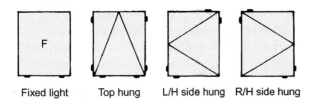

Figure 7.3 Position of setting and location blocks

Figure 7.3: If installing the double-glazed sealed units into the fixed uPVC windows yourself – which is easy enough once you know how – you must first understand that the units must be seated on plastic *setting blocks* and, if the window is an opening vent, it must have plastic *locating blocks* in various positions on the sides, as illustrated. Different thicknesses of plastic blocks may be required, but they are normally 3 mm thick, 25 mm long and should be as wide as the sealed unit's thickness (which is often 28 mm as standard nowadays).

7.3.2 Externally Beaded Glazing

Figure 7.4: With the setting and locating blocks fitted snugly in position between the glass and the uPVC casement, and the sealed unit pushed in to rest up against the upstanding inner edge, the pre-fitted shuffle beads, which were taken out to gain access for glazing, are now pivoted into the small front lip and snapped into position. Ideally, these should be pre-marked L/H, R/H, TOP and BOT, so that they relocate in their original positions.

Internal Access

On the other side of the glass now, the unit is pushed outwards, to be hard up against the external beads just fitted. This creates a small gap between the glazed unit and the grooved, upstanding edge of the window. The so-called *wedge gasket*, illustrated in position in Figure 7.4(a), is fed into this gap all round, forcing the unit forward and locking and sealing the external beading. The gasket, which has to be cut or partly snipped on its concealed face at each corner (with at least a 25 mm length-allowance added each time to ensure a well-compacted fit), is not easy to push in. To assist with the task, it would be advisable to

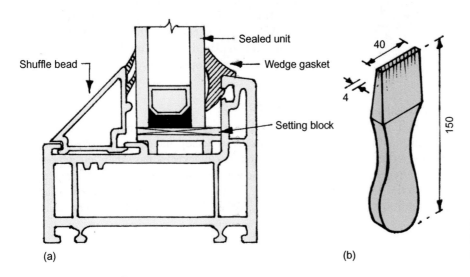

(a)

(b)

Figure 7.4 (a) Externally beaded glazing; (b) caulking tool

make a simple wooden caulking tool (Figure 7.4(b)) and, if still difficult, diluted washing-up liquid can be brushed into the gap prior to inserting the gasket.

7.3.3 Internally Beaded Glazing

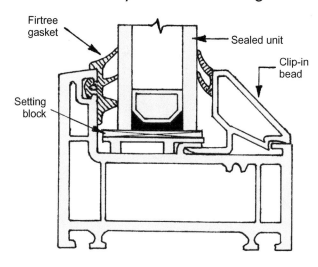

Figure 7.5 Internally beaded glazing

Figure 7.5: uPVC casements with internally beaded sealed units are mostly used nowadays, mainly for the following two reasons.

- They are thought to be more burglar-resistant.
- Being internally beaded means that the glazing operation is done from inside the building, which on certain jobs – especially those without scaffolding – can be an advantage.

On this type of window, there is an *external* upstanding edge, grooved as before, but this time to receive a so-called *firtree gasket*, as illustrated, instead of the wedge gasket. The sealed unit, still on setting blocks and, if in an opening vent, secured with locating blocks, is pushed outwards to be hard up against the firtree-gasket edge, and the internal 'clip-in' beads are pushed and clipped fairly easily into place.

7.3.4 Safety Issues: Glazing

Figure 7.6(a)(b): Related to the current Building Regulations, the 2010 edition (as amended) of Approved Document N, *Glazing – Materials and Protection*, introduced new provisions covering the safe use of glazed elements in compliance with the Workplace (Health, Safety and Welfare) Regulations 1992 (2nd edition 2013). As the title implies, these regulations are aimed at the workplace and apply mostly to non-habitable buildings such as shops, factories, offices, etc., and only Approved Document N1

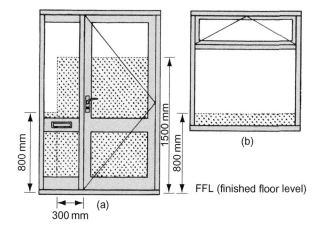

Figure 7.6 (a) Safety glass requirement indicated in shaded areas of glazed door and side panel. (b) Safety glass requirement indicated in window below 800 mm from the ffl

applies to dwellings – not Parts N2, N3 and N4. Part N1 refers to *critical locations* where people are likely to come into contact with glazing and where accidents may occur, causing cutting or piercing injuries. These critical locations, as illustrated, are identified and concern (a) the glazing of doors and side panels between the finished floor level (ffl) and 1.5 m above and (b) glass in internal or external walls and partitions between finished floor level and 0.8 m (800 mm) above. The Approved Document lists a number of alternative solutions to minimize the risk of injury in these areas. The first of these refers to the most popular solution, i.e. the installation of safety glass in any deemed critical areas. Note that five out of the six glazed units shown in the illustration would require safety glass.

7.3.5 Safety Issues: Means of Escape

Also related to safety, the Building Regulations 2010, Approved Document B1 (with 2013 amendments) covers Fire Safety in Dwelling houses, where there is reference to an upgrading of smoke alarms in *new* dwelling houses – and, in Section B2, there is reference to providing *Means of escape* via emergency egress from ground-level and upper-floor-level windows (or doors).

These provisions refer to *Escape from the ground-storey* – and *escape from upper floors not-more-than 4.5 m above ground level*. They also refer to *Escape from upper floors* **more than** *4.5 m above ground level*; but, additionally, these other categories of dwellings must also provide a fire-protected staircase and other criteria, such as *having no* **inner** *rooms above ground-level* and/ or providing an alternative escape route or a sprinkler system, depending on whether there is one floor above 4.5 m, or more than one floor.

Basically, new windows that are provided for emergency egress/means of escape, should have an unobstructed openable area of at least 0.33 m² and be at least 450 mm high × 450 mm wide. The bottom window-*ledge* of the escape-opening should be not-more-than 1.1 m (approx. 3 ft 7¼ in) above the dwelling's floor level.

The window (or door) – based on a reference to one's judgement and other criteria in B2's General Provisions in Section 2.8 – should enable a person escaping from fire to reach a safe place. Note also that there is a reference to *Approved Document K, Protection from falling*, when *guarding* might be relevant.

Note 2 of paragraph four, under section 2.8, states that: Locks (with or without removable keys) and stays may be fitted to egress windows, subject to the stay being fitted with a release catch, which may be child resistant.

And *Note 3* of paragraph five states that: Windows should be designed such that they will remain in the open position without needing to be held by a person making their escape.

7.4 WINDOW BOARDS

Windows must be fixed into position before dry-lining or plastering takes place, to enable a satisfactory abutment of the wall-lining against the window's head and sides. For the same reasons, window boards are also required to be fixed. These boards appear as a stepped extension of the sill and project beyond the plaster faces at the front and sides, like the projecting nosing of a stair tread. If made of MDF board or timber and fitting up against a wooden casement window, the back edge is usually tongued into a groove in the sill. If the window is of uPVC, the window board is square-edged and only butted and may be held with a panel adhesive such as Gripfill – or the abutment may be covered with a small plastic cloaking-fillet or quadrant held with superglue.

7.4.1 Marking and Cutting the Window Board

Figure 7.7(a): As illustrated, a portion of board about 50 mm in from each end is marked and cut to fit the window reveals, and the machined nosing shape on the front edge is returned on the ends, by hand with a smoothing plane and finished with glasspaper. Although, sadly, this traditional finishing touch seems to be omitted nowadays.

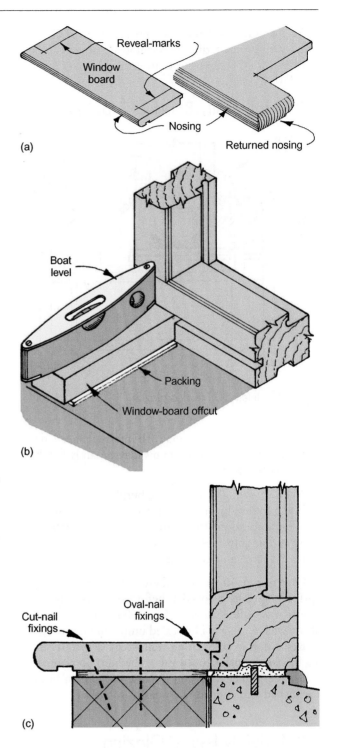

Figure 7.7 Marking and fitting a window board (traditional method)

7.4.2 Fitting and Fixing the Window Board (Traditional Method)

Figures 7.7(b) and (c): Packing is usually required between the window board and the inner skin of block-work, to level the board across its depth. Pieces of damp-proof course material, plastic shims or hardboard make ideal packings. If the board is tongued to fit grooved window sills, the board is skew-nailed to the wooden sill with 38 mm oval nails and fixed through the packings into the blockwork with either cut clasp nails or pelleted screw fixings. Packings may be established about every 450 mm prior to positioning the board, by using the end offcuts, as illustrated, with the tongue in the groove, a spirit level on top – or a try-square against the jambs – while trial packings are inserted.

7.4.3 Alternative (Modern) Window-board Fixings

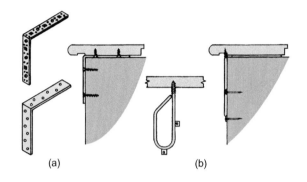

Figure 7.8 (a) Galvanized bracket-shape anchor or purpose-made, heavy-duty Fixing Band as fixings to the underside of window boards; (b) Owlett's screw ties as alternative fixings

Figure 7.8(a)(b): Most inner-skin blockwork walls nowadays use improved thermal blocks and are either too 'soft' (lightly foamed) or too 'hard' (densely foamed) to receive the traditional nail-fixings illustrated in Figure 7.7. Furthermore, the extensive use of MDF board as a replacement for softwood window boards over the last two decades or so, has added to the fixing problem. This is because MDF is a densely compacted material that does not receive nails as readily as softwood – especially such bulky fixings as cut clasp nails; the heads of which would resist being punched below the surface. For these reasons, the following alternative methods of fixing are now being used: (a) Anchors in the form of galvanized or zinc-plated, bracket-shape right-angles, or similar shaped anchors purpose-made from heavy-duty (20 gauge) galvanized-steel, multi-holed *fixing band*, are pre-screwed to the underside of the window board prior to fixing, then, once the board is refitted, they are screwed to the brickwork or blockwork face; and (b) Owlett's zinc-plated screw ties are screwed into the underside of the window board, as illustrated, prior to fixing, then, once the board is refitted, they are screwed to the wall-face through side lugs, as

illustrated. Method (a) is suitable for dry-lined walls and/or plastered walls, whereas method (b) is more suitable for plastered walls only.

7.5 BOXFRAME WINDOWS

7.5.1 Introduction

Traditional boxframe windows with double-hung, vertically-sliding sashes, sash cords, pulley wheels and cast-iron weights were extensively used many years ago and are still very much in evidence in mature and period properties. For this reason, they will still be required in their traditional form, if only as replacement windows during maintenance operations. Furthermore, these elegant windows, offering ideal top and bottom ventilation – and countless years of service *if well maintained* – appear to be making a comeback nowadays.

7.5.2 Fixing Modern-type Boxframe Windows

Figure 7.9(a)(b): However, it is unlikely that boxframe windows would be used in their precise traditional form in modern dwellings, because the present-day version dispenses with the box construction, cords, pulley wheels and weights, and has solid jambs,

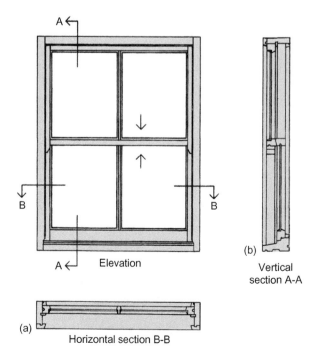

Figure 7.9 (a) Sectional view showing solid jambs of sliding-sash window; (b) sectional view of solid head

heads and sills – as illustrated – and sliding sashes hung on patent spiral-balance fittings. Also, for those who prefer less maintenance, uPVC replicas of these windows are now produced. Whether of wood or uPVC, such windows, because of their smaller jambs and head, would not need to be recessed into the brick reveals and soffit, making them ideally suitable for fitting against modern cavity walls. Being, in effect, like wooden casement windows, they would be fixed like them – or could be fitted into openings and fixed with Window-fix or Frame-fix sleeved screws.

7.5.3 Fixing (or Replacing) Traditional Boxframe Windows

Figure 7.10(a): Site measurements can be taken from the window being replaced or the following considerations should be borne in mind: The overall

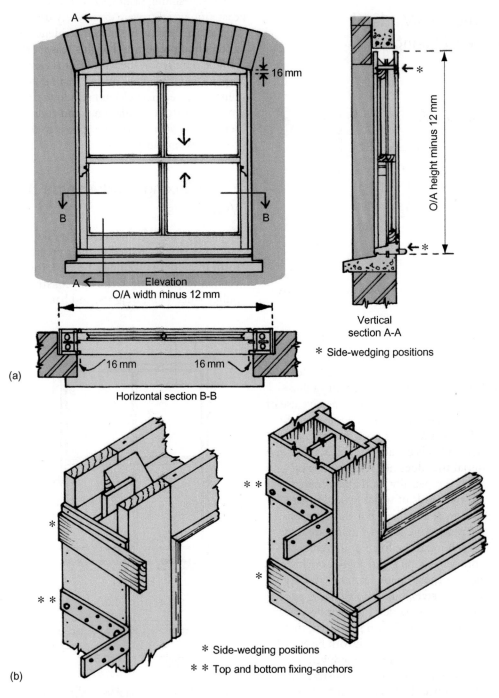

Figure 7.10 **(a)** Site measurements for box-frame windows; **(b)** top and bottom fixing-anchors and wedging positions

height and width of the boxframe should be at least 12 mm less than the internal, recessed brick opening to allow for fitting tolerances. The outer linings of the box should only project 16 mm into the window opening, with the exception of the window's head in relation to a segmental brick-arch, where the 16 mm projection is retained at the extreme ends of the springing line.

Figure 7.10(b): When positioning the window for fixing, the wooden sill should be bedded on a generous application of silicone or mastic frame sealant. Then the frame should be levelled and plumbed with hardwood wedges driven in each side, top and bottom. They should be placed immediately behind the ends of the pulley-stile head and the sill – up against the brickwork. These positions are critical, as they reduce the risk of causing the frame to bulge. Traditionally, these windows were skew-nailed through the two inner-linings' outer edges into a few wooden-plugged bed-joints of the brickwork and nailed to an internal wooden lintel above. The method nowadays, as illustrated, would be to pre-fix right-angled bracket-shape anchors, similar to those described for fixing window boards, and face-fix them to the inner brickwork or blockwork after wedging. Finally, a mastic frame-sealant or silicone should be gunned around the head and reveals on the outside and the bed-joint at the sill should be re-gunned on the face edge. The window board or nosing piece can now be fixed to complete the job.

7.6 FIXING UPVC CASEMENT WINDOWS INTO EXISTING WOODEN BOXFRAMES

7.6.1 Introduction

Traditional boxframe windows, with vertically-sliding sashes, are often replaced nowadays with uPVC casement windows that are then fitted with double-glazed sealed units. However, removing the actual boxframes exposes unacceptable, interior recesses of un-plastered rough brickwork on each side reveal and at the top. And these recesses do not readily accommodate the replacement uPVC casement windows. For this reason, it seems, most boxframes (often regardless of their degraded condition) are left *in situ* and only the sliding sashes, their sash cords, the pulley wheels and

the parting- and staff-beads are removed and done away with. The cast-iron (or lead) weights are usually left in the boxframes, after their cords have been cut to release the sashes – or they are retrieved to be sold as scrap metal.

Then the replacement uPVc unit is fitted – often wrongly, in my opinion, up against the projecting outer-linings, leaving them and a top-and-front-portion of wooden sill exposed to the elements, needing future paint-maintenance; or, one finds these boxframe projections unsatisfactorily covered with very narrow, double strips of plastic cover-mould.

Figures 7.11(a) to (h): If a boxframe is in a reasonable condition and is to be retained and have a uPVC casement window fitted, the two ways of doing this are shown here graphically (for comparison), with explanatory captions to highlight the weaknesses of the inferior method.

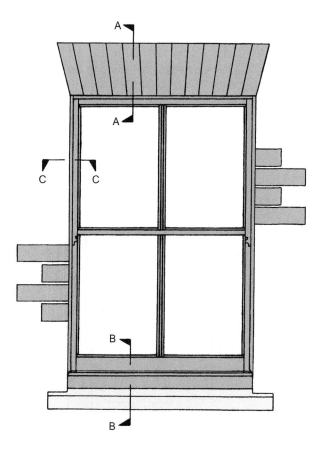

Figure 7.11(a) Sectional views A-A, B-B, and C-C show the treatment recommended to replace these typical vertically-sliding sash windows with a uPVC casement window, without removing the existing wooden boxframe.

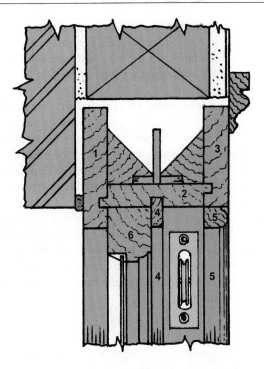

Figure 7.11(b) *Vertical section A-A* showing the head of the boxframe before conversion and installation of a new casement window. 1 = outer-lining head; 2 = pulley-stile head; 3 = inner-lining head; 4 = parting bead; 5 = staff bead; 6 = top-sash rail.

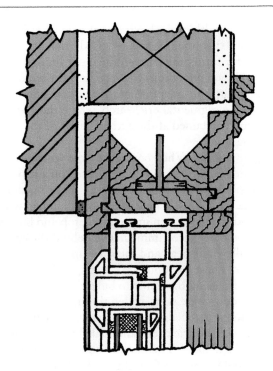

Figure 7.11(d) *Vertical section A-A* showing a uPVC casement window positioned against the protruding edges of the outer lining, with wide staff beads fixed internally. Note that this common practice leaves the protruding wooden edges of the outer linings (and the protruding edges of the wooden sill) exposed and vulnerable and requiring near-future maintenance.

Figure 7.11(c) *Vertical section A-A* showing the removal of the staff beads, the parting beads, the sliding sashes and the pulley wheels, in readiness for the uPVC casement window.

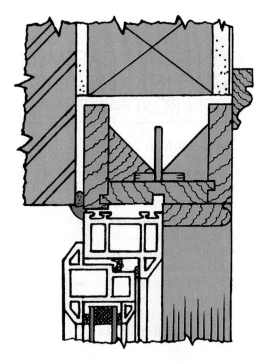

Figure 7.11(e) *Vertical section A-A* showing the protruding edges of the outer lining removed to allow the position of the uPVC casement window to be flush to the outer surface of the boxframe, with no small edges of timber requiring tedious maintenance.

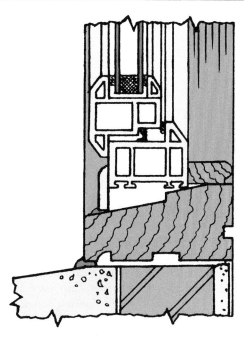

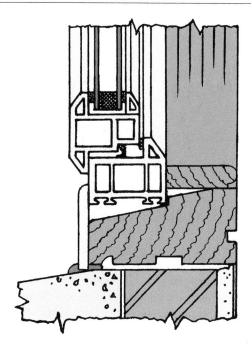

Figure 7.11(f) *Vertical section B-B* showing the situation at sill-level, when the uPVC casement window is positioned against the protruding edges of the outer linings. Note that the top- and side-edges of the wooden sill – as well as the protruding edges of the outer linings – will require painting on a maximum four-yearly cycle of maintenance.

Figure 7.11(g) *Vertical section B-B* showing the improved situation at sill-level when the protruding edges of the outer linings have been removed to allow the uPVC casement window to be flush to the outer surfaces of the boxframe, with no small edges of timber requiring tedious (often neglected) maintenance.

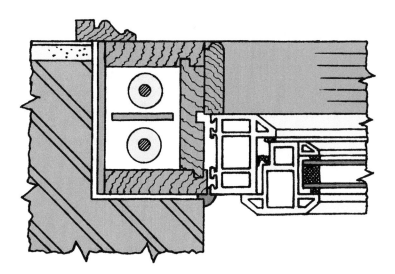

Figure 7.11(h) *Horizontal section C-C* showing the improved situation when the edges of the outer-linings have been removed to allow the uPVC window-frame to be flush to the outer face. Note that the removal of the outer-lining edges can quite easily be done by several techniques. One way is with a sharp, 25 or 32 mm firmer chisel and a mallet/hammer-side, by laying the chisel's angled, cutting-edge against and across the grain of the lining's protruding edge – and chisel-chopping deeply at 50 to 75 mm backward-spacings, so that the chopped, split fibres join up with each other. This technique is shown graphically in Chapter 18, at Figure 18.14(b); but a fussy finish is not required and, as indicated in *section C-C*, a slight protrusion of outer-lining can usually be accommodated.

8

Fixing Traditional and Modern Floor Joists and Flooring

8.1 INTRODUCTION

Although reinforced concrete floors of all kinds are used in large buildings such as blocks of flats or office blocks, timber floors are still widely used in domestic dwelling houses, especially above ground-floor level. Such floors were predominantly formed with traditional timber joists, but patented, engineered timber joists – covering wider spans – are also being used nowadays. In the UK, floors are generally referred to according to their position in relation to the ground. These range upwards from ground-floor level, first floor, second floor and so on; they may also be classified technically as single or double floors, according to the cross-formation of the structural members.

8.1.1 Single and Double Floors

Figure 8.1 (a): Suspended timber floors consist of thick board-on-edge timbers known as *joists*, spaced parallel to each other at specified centres across the

floor and, in the case of a *single floor*, resting between the extreme bearing points of the walls. In the case of a *double floor*, they rest on intermediate support(s) and the extreme bearing points of the walls. The top surface of the joists can be covered with various materials such as timber T&G floor boards, chipboard T&G flooring panels, plywood T&G flooring panels or Sterling OSB (oriented strand board) T&G flooring panels – and the underside of the joists covered with ceiling material such as Gyproc plasterboard.

8.1.2 Spacing of Joists

Figure 8.1(b): The spacing of the joists is related to the thickness of floor boarding or sheeting to be used; 400 mm centres (c/c) is required in domestic dwellings using ex. 22 mm T&G timber boarding, 18 mm chipboard T&G panels, 18 mm plywood T&G panels or 15 mm OSB T&G flooring panels. When the joists are spaced at 600 mm centres, three of these materials need to be thicker: the T&G timber boarding should be ex. 25 mm, the T&G chipboard should be 22 mm, and the OSB T&G panels should be 18 mm. Tongue and groove flooring panels are critically 2400 or 2440 mm long and 600 mm wide. The length needs to be considered when setting up the joists, because the staggered

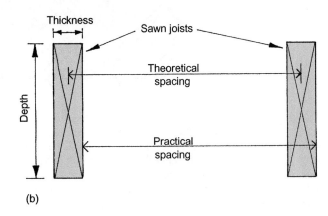

(a)

(b)

Figure 8.1 (a) Suspended floor joists; (b) spacing of joists cross-joints of these panels must bear centrally on the joists. Therefore, if the panel is a metric modular length of 2400 mm, then the joist-spacings should be 2400 ÷ 6 = 400 mm c/c, or 2400 ÷ 4 = 600 mm c/c. However, if the panel is based on an imperial length of 8 ft, converted to 2440 mm, then the joist-spacings should be 2440 ÷ 6 = 406.6 mm c/c (16 in) or 2440 ÷ 4 = 610 mm c/c (24 in). On upper floor levels, these considerations also apply to the cross-joints of the plasterboard sheets to be used later on in the ceilings below.

8.1.3 Size of Joists

The sectional size of joists is always specified and need not concern the site carpenter or builder. The subject enters into the theory of structures and mechanics and is, therefore, a separate area of study. However, for domestic dwellings, a simple rule-of-thumb calculation has existed in the trade for many years, expressed as

$$\frac{\text{span}}{2} + 2$$

but this is only an approximate method, which errs on the side of safety. In Imperial measurement, this was expressed as

$$\text{depth of joist in inches} = \frac{\text{span in feet}}{2} + 2$$

For example, if the span of the joists is 14 ft 0 in

$$\text{depth of joist} = \frac{14}{2} + 2 = 9 \text{ in}$$

In metric measurement, the formula is converted to

$$\text{depth of joist in centimetres} = \frac{\text{span in decimetres}}{2} + 2$$

For example, for a joist span of 4 m

$$\text{depth of joist} = \frac{40}{2} + 2 = 22 \text{ cm} = 220 \text{ mm}$$

The thickness of joists, by this method, is usually standardized at 50 mm. The nearest commercial size, therefore, would be 225 × 50 mm.

8.1.4 Structurally Graded Timber

By comparison, table A1 for floor joists, given in the Building Regulations' AD (approved document) A1/2, specifies joists of 220 × 38 mm section at 400 mm centres, for a maximum span of 4.43 m. However, it should be noted that SC3 (strength class 3 – now referred to as C16) structurally graded timber is specified. Such timber is now commonly relied upon for structural uses and is covered by BS 4978 – and BS 5268: Part 2: 1991. Certain standards and criteria are laid down regarding the size and position of knots, the slope of grain, etc., and the assigning of species and grade combinations to strength classes SC3 (C16) and SC4 (C24).

8.2 GROUND FLOORS

8.2.1 Suspended Timber Floor

Figure 8.2(a): The first of the various types of ground floor that involves the carpenter, is the suspended timber floor. Traditionally, joists of 100 × 50 mm

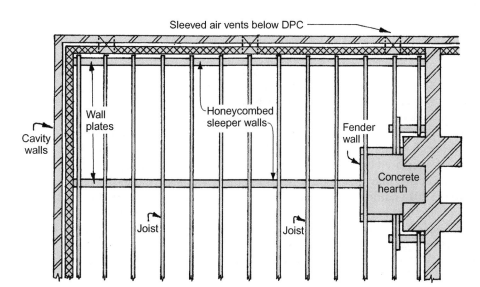

Figure 8.2 (a) Part-plan view of exposed floor

section were generally used, spaced at 400–600 mm centres. They rested on 100 × 50 mm timber wall-plates and were skew-nailed to these from each side with 100 mm round-head wire nails. The wall plates were bedded on half-brick-wide sleeper walls with a damp-proof-course material sandwiched in the mortar joint. The sleeper walls, which were honeycombed for underfloor air circulation, were traditionally built at 1.8 m centres to support the 100 × 50 mm sawn joists. Nowadays, if using structurally graded timber, designated as SC3 (structural class 3) – also known as C16 (class 16) – the walls would be built at 2.08 m centres to support the joists at 400 mm c/c, and 1.67 m centres to support the joists at 600 mm c/c. If SC4 (structural class 4) – also known as C24 (class 24) – joists were used, the walls could be built at 2.2 m centres to support the joists at 400 mm c/c, and 1.82 m centres to support the joists at 600 mm c/c. The honeycombed sleeper walls were usually built onto the concrete oversite – if adequate – rather than onto separate foundations. Of course, separate foundations were structurally better – but costlier.

8.2.2 Regulation Requirements

Since this book was first published in 1998, radical changes have been made to the Building Regulations' Part L of Schedule 1, concerning the *Conservation of fuel and power*. In an attempt to reduce CO_2 emissions and the effects of global warming, amendments to Approved Document L1 (covering dwellings) and Approved Document L2 (covering all buildings other than dwellings), came into force in April 2002, setting guidelines for higher levels of insulation and more efficient heating and lighting systems in dwellings and other buildings. ADL1 offered three methods to use for demonstrating that provision had been made to limit heat loss through the building fabric. These were (1) Elemental Method, (2) Target U-value Method and (3) Carbon Index Method. The Elemental Method was the less complex and aimed at meeting (or keeping below) the maximum U-value ratings (the rate at which heat passes through a material, or a mixture of different materials) of the individual elements – walls, floors and roofs, etc. – in a dwelling. As a guide to this, Maximum U-value tables for the building's exposed elements were given for reference.

However, further amendments have now been made and ADL1 and ADL2 have been replaced by four new Approved Documents, which came into force in April 2006 and are now in the 2013 edition, with 2016 amendments. It appears that the only changes made are to the wording; there are no technical changes. All entitled *Conservation of fuel and power*, ADL1A covers *new dwellings*, ADL1B covers *existing dwellings*, ADL2A covers *new buildings other than dwellings* and ADL2B covers *existing buildings*

other than dwellings. Under these new regulations, the methods of compliance have changed and the Elemental Method, plus the Target U-value Method are now omitted from being used in new dwellings.

The method of calculation for the energy performance is now referred to as the Target CO_2 emission rate (TER) calculations, which are applied to the whole building. In the case of existing dwellings, under ADL1B, the energy performance can be based on an Elemental Method to comply with U-value targets for the thermal elements.

As illustrated at (b) and (e), a possible way of upgrading this type of floor to meet the new regulations, would be to pack 100 mm thick mineral-wool quilt insulation (such as Crown Wool) between the joists, laid on support netting, and overlay the joists with 48 mm thick Celotex GA2000 (a rigid foam insulation slab). This would need to be supported on a sub-floor of 15 mm chipboard or OSB and overlaid with a finished floor decking.

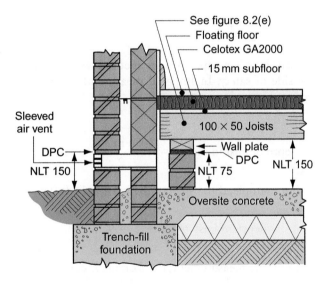

Figure 8.2 (b) Sectional view through suspended ground floor

Figure 8.2(b): Part C of the Building Regulations, which is concerned with protecting buildings from dampness, requires the site to be effectively cleared of turf and other vegetable matter and the concrete oversite to be of 100 mm minimum thickness and to a specified mix, laid on clean hardcore and finished with a trowel or spade finish. The top surface of the concrete oversite should not be lower than the highest level of the ground or paving adjoining the external walls of the building. The space above the concrete to the underside of the wall plates must not be less than 75 mm and not less than 150 mm to the underside of the joists. The space should be clear of debris (broken bricks, shavings, offcuts of timber, etc.) and be adequately through-ventilated with a ventilation area equivalent to 1500 mm² per metre run of wall in two opposite external walls.

Alternatively, the ground surface may be covered with at least 50 mm thick concrete, laid on a DPM of 1200 gauge polythene sheeting (or 1000 gauge if conforming to Agrément Certificate and PIFA standard), laid on a suitable, protective bedding material, i.e., sand. Note that if this alternative oversite covering is used, separate foundations would be required to support any honeycombed sleeper walls that may be required to support the joists.

Where external ground levels are higher than the internal levels of the oversite concrete, the oversite concrete must be laid to slope and fall to a perimeter outlet in the form of a gulley, sump or soakaway, above the lowest level of the external ground.

As illustrated, the damp-proof course in the cavity wall should be not less than 150 mm above the adjoining ground or paving – and the top of the cavity-fill should be not less than 150 mm below the level of the lowest damp-proof course (as indicated in Figure 8.9 for a *surface-battened floor*).

8.2.3 Bedding the Wall Plates

Figure 8.2 (c) Wall-plate joints

Figure 8.2(c): The first operation is to cut the wall plates to length, bearing in mind that the ends of these should be kept away from the walls by approximately 12 mm. After laying and spreading mortar on the sleeper walls, rolling out and flattening the DPC material, more mortar is laid and the wall plates are bedded and levelled into position with a spirit level. Once the first plate has been bedded and levelled, the others, as well as being levelled in length, must also be checked for level crosswise, using the first plate as a datum. If any wall plate cannot be laid in one piece, or changes direction, it should be jointed with a half-lap joint.

8.2.4 Laying and Fixing the Joists

Figure 8.2(d): When the wall plates are set, the joists can be cut to length and fixed in position by nailing or anchoring – bearing in mind that the ends of the joists should also be kept away from the walls by approximately 12 mm. The first joist is fixed parallel to the wall, with a 50 mm gap running along its wall-side face, partly to keep it away from the wall, but mainly to create easier end-of-board fixings. If timber floorboarding is to be used, the second joist can be fixed

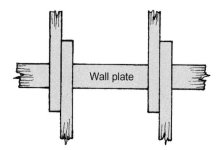

Figure 8.2 (d) Joining joists on sleeper walls

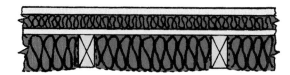

Figure 8.2 (e) Cross-section through floor at Figure 8.2(b), showing floating floor on 48 mm Celotex GA2000 slab insulation supported on a subfloor of chipboard or OSB. 100 mm thick Crown Wool packed between 100 × 50 mm joists, laid on support-netting fixed to sides of joists

at 400 mm centres from the first, but if edge-finished flooring panels are to be used, then the second joist should be fixed at 400 or 406 mm +12 mm expansion gap from the wall – *not* from the first joist. Subsequent joists are fixed at the required spacing until the opposite wall is reached. The last spacing is usually under or slightly over size. Joists joined on sleeper walls are usually overlapped, as illustrated, and side-nailed.

8.2.5 Providing a Fireplace and Hearth

Figure 8.2(f): Although omitted in recent years, there seems to be a demand for – and some return

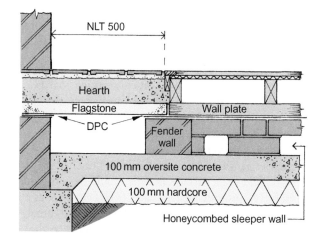

Figure 8.2 (f) Sectional view through hearth

to – traditional fireplaces able to take a gas-, electric-, or wood-burning-fire in the lounge or living room. Therefore, if a fireplace and the required concrete hearth are to protrude into the floor area, the hearth can be contained below floor level within a one-brick-thick *fender wall* built around the fireplace. The ends of the joists rest on wall plates supported by half the thickness of the fender wall. The other half supports the concrete hearth. Part J of the Building Regulations requires that no timber should come nearer to the fire opening than 500 mm from the front and 150 mm from each side (Figures 8.2(f) and (g)). Although not demanded by the Regulations, ideally the timbers should be pre-treated or treated on site with a preservative.

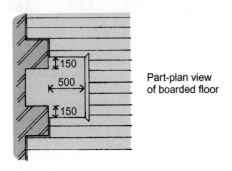

Figure 8.2 (g) Regulation size of hearth

8.2.6 Flooring Materials

The flooring material can be tongue and groove (T&G) timber boarding, flooring-quality chipboard T&G flooring panels, plywood T&G flooring panels or OSB (oriented strand board) T&G flooring panels. Their thicknesses must suit the joist-spacing, as described above.

8.3 LAYING T&G TIMBER BOARDING

Figure 8.3: When laying boarded floors, cross-joints (end grain or heading joints) should be kept to a minimum, if possible, and widely scattered. No two heading joints should line up on consecutive boards. On all sides, boards should be kept away from the walls by approximately 12 mm. This is to reduce the risk of picking up dampness from the walls and to allow for any movement across the boards due to thermal-expansion. All nails should be punched in 2–3 mm below the surface. Boards should be cramped up and fixed progressively in batches of five to six at a time. Tongues and grooves should be protected during cramping by placing offcuts of boarding between the cramps and the floor's edge. Fixings – cut floor-brads or lost-head wire nails – should be at least $2\frac{1}{2}$ times

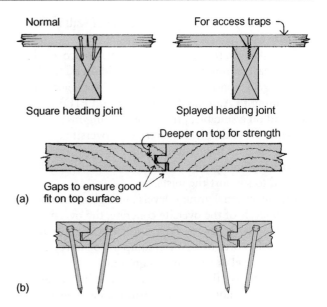

Figure 8.3 (a) The right way up for T&G boards; **(b)** Recommended dovetail-fixings

the thickness of the board, i.e. for 20 mm boards use 50 mm nails. There should be two nails to each board fixing, about 16 mm in from the edges. If using lost-head nails, each nail should be driven in at an angle to create the effect of a dovetailed fixing. Just prior to fixing, boards must be sorted and turned up the right way, as illustrated.

8.3.1 Fixing Procedure

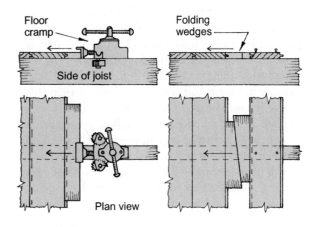

Figure 8.4 Cramping methods

Figure 8.4: Cramping can be done with patent metal floor cramps which saddle and grip the joists when wound up to exert pressure on the boards, or by using sets of folding wedges cut from tongued-and-grooved offcuts, 200–300 mm long. The first board is nailed down about 12 mm away from the wall, with small wedges inserted to retain the gap during cramping. Five or six more boards are cut to length and laid.

When using wedges as cramps, a seventh board is cut and partially nailed, set away from the laid boards at a distance equal to the least width of the pre-cut folding wedges. The wedges are inserted at about 1–1.5 m centres and driven in to a tight fit. The boards, having been marked over the centre of the joists, are then nailed. When complete, the wedges and the seventh board are released. This board becomes the first of another batch of boards to be laid and the sequence is repeated until the other wall is reached. The final batch of boards are levered and wedged from the wall with a wrecking bar or flooring chisel, then nailed; the last board having been checked and ripped to width to ensure a 12 mm gap from the wall on completion.

8.4 FLOATING FLOOR (WITH CONTINUOUS SUPPORT)

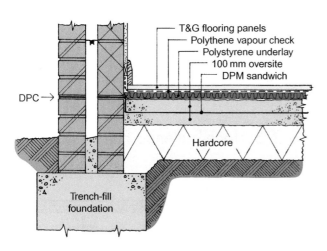

Figure 8.5 Floating floor

Figure 8.5: This type of floor, consisting of manufactured T&G flooring panels, laid and floating on close-butted 100 mm thick rigid polystyrene sheets, such as Jabfloor Type 70, and held down by its own weight and the perimeter skirting, turns a cold unyielding concrete base into a warm resilient floor at relatively low cost. The flooring panels used may be 18 mm T&G chipboard (preferably moisture-resistant grade), or 18 mm T&G plywood, with laid-sizes of 2400 × 600 mm or 2440 × 600 mm. Sterling OSB T&G flooring panels are not recommended for use on floating floors.

8.4.1 Laying Procedure

After the closely butted sheets of Jabfloor insulation have been laid over the concrete floor, a roll of 1000 gauge minimum thickness polythene sheeting is rolled out and laid over it to act as a vapour check. This should be turned up at least the flooring-thickness at the wall

edges and all joints should be lapped by 300 mm and sealed with waterproof adhesive tape, such as Sellotape 1408. Next, the tongued-and-grooved panels are laid, taking care not to damage the polythene and leaving an expansion gap of 10–12 mm around all walls and other abutments. The cross-joints must be staggered to form a stretcher-bond pattern and all joints should be glued with polyvinyl acetate adhesive, such as Febond. Laying is started against the wall from one corner and when the other corner is reached, any reasonably sized offcut can be returned to start the next row. Otherwise half a sheet should be cut to do this. Temporary wedges should be inserted around perimeter gaps until the glued joints set. A protective batten or flooring-offcut should be held against the flooring's edge when hammering panels into position. Finally, before fixing the skirting, the perimeter gap should be checked and cleaned out, if necessary.

8.4.2 Other Points to Note

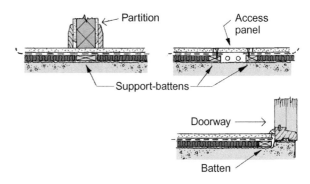

Figure 8.6 Battens required within the insulation

Figure 8.6: Any pipes or conduits to be accommodated within the insulation material must be securely fixed to the oversite concrete and the thickness of insulation material may need increasing to exceed the diameter of the largest pipe. Hot-water pipes should not come in direct contact with any polystyrene underlay. Preservative-treated battens should be fixed to the slab to give support to any concentrated load on the floor, such as a partition or the foot of a staircase, etc. Such battens are also required where an access panel is to be formed and where the floor abuts a doorway, as illustrated. As well as being stored carefully on site, flooring panels should be *conditioned* by being laid loosely in the area to be floored for at least 24 hours before fixing.

8.5 FLOATING FLOOR (WITH DISCONTINUOUS SUPPORT)

Figure 8.7: This type of floor is referred to as a *battened floating floor*. Staggered flooring panels are laid and fixed to a framework of 50 × 50 mm battens

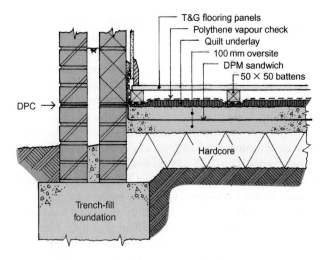

Figure 8.7 Battened floating floor

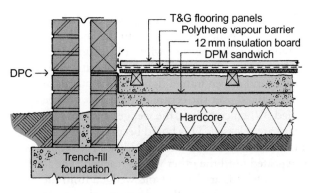

Figure 8.8 Embedded-fillet floor

(the final notes on upper floors give fixing details of panels on joists, which also apply here in fixing panels to battens). The battens are laid unfixed on 48 mm thick Celotex GA2000 rigid-insulation slabs, resting on a dry and damp-proofed concrete oversite and covered with a vapour barrier as described previously. When fixed to the battens, the floor is floating on the insulation and is held down by its own weight and the perimeter skirting.

8.6 FILLET OR BATTENED FLOORS

Another type of floor to be considered consists of 50 × 50 mm sawn fillets or battens, spaced at 400 mm or 600 mm centres and either embedded in the concrete over-site or fixed on top. Both of these applications are covered by Part C of the Building Regulations' Approved Document Regulation 7, and the detailed examples given here are aimed at meeting the requirements for timber in *floors supported directly by the ground*.

8.6.1 Embedded-fillet Floor

Figure 8.8: The timber fillets are splayed to a dovetail shape and must be pressure-treated with preservative in accordance with BS 1282: 1975, *Guide to the choice, use and application of wood preservatives*, prior to being inserted in the floor. The concrete oversite must incorporate a damp-proof sandwich membrane consisting of a continuous layer of hot-applied soft bitumen of coal tar pitch not less than 3 mm thick, or at least three coats of bitumen solution, bitumen/rubber emulsion

or tar/rubber emulsion. After the DPM has been applied and has set, the splayed fillets, having been cut to length and the cut-ends resealed with preservative, are bedded in position at the required centres and levelled. This can be done by placing small deposits of concrete in which the fillets are laid and tamped to level positions. When set, the top half of the concrete sandwich is laid, using the fillets as screeding rules. Again, 48 mm Celotex GA2000 rigid foam insulation can be used.

8.6.2 Surface-battened Floor

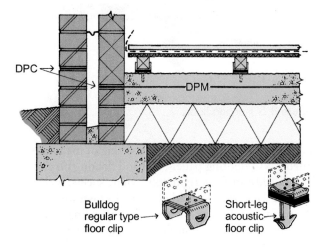

Figure 8.9 Surface-battened floor

Figure 8.9: A damp-proof sandwich membrane, as described above, is required and must be joined to the damp-proof course in the walls. Standard or acoustic Bulldog floor clips can be used to hold the battens in position at the required spacing. The clips are pushed into the plastic concrete at 600 mm centres within 30 minutes of laying and levelling. A raised plank is placed across the concrete to support the operative and a batten marked with the clip-centres is used to act as a guide for spacing and aligning the clips. When laying the battened floor, after the concrete is set and thoroughly dry, the ears of the clips are raised with the claw hammer, the battens are inserted and fixed with special friction-tight nails supplied with the clips. As illustrated, both floors may be insulated with 48 mm Celotex GA2000, covered with a polythene vapour barrier prior to being floored with T&G chipboard, plywood or OSB flooring panels.

8.7 BEAM-AND-BLOCK FLOORS

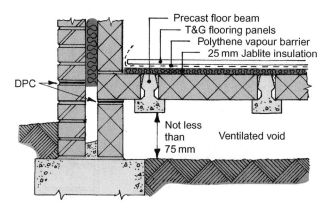

Figure 8.10 Floating floor on beam-and-block construction

Figure 8.10: This is a modern construction used at ground floor level on dwelling houses. The precast reinforced concrete floor-beams have to be ordered to the lengths and number required. They take their bearings on the inner-skin of blockwork and, as illustrated, must be at least 75 mm above ground level to the underside of the beams. They are spaced out to take the length of 440 × 215 × 100 mm standard wall blocks, laid edge to edge and resting on the beams' protruding bottom-edges. A row of cut blocks is inevitably required along one side, as illustrated. The gaps between blocks and beams are filled with a brushed-over sand-and-cement grout. The void beneath the precast beams should be provided with through-draught ventilation, continuous through any intermediate cross sleeper-walls, with an actual ventilation area equivalent to 600 mm^2 per metre run of wall. The floating floor laid on this sub-floor construction will be as described previously under 'Laying Procedure'. Note: By using AD (Approved Document) L1B's Elemental Method of calculating energy performance, the insulation shown above the beam-and-block floor would need to be either 100 mm (or two layers of 50 mm) Jabfloor Type 70, or 48 mm slabs of Celotex GA2000.

8.8 ENGINEERED-TIMBER FLOORS

8.8.1 Introduction

Engineered-timber products such as beams and joists have been around for many years now – in various forms – but have previously been mostly used in buildings of timber frame construction. However, they are now also used in buildings and dwelling houses of traditional masonry construction. Apart from the ecological benefit gained by these products using less of the limited forest resource compared with traditional solid-timber beams and joists, they can be manufactured to bridge larger spans between bearing points.

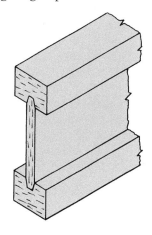

Figure 8.11 (a) Sectional shape of a TJI® joist

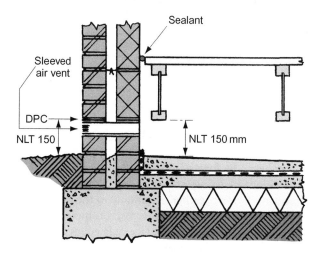

Figure 8.11 (b) TJI® joists to ground floor level. Insulation omitted for clarity (see Figure 8.11(c))

Figure 8.11(a)(b): The details illustrated here represent a suspended timber floor at ground floor level using engineered timber joists manufactured by the

Trus Joist company. The floor is referred to as the *Silent Floor*® *System* and the joists as *TJI*® *joists*; TJ obviously being the manufacturer's initials and the 'I' being a trade reference to an 'I Beam': the joists' sectional shape, as illustrated at (a). The joists are described as being made of continuous Microllam® laminated veneer lumber (LVL) flanges, which are groove-routed to house a central web of Performance Plus™ OSB (oriented strand board). In manufacture, the three components are joined together by a high-speed bonding process using a waterproof, synthetic resin adhesive and radio-frequency heating whilst under pressure.

Trus Joist's detailed illustrations in their technical literature show the TJI® joists built in to the inner-skin blockwork and state that where the external ground levels are higher than the internal subfloor levels, the internal subfloor surface (oversite concrete) should be sloped to fall to a perimeter drain, gulley, sump or soakaway – in compliance with Part C of the Building Regulations, as already detailed in this chapter (Regulation Requirements). If required – to keep the joists to a shallower depth over areas with large spans – intermediate, honeycombed sleeper walls can be used in this system.

8.8.2 Insulation Details

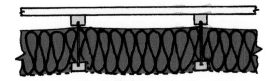

Figure 8.11 (c) 200 mm thick Crown Wool packed between TJI® joists, laid on support-netting

Figure 8.11(c): Typical insulation details for these floors are shown in the Trus Joist literature and are represented here. Reference is made to a mineral-wool quilt-type insulation being packed tightly between the webs of the TJI® joists. To achieve this, a trough or tray has to be formed at the bottom of the joists. This is done by fixing (stapling) galvanized wire, plastic mesh or a breather membrane to the bottom of the joists. In practical terms, this membrane would have to be fixed to the face-side edges of the bottom flanges, or laid snake-fashion over the joists. Alternatively, a slab-type, solid foam insulation may be used without the need for supporting mesh. Note that to comply with the amended regulations and guide details in Approved Document L1A, for new dwellings (outlined under Regulation Requirements in this chapter), the insulation of this floor would be subject to so-called Target CO_2 emission-rate (TER)

calculations for the energy performance of the whole building. However, if the floor only needed to comply with Approved Document L1B, for existing dwellings, the Elemental Method of complying with the U-value targets could be used. By this method, the insulation needs to be at least 200 mm thick between the webs – and packed into the gaps between the joists and the wall.

8.9 UPPER FLOORS

8.9.1 Timber Joists

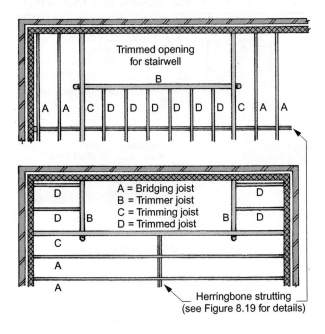

Figure 8.12 (a) Part-plan views of alternative joist-arrangements around trimmed opening

Figure 8.12 (a): In dwelling houses, suspended timber floors at first-floor level and above are usually single floors comprising a series of joists supported only by the extreme bearing points of the structural walls. These joists are called *bridging joists*, but any joists that are affected by an opening in the floor, such as for a stairwell or a concrete hearth in front of a chimney-breast opening are called *trimmer, trimming* and *trimmed joists*, as illustrated. Because the trimmer carries the trimmed joists and transfers this load to the trimming joist(s), both the trimmer and the trimming joists are made thicker than the bridging joists by 12.5–25 mm, or double joists, bolted together, are used. The depth of the joists, as mentioned in the opening pages of this chapter, does not usually concern the site carpenter or builder, such structural detail being the responsibility of the architectural team. However, if ever needed, the size of joists relevant to the span and joist-spacing, can be gained easily enough by reference to the *Span Tables* for floor joists, published by TRADA (Timber Research and Development Association) and available from HMSO Publications Centre or bookshops.

8.9.2 Framing Joints

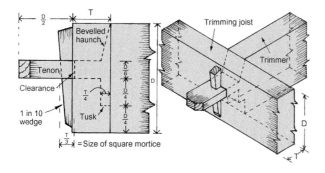

Figure 8.12 (b) Traditional tusk tenon joint (T = thickness; D = depth)

Figure 8.12 (b): Traditionally, a tusk tenon joint was used between the trimming joist and the trimmer – and is given here for reference only. This joint was proportioned as shown and was set out and cut on site with the aid of hand tools. The wedge was cut to a shallow angle of about 1 in 10 ratio to inhibit rejection, made as long as possible and, upon assembly, was driven into an offset draw-bore mortice in the tenon. The offset clearance that was needed to effect the drawn-tight fit between the two structural members, is indicated in the illustration. The slope on the bottom of the wedge was to facilitate entry and the top slope lent itself to the angle of the hammer blow with less risk of shearing the short grain. When jointing, particular care was taken to ensure that the bearing surfaces of the tusk and tenon were not slack against the stopped housing and the mortice.

8.9.3 Blind Tenon and Housing Joints

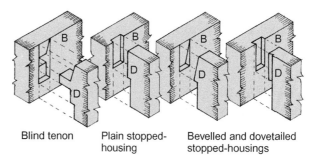

Blind tenon   Plain stopped-housing   Bevelled and dovetailed stopped-housings

Figure 8.12 (c) Traditional framing joints

Figure 8.12 (c): Traditional joints used between trimmed joists (D) and trimmer joist (B), varied between a *blind tenon* and a *plain stopped-housing*. Other joints, seen more in textbooks than in practice, included a *bevelled stopped-housing* and a *dovetailed stopped-housing*. The blind tenon joint was made to the same proportions as the tusk tenon, but did not have a wedge or projecting tenon. The plain stopped-housing joint was set out and gauged to cut into the *in situ* trimmer 12.5 mm on the top edge and half the joist-depth on the side. It was quickly formed by making three diagonal saw cuts across the grain (two on the waste side of the lines, one in the mid-area), chopping a relief slot at the bottom of the housing and chisel-paring from above.

8.9.4 Modern Framing Anchors

Figure 8.13: Metal timber-connectors are now extensively used to replace the above-mentioned joints, in the form of metal framing anchors and, more commonly, timber-to-timber joist hangers. The advantages to be gained in using these connectors are a saving of labour hours and, in the case of the hangers, more

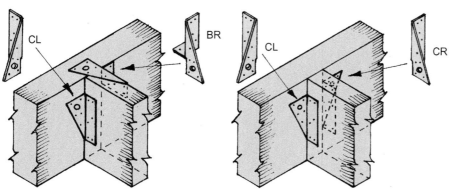

Trip-L-Grip BR and CL type   Trip-L-Grip CL and CR type

Figure 8.13 Modern framing anchors

effective support of the trimmer or trimmed joists, by the bearing being at the bottom of the load. However, it must be mentioned that traditional framing joints have held up to the test of time in houses of several hundred years of age. When using sherardized framing anchors, such as MAFCO Trip-L-Grip, for floor joists, the loads to be carried are such that each trimmed joint should comprise both a B type and a C or two C type anchors. When using two C types (CL and CR), one on each side of the joist, they should be slightly staggered to avoid nail-lines clashing. The anchors are recommended to be fixed with 3 mm diameter by 30 mm sherardized wire nails.

8.9.5 Timber-to-timber Joist Hangers

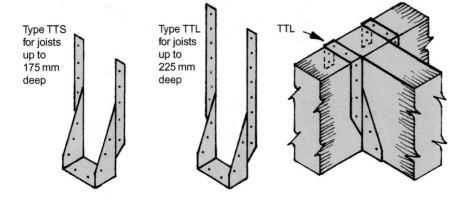

Figure 8.14 Timber-to-timber joist hangers

Figure 8.14: Steel joist hangers, such as those manufactured by Catnic–Holstran, type TT (timber to timber), S and L (short and long), are made from 1 mm galvanized steel with pre-punched nail holes. With the aid of a hammer, the straps are easily bent over the joists, as illustrated, and fixed with 32 mm galvanized nails. Another advantage of the thin-gauge metal is that hangers do not require housing into the top or bottom edges of the joists. Although perhaps not structurally necessary, it is advisable to place nail-fixings in all the available fixing holes.

8.9.6 Double Floors

Figure 8.15 (a)(b): The sensible structural rule of timber joisted floors, is that the joists should always bridge across the shortest span of an area, unless a double floor is required, whereby a steel beam (or beams) bridges the shortest span and the timber joists run the longest span, bearing on the intermediate beam(s), as illustrated. The protruding beam at ceiling level is encased in several ways to achieve fire-resistance and a visual finish. One way of doing this is to make a quantity of U-shaped frames – known as *cradles* or *cradling* – using 50 × 50 mm or 38 × 38 mm timber, with lapped or half-lapped, clench-nailed joints at each corner, and fix them, one to each joist-side, as indicated at Figure 8.15(b), close to the beam's bottom and

flange-edges. These frames are then clad with 12.5 mm plasterboard or other such non-combustible material – or clad as directed in the Works Specification.

8.9.7 Solid-wall Bearings

Figure 8.16: The old practice of building the ends of joists into solid (non-cavity) walls is now frowned upon, because of the increased risk of timber decay

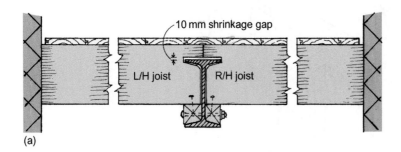

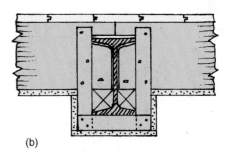

Figure 8.15 (a) Double floor; (b) cradling

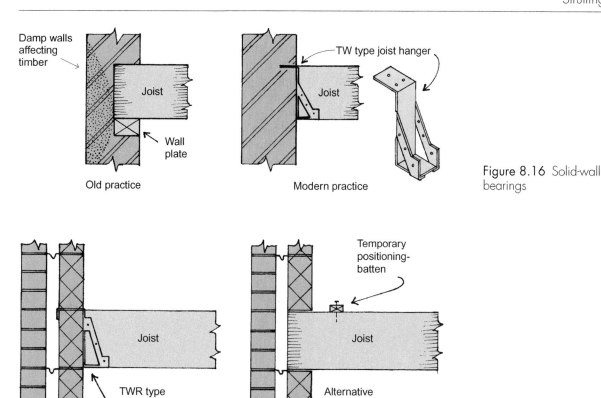

Figure 8.16 Solid-wall bearings

Figure 8.17 Cavity-wall bearings

through lateral damp-penetration. As illustrated, the modern practice is to use steel joist hangers, such as those manufactured by Catnic–Holstran, type TW (timber to wall), made from 2.5 mm galvanized steel. When fixing, 32 mm galvanized nails are recommended. Owing to a double metal-flange on the bottom, equalling a thickness of 5 mm, the bottom edge of the joists require notching out to achieve a flat surface for the plasterboard ceiling.

8.9.8 Cavity-wall Bearings

Figure 8.17: TW (timber to wall), or, as illustrated, TWR (timber to wall return) joist hangers with a turndown top flange to ensure correct and safe anchorage, especially when there is insufficient weight above, may be used for cavity walls. Alternatively, the ends of joists, which should be treated with timber preservative, are positioned, levelled up and built into the inner skin of the cavity wall. Care must be taken to ensure that the joists do not protrude past the inner face of the bearing-wall, into the cavity. The temporary positioning-batten, illustrated, should be attached to the scaffold or return wall at its end(s) to create stability and to stop the joists toppling sideways until they are built in.

8.10 STRUTTING

8.10.1 Introduction

Strutting in suspended timber floors is used to give additional strength by interconnection between joists. This removes the unwanted individuality of each joist and effects equal distribution of the weight and prevents joists bending sideways. Struts should be used where spans exceed 50 times the joist thickness. Therefore, with 50 mm thick joists, a single row of central struts should be used when the span exceeds 2.5 m and two rows are required for spans over 5 m and up to 7.5 m.

8.10.2 Solid Strutting

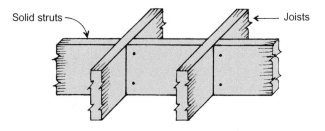

Figure 8.18 Common practice of strutting

Figure 8.18: The common practice of strutting a floor with solid struts or noggings (not *noggins*, the present-day spelling, which means a small glass of spirits!) was frowned upon traditionally as adding unnecessary weight and creating an inflexible floor, but nowadays, there are at least two exceptions where it is recommended: (1) Section 5 of the New Build Policy's Technical Manuals recommends that solid strutting should be used instead of herringbone strutting where the distance between joists is greater than three times the depth of the joists; and (2) the TRADA (Timber Research And Development Association) document: 'Span tables for solid timber members in floors, ceilings and roofs for dwellings' recommends that solid strutting should be used for certain long-span domestic floors. Note that the New Build Policy's Technical Manuals mentioned above are registered-builders' guidance notes supplied by Zurich Municipal Insurance, whose surveyors monitor the building work during construction and issue a certificate on completion, guaranteeing the building's fitness for a period of ten years. Such schemes nowadays are in competition with some local Building Regulations Control Departments, who have their own, similar schemes that offer LABC (Local Authority Building Control) certificates and guarantees on completion.

8.10.3 Herringbone Strutting

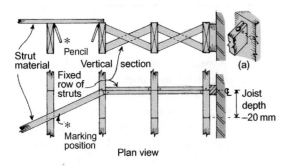

Figure 8.19 Plan and sectional view of herringbone strutting

Figure 8.19: This traditional method of strutting, using 38 × 38 mm or 50 × 38 mm sawn timber, although still effective and occasionally used, nowadays has to compete for speed with struts made of steel. The method of fixing involves marking a chalk line across the joists, usually in the centre of the floor as stipulated in the introductory notes. From this, marks are squared down the sides of the joists and – in the case of timber strutting – another line is struck on top with the chalk line, parallel to the first and set apart by the joist-depth minus 20–25 mm. As illustrated, the strutting material

is laid diagonally within these lines and marked from below to produce the required plumb-cuts (vertical faces of an angle). Cutting and fixing the struts is done in a kneeling position from above, using 50–63 mm round-head wire nails. Prior to fixing the struts, the joists running along each opposite wall should be packed – with wedges, technically – and nailed immediately behind the line of struts. As indicated at (a), a sawcut can be made at the end of each strut to receive the nail fixing and eliminate splitting.

8.10.4 Catnic Steel Joist-struts

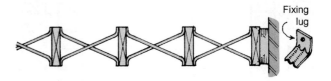

Figure 8.20 Catnic steel joist-struts

Figure 8.20: As illustrated, at least two types of galvanized steel herringbone struts are produced to compete with wooden strutting. The first type, by Catnic–Holstran, have up-turned and down-turned lugs for fixings with minimum 38 mm round-head wire nails. As before, fixing is done from above.

8.10.5 Batjam Steel Joist-struts

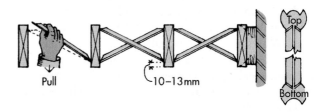

Figure 8.21 Batjam steel joist-struts

Figure 8.21: The second type, with the 'BAT' trademark, has forked ends which simply bed themselves into the joists when forced in at the bottom and pulled down firmly at the top. This time, fixing is done (more safely) from below. One disadvantage with steel strutting, which is made to suit joist centres of 400, 450 and 600 mm, is that there are always one or two places in most floors that do not conform to size and require reduced-size struts. When this occurs, it is necessary to use a few wooden struts in these areas.

8.10.6 Fixing-Band Strutting

Figure 8.22: Another metal fixing-device, sometimes used for strutting, is *Fixing Band*. This is composed of

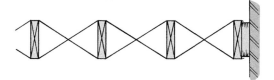

Figure 8.22 Fixing-Band strutting

a 20 gauge × 20 mm wide galvanized-steel band, usually in 10 m rolls. The band is perforated with continuous fixing holes and is easily bent to any required shape and cut to length with a hacksaw. As indicated in the illustration, the band is fixed to form a continuous, taut zigzag pattern over the tops and bottoms of the joists. This is best done up against a line marked over and squared down the face of each joist. Then another band is fixed across the joists, against the other side of the line. This second band is fixed on the alternate tops and bottoms missed by the first band. The downside with Fixing-band strutting, is that the joists near each opposite wall would need to be fixed firmly to the walls, through the 50 mm packings, to offset the pulling effect of the taut banding. Alternatively, the band could be turned up (or down) against the wall at each end and fixed firmly into the blockwork. But this is not a very practical solution.

8.10.7 Horizontal Restraint Straps

Figure 8.23: Modern construction methods involving lighter-weight materials in roofs and walls, has led to the need for anchoring straps, referred to in *The Building Regulations' Approved Document A*, to restrict the possible movement of roofs and walls

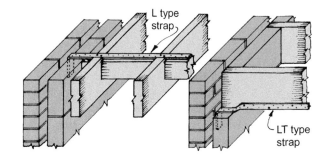

Figure 8.23 30 × 5 mm restraint straps for joists parallel or at right-angles to wall

likely to be affected by wind pressure. Such straps are made from galvanized mild-steel strip, 5 mm thick for horizontal restraint and 2.5 mm thick for vertical restraint. The straps are 30 mm wide and up to 1.6 m in length. Holes are punched along the length at 15 mm staggered centres.

8.10.8 Joists Parallel or at Right Angles to Wall

As illustrated, the straps, which may be on top or on the underside, require notching-in when the floor joists run parallel to the supported wall, so as not to clash with the seating of the flooring or ceiling materials – but only require surface-fixing on the sides when the joists are at right-angles to the supported wall. The walls should be anchored to the floor joists at centres of not more than 2 m on any wall exceeding 3 m in length, including internal load-bearing walls, irrespective of length. When joists run

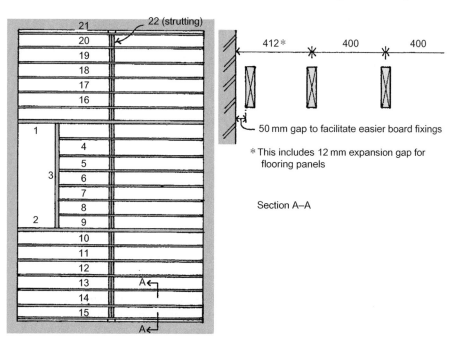

Figure 8.24 Sequence of fixing joists

parallel to the wall to be supported, the straps should bridge across at least three joists and have noggings and end-packing tightly fitted and fixed between the bridged spaces. The noggings should be at least 38 mm wide by half the depth of the joists. If the straps are fixed on the underside, the noggings and packing should be equal to the depth of the joists. There should be at least four 38 mm × 4.0-gauge screw-fixings in each strap.

8.11 FITTING AND FIXING TIMBER JOISTS

8.11.1 Joists Built-in to Cavity Walls

When the load-bearing walls have been built up to storey height and allowed to set, the joists may be fitted. After cutting to length and sealing or re-sealing the ends with preservative, the joists – with any cambered edges turned upwards – are then spaced out to form the skeleton floor and temporary battens are fixed near each end to hold the joists securely in position. The side-fixed restraint straps are then fixed and the joists are built-in by one course of blocks being laid all round. When set, the notches may be cut for those restraint straps that run across the joist tops, the 50 mm packing and the noggings cut and fixed, then the straps are screwed into position – and the blockwork may be continued.

8.11.2 Sequence of Fixing Joists

Figure 8.24: Normally, the first consideration is to position the trimming joists and trimmer of any intended opening, then, from this formation, the trimmed and bridging joists are spaced out. Joists should be checked for alignment with a straightedge or line and, if necessary, packed up with offcuts of thin material such as roofing-felt or oil-tempered hardboard – or lowered by minimal paring of the joist-bearing area. However, if regularized joists are used, the need for vertical adjustments should be eliminated. Herringbone strutting is fixed later, after the bricklayers have finished their work.

8.11.3 Joists Bearing on Joist Hangers

Figure 8.25: When joist hangers are to be used as an alternative to the joists being built-into the inner skin of blockwork, the blockwork is built to the top of the floor-joist level and the joists are cut to length and positioned at the same time as placing the joist

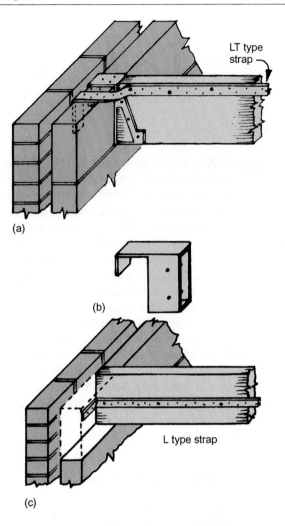

(a)

(b)

(c)

Figure 8.25 (a) Restraint strap in relation to joist hanger; **(b)** Restraint-type joist hanger; **(c)** Restraint-strap position when joist has a wall-bearing

hangers. The joists must be cut carefully to the correct length, to hold the hangers snugly against the walls, yet not push the blockwork from its set position. As illustrated at (a), the restraint straps are fixed to the upper edge on the side, when the joists are at right-angles to the supported wall, or, alternatively, restraint-type joist hangers can be used (Figure 8.25(b)) where required.

8.12 FIXING FLOORING PANELS ON JOISTS

8.12.1 Introduction

Tongue and groove (T&G) flooring panels of chipboard, plywood or Sterling OSB, machined on all four edges with compatible tongue and groove profiles and with laid-measure dimensions of 2400 or 2440 mm × 600 mm, can be laid on suspended timber floors in the following thicknesses related to joist-spacings: 18 mm chipboard, 18 mm plywood and 15 mm Sterling OSB for joists at 400 mm centres – and 22 mm chipboard, 18 mm plywood (as before) and 18 mm Sterling OSB for joists at 600 mm centres.

8.12.2 Fixing Procedure for T&G Panels

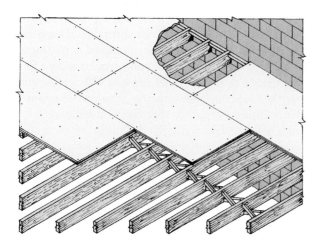

Figure 8.26 Fixing T&G flooring panels

Figure 8.26: The T&G boards, as illustrated, are laid with the long edges across the joists. The short edges bear centrally on the joists and *only* the long edges against the walls must be supported by noggings of at least 38 mm width – but preferably of 50 mm width and 75 mm depth. The boards should be nailed with three or four nails to each joist, two at about 25 mm from each edge and one or two nails equidistant between. The nails should be 45 mm annular-ring shank type for floor thicknesses up to 18 mm and 56 mm for floors of 22 mm thickness. All joints should be glued with polyvinyl acetate (PVA) adhesive, preferably the waterproof type. Gluing of joints, which is often skimped, is important to eliminate an aggravating squeaky floor.

8.12.3 Fixing Square-edged Panels

Figure 8.27: These boards, as illustrated, are laid with the long edges bearing centrally on the joists. *All* short edges, including the edges against the walls, must be supported with, preferably, 75 × 50 mm noggings tightly fitted and fixed between the joists. As before, the boards should be fixed with 45 mm or 56 mm

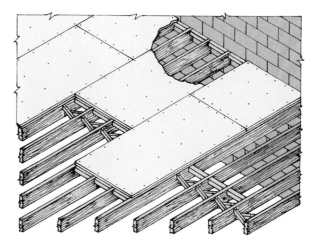

Figure 8.27 Fixing square-edged flooring panels

annular-ring shank nails at 300 mm centres around the edges and at 400–500 mm centres on intermediate joists. Nail fixings should be at least 9 mm in from the edge of the boards.

Cross joints on both types of board must be staggered and expansion gaps of 10–12 mm allowed around the perimeter of walls and any abutment. Sterling OSB square-edged panels are recommended to have a 3 mm expansion gap between boards in addition to the perimeter gap. Traps in the floor must be supported on all four edges and fixed with 45 mm × 4.0 gauge countersunk screws. As stated on floating floors, all panels should be *conditioned* by laying loosely in the area to be floored for 24 hours before being fixed.

8.13 FITTING AND FIXING ENGINEERED JOISTS

8.13.1 Introduction

As mentioned in the ground floor details, engineered joists are not only used in buildings of timber frame construction, but are now used in buildings and dwelling houses of traditional masonry construction – especially above ground-floor level. The details and illustrations in this section, therefore, cover the fitting and fixing of a Silent Floor® System of Trus Joist's TJI® joists at or above first-floor level.

8.13.2 Main Structural Components

Figure 8.28(a)(b)(c): The three main components in Trus Joist's Silent Floor® System for residential buildings of masonry construction are: **(a) TJI® joists** – as

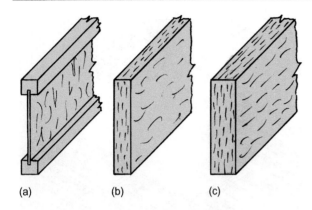

Figure 8.28 (a) TJI® joist; (b) TimberStrand™ LSL joist or beam; (c) Parallam® PSL joist or beam

previously described in the paragraph headed *Figure 8.11(a)(b)* – of varying depths and material sizes according to the floor span; **(b) Timber Strand™ LSL** (laminated strand lumber), which is a manufactured resin-bonded timber product of great strength. It is produced in wide slabs or billets and cut up into rectangular standard sections to be used as trimming or trimmer joists and beams or lintels for short and intermediate spans; **(c) Parallam® PSL** (parallel strand lumber) is also a manufactured resin-bonded timber product, but with superior strength, stiffness and dimensional stability. It is used for carrying greater loads and/or achieving greater spans in the Silent Floor® System. Parallam® PSL is also claimed to be aesthetically pleasing, if its natural, laminated appearance is left exposed.

8.13.3 TJI® Standard Sections

Figure 8.29(a)(b)(c)(d): The standard TJI® joist sections available in the Silent Floor® System are:
(a) Series reference number 150, with a 9.5 mm web thickness, 38 × 38 mm top and bottom flanges and a depth (height) of 241 mm or 302 mm;
(b) Series number 250, with a 9.5 mm web, 45 mm wide × 38 mm deep flanges and a depth of 200 mm, 241 mm, 302 mm, 356 mm or 406 mm; **(c)** Series 350,

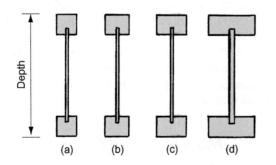

Figure 8.29 (a) Series 150 joist; (b) Series 250 joist; (c) Series 350 joist; (d) Series 550 joist

again with a 9.5 mm web, but top and bottom flanges of 58 mm width × 38 mm depth and with joist-depths equal to Series 250; **(d)** Series 550, with a 11.1 mm web, 89 mm wide × 38 mm deep flanges and joist-depths equal to Series 250 and 350.

8.13.4 Joist-size, Span and Spacing

Trus Joist's Technical Guide notes for the UK and Ireland contain span tables giving joist-size and joist-spacing related to spans for residential applications. The spacing centres used are 300 mm, 400 mm, 480 mm and 600 mm. The tables refer to clear spans for uniformly loaded joists and they assume the provision of a 22 mm chipboard floor and a directly applied plasterboard-type ceiling for joists spaced at 600 mm – and a 18 mm chipboard floor and a similar ceiling for joists spaced at 300 mm, 400 mm and 480 mm. Apart from the Guide information, the Trus Joist company offer technical support to specifiers and builders throughout the UK and Ireland. This includes training in the provision of specification and installation. Also, leading-edge automation tools are available and include: **TJ-Beam**® software, which produces single-member sizing options in floors, and **TJ-Xpert**® software, which automatically tracks loads throughout the structure and develops sizing solutions, material lists, framing plans and installation details.

8.13.5 Typical Upper-floor Layout

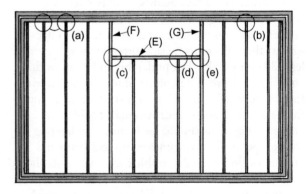

Figure 8.30 Typical upper-floor layout, indicating the various Trus Joist framing arrangements, (a) to (e), detailed in Figure 8.31

Figure 8.30: The illustration shows a simple, yet typical upper-floor layout of TJI® joists bridging across the shortest span, either bearing on the inner-skin block wall or, alternatively, bearing on joist hangers with their top flanges bearing on the block wall. A stairwell opening is shown typically against a structural wall,

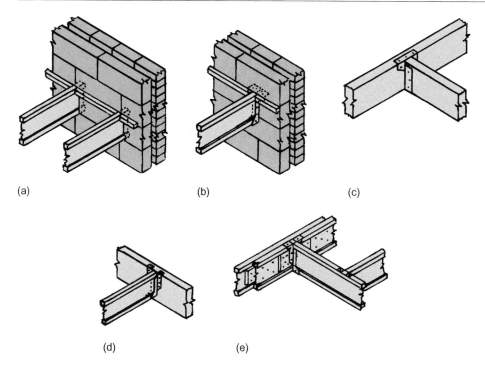

(a)   (b)   (c)

(d)   (e)

Figure 8.31 (a–e) Trus Joist framing details

but may of course be anywhere in the mid area; the main point being that the opening creates a transference of load to the joist marked (E), which in turn will transfer the imposed load to the joists marked (F) and (G). Traditionally, joist (E) would be referred to as a trimmer joist and joists (F) and (G) as trimming joists. The short bridging joists would be referred to as trimmed joists. In Trus Joist's Silent Floor® System, the components shown here acting as trimmer and trimming joists would be either TimberStrand™ LSL or Parallam® PSL beams. Alternatively, as illustrated at (e) in the framing details in Figure 8.31, TJI® joists may be joined together to carry an additional load. Finally, Trus Joist's Guide notes state that herringbone strutting (referred to by them as bridging) or solid strutting (mid-span blocking) is not required, but may enhance the floor's performance if properly installed.

8.13.6 Framing Details

Figure 8.31(a)(b)(c)(d)(e): Although Trus Joist's Guide notes show a wide range of framing details, the five examples given here only represent those that could be used in the typical upper-floor layout illustrated in Figure 8.30. Detail (a) shows TJI® joists bearing on the inner-skin blocks of an external cavity wall; (b) shows a TJI® joist supported by a Simpson's JHMI single-sized joist-hanger. Note that timber web-stiffeners will be needed with these joist hangers to eliminate lateral movement at the top of the web, within the hanger, and to enable the hangers to be fixed through their side holes. Also, joists supported by

hangers built into walls, should have at least 675 mm of cured (set) masonry built above the hangers' flanges before the floor has any construction materials (loads) placed upon it; (c) shows a TimberStrand™ LSL trimmer beam to trimming beam arrangement, supported by a Simpson's WPI timber-to-timber top-flange joist hanger; (d) shows a TJI® joist supported by a TimberStrand™ LSL trimmer beam, via a Simpson's ITT timber-to-timber, single-sized joist hanger; (e) shows a TJI® joist to a double TJI® joist trimmer beam arrangement using – as in the framing detail at (d) – an ITT timber-to-timber hanger. Note that there are side lugs on the ITT hangers that must be bent over and nailed to the top edges of the bottom flanges with 3.75 × 38 mm nails.

8.13.7 Restraint Strap Details

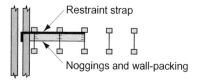

Figure 8.32 Restraint strap slotted through TJI® webs

Figure 8.32: The need for restraint straps and their legal requirement and application to floors above ground-floor level, is covered in detail under section 8.10.7 (Horizontal Restraint Straps) of this chapter. However, as illustrated, when TJI® joists are parallel to the structural wall, the straps cannot be notched-in to the flanges (as is the practice with solid timber joists) and need to be inserted through slots cut in the webs, 38 mm wide × strap-thickness + tolerance deep. The slots should be close to the underside of the top flanges, with not less than 3 mm of web remaining. The 38 mm × half joist-depth noggings and the timber wall-packing should be of C16 structurally graded timber. The Trus Joist Guide notes do not show a detail of wall restraint when the TJI® joists are at right angles to the wall (this may be in their software), but it seems feasible to either fix them to the vertical face of the top or bottom flanges, or stiffen-up the web thickness with plywood or OSB and fix the straps to the top or bottom area of the web. Finally, note that the Trus Joist details indicate a minimum 25 mm gap between the wall and the TJI® joist parallel to it.

8.13.8 Perimeter Noggings

Where the TJI® joists are at right angles to a structural wall, noggings are fitted between the joists and against the wall and are skew-nailed with 65 mm or 75 mm round-head wire nails. Note that these perimeter noggings provide lateral restraint when the joists are on hangers, but otherwise are only needed when using sheet decking material such as chipboard, OSB or plywood, to provide edge-support; they can be omitted when using ex. 25 mm (20 to 21 mm finish) T&G boarding for decking, providing the joists are built-in to the wall. However, noggings can be used solely to achieve lateral joist-support and positional spacing. This could particularly improve the working stability of a joisted-layout, seated on block walls, waiting to be built-in. Because skew-nailing would be impractical in this situation, Simpson's Z35N clips could be used. Noggings are usually of sawn finish and may be cut from timber sizes of 38 × 50 mm, 50 × 50 mm, or 50 × 75 mm.

8.13.9 Permissible Holes in Joists and Beams

Round holes and square or rectangular holes can be drilled or cut into the webs of TJI® joists, but only in specific places and of certain sizes as specified in Trus Joist's Hole Charts (too detailed to reproduce here). The holes may be drilled or cut anywhere vertically in the webs, but not less than 3 mm of web must be left at the top and/or bottom – and joist-flanges must not be cut at all. Note that TJI® joists are manufactured with 38 mm Ø perforated knockout holes in the web, at approximately 300 mm centres along the joist's length. These do not affect other holes that may be required, providing they conform

to the charts' criteria. Rectangular or square holes are not allowed in uniformly loaded beams comprised of TimberStrand™ LSL or Parallam® PSL. Furthermore, round holes above 50 mm Ø are not allowed – and holes that may be drilled at or below this diameter must conform to the chart's criterion regarding position. Finally, note that cuts or notches are not allowed in the bottom or top edges (chords) of beams, or in the bottom or top flanges of TJI® joists.

8.13.10 Safety Bracing

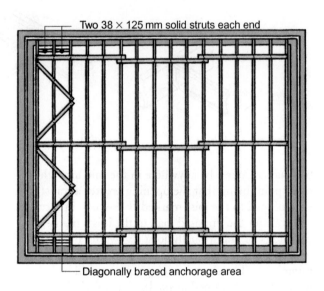

Figure 8.33 Safety bracing to newly joisted TJI® floor

Figure 8.33: For safety's sake, it must be realized that newly joisted floors are unstable until all the end bearings are complete and the 'green' masonry alongside built-in joists or above built-in joist-hangers has cured (set). Workers must not be allowed to walk on the joists, and construction loads must not be laid on the joists, until longitudinal safety bracing (with a recommended minimum 19 × 89 mm section) is attached to the tops of joists, as illustrated. Also, at least two 38 × 125 mm timber blocks – acting as solid strutting – must be fixed at each opposite end of the layout (two blocks each end), between the webs of the joists, to create an anchorage area of lateral restraint. The top bracing battens, starting from above the completed anchorage-area, are recommended to be fixed at every joist with two 3.35 × 65 mm nails (left with their heads protruding). Longitudinal bracing should be positioned near the bearings and in mid-positions between 1.5 mm and 2.4 mm, according to the Series' number of the TJI® joists used. Safety bracing should only be removed at the decking stage, progressively as the decking is laid – not all at once.

8.14 POSI-JOIST™ STEEL-WEB SYSTEM

8.14.1 Introduction

This section covers the fitting and fixing of the **Posi-Joist™** system of engineered floor-joists at or above first-floor level (although they can also be used at ground-floor level, if in compliance with Part C of the Building Regulations, as described in section 8.2.2 (Regulation Requirements) of this chapter). Although Posi-Joists are also used as roof components and in buildings using a structural inner-skin of timber-frame construction, the references here are only intended for floors in dwelling houses of masonry construction. Furthermore, general floor-joisting principles and other details – including safety bracing – already covered in this chapter, will not be repeated here.

8.14.2 Component Details and Advantages

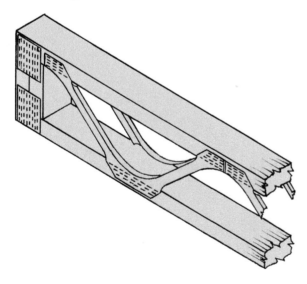

Figure 8.34 Part view of a typical Posi-Joist

Figure 8.34: Posi-Joists are a combination of stress-graded structural timber flanges, joined by the **MiTek Posi-Strut™** galvanized-steel web system. The timber flanges for the different depths of standard joists, vary only in width. The depth of the flanges is consistent at 47 mm finish. The width is either 72 mm or 97 mm finish, determined by joist-spacing related to span. There are four standard Posi-Joist™ depths, which are referenced by a **PS** (Posi-Strut) number. **PS8** = 202 mm deep, **PS9** = 223 mm, **PS10** = 254 mm, and **PS210** = 302 mm deep. An overview of how these

referenced joist-depths relate to span, is shown in MiTek's chart reproduced at Figure 8.35.

These relatively lightweight, easily manageable joists' main advantage is their unique open-web design, which allows electrical cables, plumbing pipes and venting ducts to be laid quickly underfloor in any direction. Other benefits include achieving greater spans without the need for intermediate load-bearing walls (note that if the span exceeds 4 m, a strong-back is installed at mid-span, as shown here in the separate, illustrated details); even over long spans, no herring-bone strutting is required; Posi-Joists can be ordered with solid-block trimmable ends, for on-site reductions of up to 150 mm each end.

8.14.3 Typical Floor Spans

Joist reference	Spacings	TR26 Flange size	Max span
PS8	400	72 × 47	5.100
		97 × 47	5.560
	600	72 × 47	4.310
		97 × 47	4.720
PS9	400	72 × 47	5.220
		97 × 47	5.710
	600	72 × 47	4.330
		97 × 47	4.730
PS10	400	72 × 47	5.980
		97 × 47	6.550
	600	72 × 47	4.760
		97 × 47	5.520
PS210	400	72 × 47	6.920
		97 × 47	7.800
	600	72 × 47	5.550
		97 × 47	6.640

Figure 8.35 Floor spans over bearings

Figure 8.35: As mentioned above, MiTek's chart of Typical Floor Spans is a simple overview of spans measured over bearings, which will give an indication of the particular joists required. Full load and clearly detailed information is available from local Posi-Joist™ manufacturers or from MiTek Industries.

8.14.4 Posi-Joist™ Trim-It™ Floors

Figure 8.36: Another innovative engineered floor-system pioneered by MiTek®, is the Posi-Joist™ Trim-It™ floor, with all the advantages of its open-web configuration and a unique solid-end insert

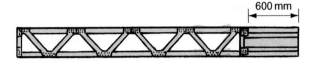

Figure 8.36 Trim-It joist with 600 mm solid trimmable end

allowing up to 600 mm to be trimmed on site from one end of each joist – if necessary. This system offers a range of five spans at a joist-depth of 253 mm, for standard floor loads. The available spans are 3.3 m, 3.9 m, 4.5 m, 5.1 m and 5.7 m. Bearing in mind that each joist-span has the capability of multiple reductions of the 600 mm Trim-It end, the range of possible spans available from the stocked range of five, runs into high double figures.

8.14.5 Posi-Joist™ Standard Details

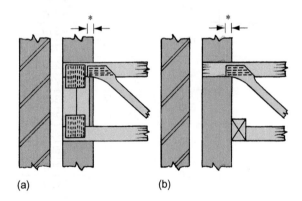

(a)   (b)

Figure 8.37 (a) Bottom-chord bearing; (b) top-chord bearing
* Posi-web must be 15 mm minimum over the bearing

Figure 8.37(a)(b): One of the prime considerations of any joist-arrangement is the wall-bearing detail. The first option shown above is referred to in MiTek's literature as a bottom-chord bearing into masonry (the inner-skin (blockwork) of a cavity wall). As illustrated, this is where the square end of the Posi-Joist™ is seated on the wall, flush to the rear face of the blockwork, and built-in at the sides and above. The second option above is referred to as a top-chord bearing. The top chord (flange) only is seated on the wall and built-in and the bottom chord (flange) butts up against a continuous 47 × 72 mm so-called ledger fixed to the wall. The end of the chord should touch the ledger, but is allowed a maximum gap of 6 mm. The metal Posi-web, as illustrated, must overrun the wall-bearing by at least 15 mm.

8.14.6 Joist Hanger to Inner Cavity-wall

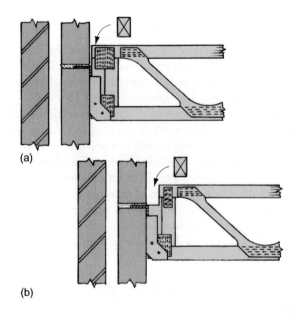

(a)

(b)

Figure 8.38 (a) Posi-Joists bearing on wall hangers with 47 × 72 mm noggings between top chords; (b) special notched ends of Posi-Joists to allow a continuous 47 × 72 mm restraint ledger to be inserted

Figure 8.38(a)(b): The third and fourth wall-bearing options are shown above. As illustrated, the ends of the Posi-Joists bear into wall hangers, which are fixed to the joists with 3.75 × 32 mm galvanized plasterboard nails. The choice of hanger and minimum bearing is determined by the floor-design, the load, the joist-width (72 or 97 mm) and the blockwork level for the hanger's bearing flange. Option (a) above, has 47 × 72 mm restraint-noggings fixed between the top chords, but option (b) shows the advantage (by being more easily fixed) of a continuous restraint ledger, facilitated by the manufacturer's special end-blocking.

8.14.7 Staircase Openings

Figure 8.39(a): The stair-opening shown below uses so-called two-ply joists as trimming joists. These are formed on site by fixing two Posi-Joists together with the aid of Cullen UZ47 clips, or similar, as specified by the design. The single Posi-Joist™ shown here, attached to the trimming joists with Simpson LBV or Cullen OWF/OWT hangers, is acting as a trimmer joist, carrying the trimmed joists. If carrying above a certain number of trimmed joists, it too might have to be of two-ply construction. The trimmed joists are attached to the trimmer with similar hangers. Note

Overlay Flooring 107

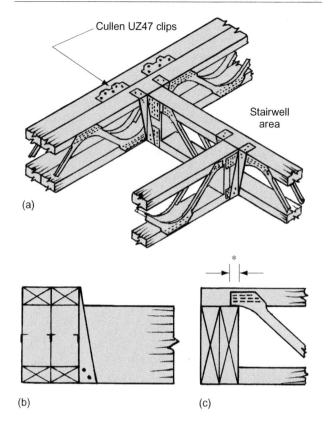

(a)

(b)   (c)

Figure 8.39 (a) Posi-Joist arrangement around stairwell; (b) two-ply trimming joist with solid trimmer on joist hanger; (c) sectional view through solid trimmer showing trimmed joist bearing upon it.
* Note that Posi-web must be 15 mm minimum over the bearing

that the bottom flanges of all hangers need to be notched into the Posi-Joists.

Figure 8.39(b)(c): An alternative stair-opening detailed above uses similar two-ply trimming joists, but solid timber or LVL (laminated, engineered timber) as a trimmer joist. The trimmer is fixed to similar Simpson or Cullen joist hangers, but the lowered, top surface of this trimmer acts as a bearing for the top chords of the trimmed joists nailed to it. The bottom chords should butt up against the trimmer. Note that the metal Posi-web must overrun the trimmer-bearing by at least 15 mm and, as before, the bottom flanges of the hangers need to be notched into the trimmer joist.

8.14.8 Horizontal Wall-Restraint

As covered already in this chapter, where horizontal wall-restraint is needed to meet the requirements of *The Building Regulations' Approved Document A*, horizontal restraint straps are fixed. When Posi-Joists are parallel to the wall, the straps can be hooked through the blockwork and fixed to the side of a timber plate fixed across the topside of the bottom chords, at

right-angles to the wall. The plate should be of at least 35 × 97 mm section and of Class C16 timber, twice fixed to each of three joists with 3.1 × 75 mm galvanized wire nails. The strap should be fixed with a minimum of four No 4 gauge × $1\frac{1}{2}$″ (4.0 × 38 mm) screws, one at each end and others equidistant between. If restraint straps are required when Posi-Joists are at right-angles to a wall, they can be screwed to the side-edges of the joists' flanges.

8.14.9 Strongbacks

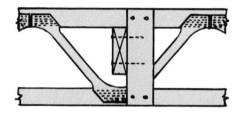

Figure 8.40 Sectional view of a 35 × 150 mm 'strongback' fixed to a vertical block

Figure 8.40: If the joist-span exceeds 4.0 m, a strongback strut is required in mid-span to stiffen the floor. If more than one strongback is required, they should be fixed at a maximum 4.0 m centres. The size, detail and position of strongbacks, is usually specified by the manufacturer. The illustration above shows the simplest of the strongbacks running through the open webs at right-angles to the joists, twice nailed to 38 × 75 mm blocks. The blocks are also twice nailed to the top and bottom chords. The nails used are 3.1 × 75 mm galvanized wires. As illustrated, the strongback must be kept tight to the underside of the top chord. Note that some strongbacks are fixed to built-in soldier pieces formed within the webs, according to the span and design.

8.15 OVERLAY FLOORING

8.15.1 Introduction

Overlay floors of hardwood, or laminate flooring with a hardwood effect, are very popular nowadays. However, hardwood floors and overlay floors are not new. Wide, square-edged oak boards were used centuries ago in certain properties and hardwood-*strip* floors, in the form of narrow strips of T&G boarding, about 60 mm wide, have been used for overlaying sub-floors for many decades. Traditionally, they were mostly 'secretly nailed' through the inner-edges of the tongues, whereas nowadays they are more likely to be glued together and laid as a 'floating floor'. Thin plastic-laminate floors – with no alternative other than being laid to float – use self-locking click joints without glue. Today's seemingly high demand for hardwood floors is no doubt because more people can afford them – and the popularity of plastic-laminate flooring perhaps indicates those who have to (or choose to) compromise.

8.15.2 Hardwood Strip Flooring

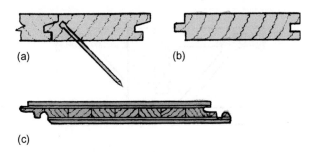

Figure 8.41 (a) T&G hardwood strip flooring for secret-nailing technique; (b) T&G hardwood strip for floating floor; (c) hardwood veneered, ply-laminate panel for floating floor

Figure 8.41 (a): If hardwood strip floors are to be nailed, the tongue and groove detail should be as illustrated, with a sloping tongue and an offset shoulder to facilitate and accommodate secret-nailing. By this method, lost-head nails (of 65 mm length) are driven-in at 45° angles, at maximum 600 mm centres. These angles and the final punching in of the nails, effectively cramps up each strip as the floor proceeds. Alternatively, the boards can be fixed more speedily with an angled brad-nailer gun, such as the Paslode Impulse IM65A (covered in Chapter 2), or with a semi-automatic flooring-nailer, such as the Primatech, that drives in so-called 'L Cleats' at 45° angles.

Figure 8.41(b): The standard T&G profile illustrated here, lends itself to being glued and cramped together when laid as a floating floor. It can be secretly nailed without being glued, but splitting and bulging of the tongue and lower edge is almost unavoidable in places. If this occurs, the holding-power of the fixing is impaired and the next row of boards might not cramp up effectively, leaving unsightly gaps on view. The header joint used for both types of tongue-and-groove board should be square and neatly butted. They must be strategically placed to avoid closeness to each other and no header joint in any following row should be closer than 300 mm to any header joint in the preceding row. When being glued, it is only necessary to run a small bead of PVA adhesive – from a squeezable plastic container – along and within the groove. Although

cramping will slow down the job, it is necessary to cramp each board for a short period of time, until the glue reaches an *initial-set* stage – which might be only a few minutes. Adjustable-strap type cramps are available for this, or a traditional wedging technique can be used.

Figure 8.41(c): The third type of hardwood floor consists of interlocking, laminated floor panels made of real wood plies (plywood). The core plies are of softwood and the surface ply – or veneer – is of hardwood, about 3 mm thick. Typical panel sizes are 145 mm wide × 1220 mm long × 13.5 mm thick. The panels are laid as a floating floor and they have the advantage of glueless, self-locking click joints on all four edges. Also, by being an engineered-timber product, they retain the benefits of natural wood, without the inherent weaknesses of twisting, cupping, shrinking or splitting. As before, the header joints must be staggered. Ideally, half a panel's length should start the second row, but this rarely works out in practice.

8.15.3 Hardwood-effect Laminate Flooring

Figure 8.42 Plastic-laminate floating-floor panel

Figure 8.42: Plastic-laminate flooring – usually just referred to as laminate flooring – is composed of a hard-wearing, wood-grain-patterned plastic membrane on the face-side, which has been fused (under great heat and pressure) to a high-density fibreboard (HDF) core of 7 to 8 mm thickness. The 'boards' are about 150 mm longer and 50 mm wider than the engineered-timber boards described above and each panel usually displays a three-row, strip-floor effect. This type of flooring is produced to varying grades to suit different levels of use, and can only be laid as a floating floor. In recent years, these panels had standard T&G joints that were glued together, but they now have glueless, self-locking click joints on all four edges.

8.15.4 General Considerations

The following points apply to either fixed or floating overlay-floors:

● Follow the manufacturer's instructions regarding floor preparation and the types of underlay required. If not familiar with floor-laying techniques, study the excellent detailed illustrations usually supplied by the manufacturers.

- Consider in what direction the boards should run. Running in the direction of the longest wall will reduce end cuts, but often the visual effect is improved if boards run in a forward direction from an entrance doorway.

- Once the direction is determined, it is wise to divide the length of the room by the precise width-measurement of a board or panel, to find out how it will meet the other side. Often it is discovered that a very narrow, awkward rip will be required. If so, create a better balance by reducing the width of the first row laid. In a more complex situation, forethought should also apply to any projecting or receding walls that might be met before the far side is reached – as in, for example, an L-shaped room. When calculating, take the required expansion gaps into account.

- By using temporary plastic or wooden folding wedges, always establish an expansion gap of approximately 12 mm between the edges of hard-wood flooring and the perimeter of the room – and 10 mm in the case of plastic-laminate flooring.

- Additionally, in large rooms, expansion joints will be needed within the visual area of the floor when the length or breadth of a room is in excess of 10 linear metres. If required, plan these joints strategically to be symmetrical or obscure within the room. Cover the joint-gaps with co-ordinating T Bars/strips.

- Floating floors are held down by the perimeter skirting, which – if necessary – should be carefully scribed to the floor. If skirting boards already exist in a room, remove them to achieve a professional finish and re-fix or renew them afterwards. Alternative skirting beads or quadrants always look amateurish, in my opinion.

- Floors should always fit under door-linings, frames and architraves, etc. To achieve this, lay an offcut piece of flooring face down against the architrave, etc., and run a fine handsaw against it to remove the obstruction. Keep the saw pressed down flat whilst sawing, by applying light finger pressure from your idle hand onto the blade and flexing the saw slightly. Alternatively, this can be done with a corded or cordless oscillating *Multi-Cutter Tool*, fitted with a segmented cutter blade.

- When a hole is drilled near the edge of a laminate panel to accommodate a radiator pipe, to remove the back-piece (and be able to reuse it), make two V-shaped cuts towards the hole with a fine saw (gent's saw or tenon saw) at 45° in plan – and at about 60° vertically, leaving only a semi-circle in the board when the back-piece is removed. When the panel is in position, the twice V-shaped back-piece can be glued in flush to the surface.

8.16 NOTCHING AND DRILLING TRADITIONAL FLOOR/CEILING JOISTS

Traditional floor joists, commonly with a 225 mm depth for upper floors/ceilings and 100 mm depth for suspended ground-floors, can often be found to have suffered much structural abuse by having been thoughtlessly and excessively notched and drilled to receive electrical cables and small-bore service pipes. Ill-considered and excessive notching and drilling can weaken a floor and cause varying degrees of structural failure. Therefore, any notching and drilling of joists should comply with the following requirements given in the TRADA (Timber Research and Development Association) span tables and interpreted here graphically in **Figure 8.43** below, with textual explanations of the capital-lettered references (A), (B), (C), (D), (E), (F), (G), (H), given here separately in the final two penultimate paragraphs below. Note that capital D, illustrated on the left-hand side of the joist, represents *depth* – and D/4, therefore (displayed in the central, top area of the joist), represents ¼ of the joist-depth as being the maximum diameter of drilled-holes in the indicated areas. Finally, note that the dot-dash centre line (shown as C/L) drawn horizontally through the joist, represents the structurally important *neutral axis*, above which the fibres of the joists are *in compression* (being compressed) – and, below which the fibres are *in tension* (being stretched). These two forces change from one to the other in the centre-line area, hence the term *neutral (neutralized-stress) axis*.

Figure 8.43(A)(B)(C): **Notches**, as indicated, should not be deeper than 0.125 (one-eighth) of the joist's depth and should only be positioned between **(A)** 0.07 × span and **(B)** 0.25 × span from either end. They may only be cut into the tops of the joists (never on the underside) and **(C)** they should be kept at least 100 mm apart, as illustrated. And the maximum depth of a notch in any depth of joist (regardless of the formula) is 35 mm.

Figures 8.43 (E)(F)(G)(H): **Holes** may only be drilled at each end of the joists, through the designated zones shown here with broken lines on the neutral axis (the centre-line), which are indicated at **(G)** as being not-less-than 0.25 × span from the joist-ends and at **(H)** as being not-more-than 0.40 × span from the joist-ends. The maximum diameter of the holes should be 0.25 × depth of joists. The vertical centre-lines of holes in the permitted zones, indicated at **(E)**, must be no nearer to each other than three times the hole's diameter. And, as indicated at **(F)**, the centre of a hole should not be less than 100 mm from

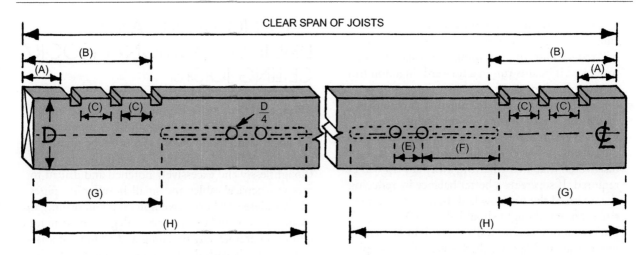

Figure 8.43 This illustration represents an elevational image of a *simply supported* joist (meaning that the joist's ends are not built-in/not restrained), with no bearings shown (i.e., the *clear span*), displaying capital-lettered **references (A)(B)(C)(D)(E)(F)(G)(H)**, related to TRADA's calculated notching-and-drilling requirements outlined above.

the side-cut of a notch. Finally, the maximum diameter of holes in any depth of joist (regardless of the formula) should be 65 mm.

Note that TRADA's notching-and-drilling rules are for simply-supported joists with a maximum depth of 250 mm.

9

Fixing Interior and Exterior Timber Grounds

9.1 INTRODUCTION

Timber grounds are either sawn or prepared battens, fixed to walls or steel sections, to create a true and/or receptive fixing surface. Depending on the material the grounds are to be attached to, they may be *fired* onto steel flanges or webs, brickwork and concrete with a cartridge powered tool, fixed to lightweight aerated blocks with screws and plugs, or fixed to blockwork and brickwork with Fischer type Hammer-fix screws.

9.2 SKIRTING GROUNDS

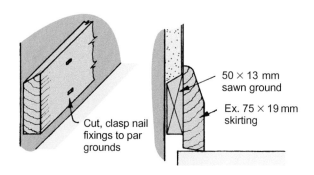

Figure 9.1 **(a)** Skirting grounds

Figure 9.1(a): Although grounds were traditionally only used on good class work to promote truer plastered surfaces and provide a means of fixing for the skirting, they are included here because research has shown that a small percentage of dwellings being built still use wet-plastering techniques and therefore may use grounds. The grounds are bevelled on their top edge to retain the bottom edge of the plaster and must, of course, be equal to the required plaster thickness. As packing pieces (off-cuts of damp-proof course material and plastic shims are ideal) are often required on uneven walls, the grounds should be slightly less thick than the required plaster thickness.

9.2.1 Fixing Technique

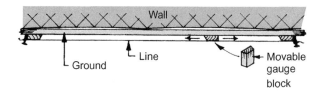

Figure 9.1 **(b)** Checking straightness of ground with stringline and gauge blocks

Figure 9.1(b): The top of the grounds should be levelled (or parallel to the finished floor) and set up to finish between 6 and 10 mm below the anticipated skirting height. Long grounds should be fixed at each end and have a string line pulled taut along the face. Two pieces of offcut ground, one at each end, are pushed in between the line and ground, while a third piece of offcut ground is tried between the taut line and the unfixed ground at 600–900 mm intervals, packed if necessary and fixed. Shorter grounds may be checked for straightness with a timber or aluminium straightedge. Internal and external angles are butt jointed – not mitred. On external angles, run the first ground about 50 mm past the corner, butt the end of the second ground up to this and when fixed, cut off the first projection flush to the second ground's face.

9.2.2 Deep and Built-up Skirtings

Figure 9.1(c): Grounds for deep or built-up skirtings may be required on refurbishment, maintenance and repair work. Such grounds, as illustrated, have a longitudinal top ground and vertical, face-plumbed soldier pieces of ground fixed at 600–900 mm centres. Depending upon the particular skirting design and height, additional stepped soldiers may be required to be fixed onto the first row.

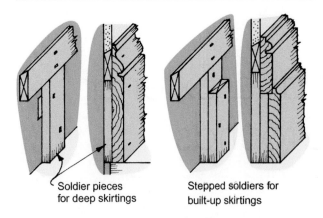

Soldier pieces
for deep skirtings

Stepped soldiers for
built-up skirtings

Figure 9.1 (c) Soldier pieces for deep skirtings and stepped soldiers for built-up skirtings

9.3 ARCHITRAVE GROUNDS

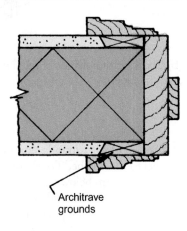

Architrave
grounds

Figure 9.2 Architrave grounds

Figure 9.2: Because the modern architrave section is relatively narrow, the traditional use of grounds in these situations is rarely required nowadays. Similar to skirting grounds, the edge against the plaster was bevelled and had to be concealed under the outer-architrave edge by about 6–10 mm. These grounds helped to keep the wet plaster away from the lining's edges and provided a true and receptive fixing surface for the outer-architrave edges.

9.4 APRON-LINING GROUNDS

Figure 9.3: Grounds are often required behind the apron lining around the edge of a trimmed stairwell. This is to bring the face of the lining to a position equal to the centre of the newel post and to further support the projecting landing-nosing. The grounds may be longitudinal if being faced with an MDF or

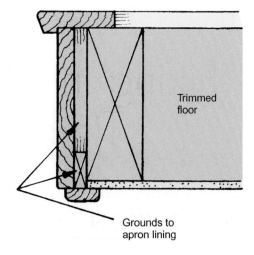

Trimmed
floor

Grounds to
apron lining

Figure 9.3 Grounds to apron lining

plywood apron lining, or in the form of vertical soldier pieces if a timber lining is being used. The soldier pieces give better support by being across the grain to offset any cupping of the lining.

9.5 WALL-PANELLING GROUNDS

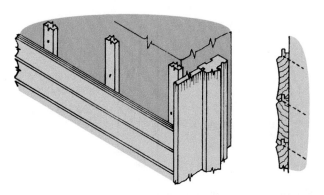

Figure 9.4 Horizontal cladding secret-nailed to vertical grounds

Figure 9.4: Grounds – without bevelled edges and sized about 50 × 25 mm – may be fixed horizontally across a wall and spaced at 600 mm centres from floor to ceiling, acting as a straight and plumb fixing medium for vertical boarding. Alternatively they may be fixed vertically at similar centres across the wall, acting again as a true surface and fixing medium for horizontal boarding. The technique in these situations is, having fixed two extreme grounds straight and true (one horizontally near the floor, one near the ceiling – or one vertically on the extreme left, one on the extreme right), they are used as a fixed datum for all the in between grounds to relate to, by means of a straightedge or string line.

9.6 FRAMED GROUNDS

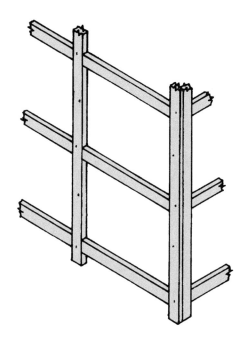

Figure 9.5 Framed grounds

Figure 9.5: Occasionally, on traditional forms of wall panelling, framed grounds are still used. Basically, these consist of either an arrangement of grooved uprights and tongued cross-rails of about 50 × 25 mm section, or morticed uprights and stub-tenoned cross-rails of a similar section. The fixing technique for framed grounds is as explained above for wall-panelling grounds.

9.7 EXTERNAL GROUNDS

Fixing timber grounds for external work such as for timber or plastic cladding will again involve a fixing technique similar to that described above for wall-panelling grounds. The main difference will be that the grounds should be tanalized or protimized, or protected with a similar acceptable preservative treatment.

10

Fixing Stairs and Balustrades

10.1 INTRODUCTION

Figure 10.1: Traditionally, a series or flight of steps, rising from one level to another, whether it be a floor to a landing or vice versa, was known as a *stair*, but is now more commonly referred to as *stairs* or a *staircase*. Originally, *stairs* was the plural of stair, meaning more than one flight of steps and the word *staircase* meant the space within which a stair was built. This space is now called a *stair well*. These more-recent, modern terms are used here.

10.1.1 Manoeuvrability

For reasons of easier transportation, manoeuvrability through doorways, and practical issues involved in the fitting and fixing, staircases usually arrive on site separated from the newel posts and balustrade, the bottom step (if such protrudes beyond the newel post, as with a bullnose step), the top riser board, the landing nosings and the apron linings.

10.1.2 When to Fix a Stair

From a carpenter's point of view, fixing is best done before dry lining or traditional plastering takes place, soon after the shell of the building is formed and the roof completed. This sequence allows the staircase to be fixed to the bare wall, so ensuring a better finish by the plasterer or dry lining being seated on the edges of the wall-string board, sealing any gaps that otherwise might appear if the stair-string were fixed to the dry-lined or plastered surface.

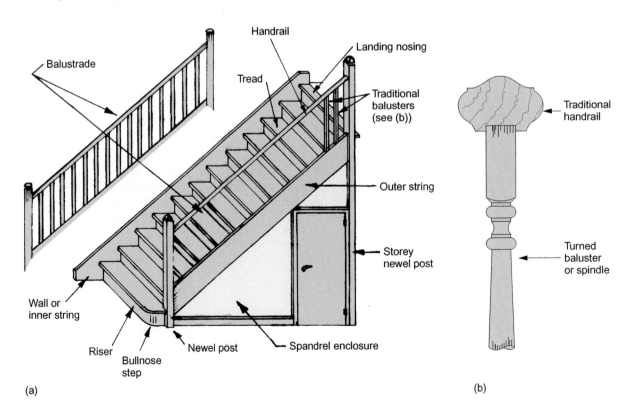

(a)

(b)

Figure 10.1 Stair terminology

10.1.3 Other Considerations

Fixing at this stage also effectively reduces the disproportionately thick-edge appearance of the wall string and, if worked out, perhaps by packing the wall string when fixing, it can be gauged so that the remaining thickness of wall string equals the thickness of the skirting board that will eventually abut its ends at the top and bottom of the staircase. This is an important point on good quality work, because abutting skirting boards ought to be flush with the string-face.

Another reason for installation at this stage is to allow building operatives quick and easy access to the upper floor(s). However, from a plasterer's/dry-liner's point of view, it seems likely that they might prefer to dry-line the staircase wall(s) first, to avoid being involved (time-wise) with angled cuts to suit the staircase-pitch. The following steps outline the operations involved in fitting and fixing a straight flight of stairs.

10.2 INSTALLATION PROCEDURE

10.2.1 Checking Floor Levels

Figure 10.2(a): Check whether the existing floors (upper and lower) are finished levels. In the case of a boarded or ply/chipboard/OSB sheeted floor, these are usually the levels to work to – as any additional floor covering can be assumed to cover the steps as well, thereby retaining equal rises to all steps. If, however, the ground floor is of concrete (ground-slab form or a beam-and-block floor) and has yet to receive a finishing material such as a 50 mm sand-and-cement screed, or a floating floor of polystyrene sheets (Jablite) and tongued and grooved chipboard panels, or sand/cement screed and wood parquet blocks, then the finished floor level (ffl) must be known or found out and established – and packing blocks prepared to fit under the bottom step.

10.2.2 Establishing the Finished Floor Level

Figure 10.2(b): At this stage of the job, the ffl has usually been established and may be found, ready to transfer from the bottom of door linings or the sills of external door frames. Alternatively, a bench mark above the site datum can be levelled across to the stair area, marked on the wall and measured down the set amount to the ffl (as outlined in Chapter 22).

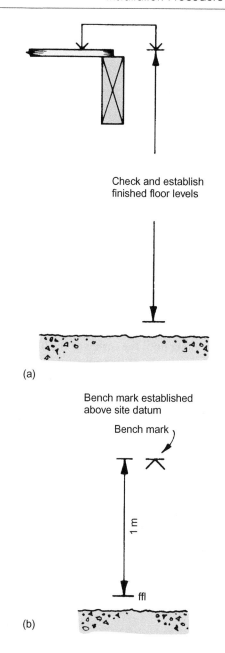

Check and establish finished floor levels

(a)

Bench mark established above site datum

Bench mark

1 m

ffl

(b)

Figure 10.2 **(a)** Establish finished floor levels; **(b)** Establish bench mark

10.2.3 Cuts for Floor and Skirting Abutments

Figure 10.2(c): Next, cut the wall string at the bottom to fit the ffl (even if the finished floor is yet to be laid). If not already cut or marked during manufacture, then simply measure down the depth-of-rise from the top of the first tread-housing (if such exists, as in the case of a bottom step left out for site-fixing), or measure down from the tread and mark a line through this point at right-angles to the face of the first riser-housing or riser board. Cut carefully on the waste side with

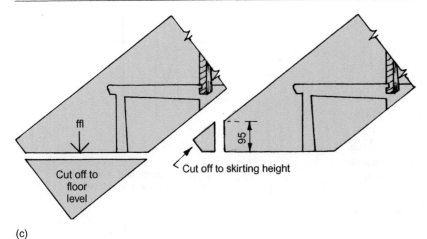

(c)

Figure 10.2 (c) Wall-string cuts at bottom

a sharp saw (to produce a clean cut). Then mark the plumb cut to form the abutment joint between the string and skirting board, as indicated. To do this, set the skirting height, say 95 mm, on the blade of the combination mitre-square and square-up from the ffl cut, sliding along until the corner of the blade touches the edge of the string. At this point, mark the plumb line and cut with a panel or fine hardpoint saw.

10.2.4 Cutting to Fit Trimmer and Skirting Abutment

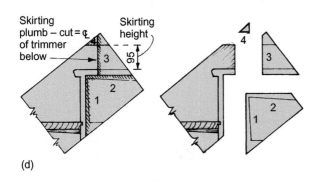

(d)

Figure 10.2 (d) Wall-string cuts at top

Figure 10.2(d): At the top end of the wall string, more elaborate marking out and cutting is required to enable the staircase to fit against the landing/floor trimmer or trimming joist. This also includes preparation of the string to meet the skirting. As indicated, the cuts are made in four places:

1. within the riser housing, on a line equal to the back of the riser;
2. within the tread housing, on a line equal to the underside of the flooring;
3. on a plumb line equal approximately to the centre of the landing trimmer (this is for the skirting abutment);

4. a horizontal cut at the very top of the string, equal to the skirting height.

This last cut should be planed to a smooth finish, as it becomes a visual edge of the string margin.

10.2.5 Offering Up and Checking

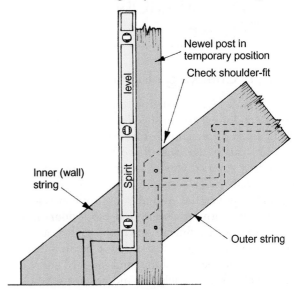

Figure 10.3 Checking for error in total rise

Figure 10.3: Now offer the staircase up into position, resting against the landing and packed up at the bottom, if necessary. Check the treads across the width and depth with a spirit level. Any inaccuracies registering in the depth of the tread will infer that either a fundamental error has been made in the mathematical division of the total rise of the staircase, or that the floor-to-floor storey height is not what it should be. A more positive way of confirming this will be to position the bottom newel post temporarily onto the outer-string tenons, making sure that the shoulder of the bare-faced tenon fits snugly against the newel, and checking for plumb with the spirit level, as illustrated.

10.2.6 Dealing with Inaccuracies

If inaccuracies are confirmed and they are only minor, they may have to be suffered, as very little can be done – short of shoddy tactics such as adjusting the shoulder of the string-tenons to improve the plumb appearance of the newel posts. If inaccuracies in level and plumb are more serious, then measure the rise of one step carefully, multiply it by the total number of steps in the staircase and compare this figure with the actual measurement of the storey height from ffl below to ffl above. Armed with this information, it would be wise to confer with the site foreman or builder's agent before proceeding.

10.2.7 Fixing the Wall String

Figure 10.4: After minor adjustments, if any, to the normal correctly fitting staircase, the next operation to

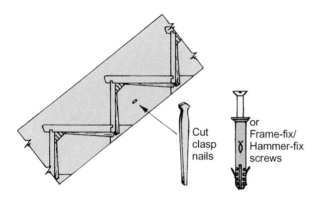

Cut clasp nails

or Frame-fix/ Hammer-fix screws

Figure 10.4 Fixing the wall string

consider is the fixing of the inner string to the wall. If the wall, being the inner-skin of cavity construction, is built of aerated lightweight material such as Celcon or Thermalite blocks, then nailing with 100 mm cut clasp nails will be satisfactory. These fixings are driven through the string on the underside of the treads, within the triangular area of every third or fourth step. However, if the wall is built with bricks or concrete blocks, and is not receptive to direct nailing, the wall string/wall will have to be drilled through in one operation to receive Fischer-type nylon-sleeved Frame-fix or Hammer-fix screws of 100 mm length. These may also be used, of course, if the first-mentioned aerated lightweight blocks do not prove to be dense enough to grip the cut clasp nail.

10.2.8 Preparing the Bottom Newel Post

Figure 10.5: Having decided on the method best suited to fixing the wall string, the next job is to prepare the bottom newel post to meet the floor level. The post is usually left longer at its lower end to allow for site treatment in various ways, according to the construction of the floor. Unless specified, the carpenter will decide exactly which way is suitable for a particular floor. The various methods of treatment at floor level are now described.

Newel Rests on Concrete

Figure 10.5(a): On concrete floors (slab form or beams and blocks), the newel post can be cut to rest on the

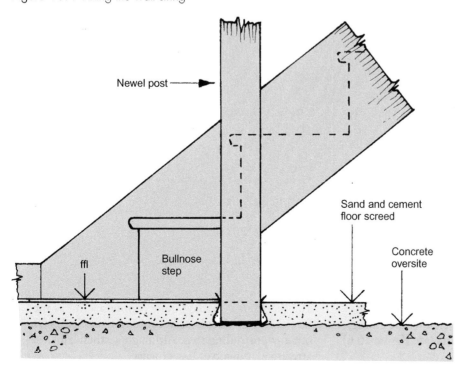

Newel post

Sand and cement floor screed

Concrete oversite

ffl

Bullnose step

Figure 10.5 (a) Screed bedded around newel

concrete – although the end should be sealed with preservative and/or wrapped with a piece of polythene sheet. When, after installation of the staircase, the sand-and-cement floor screed is bedded and set around the post, a further degree of rigidity is achieved.

Secure with a Metal Dowel

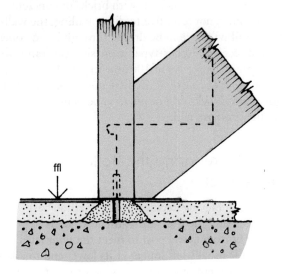

Figure 10.5 (b) Screed bedded around metal dowel

Figure 10.5(b): Alternatively, as illustrated, the newel post can be cut off at floor level, be drilled up into the end grain and have a metal dowel inserted. The dowel, which can be cut from 18 mm diameter galvanized pipe, should be inserted for at least half its length and protrude to rest on the concrete. Separate, localized bedding, with a strong mix of sand and cement around the dowel, is recommended before the main floor screed is laid.

Housed into Floor

Figure 10.5(c): The position of the newel post is marked on the wooden or chipboard floor and chopped out to form a shallow housing, equal to about one-third of the floor thickness. The post should fit this snugly and be skew-nailed into position.

Skew-nailed

Figure 10.5(d): Alternatively, on wooden or chipboard floors, bottom newel posts are quite commonly cut off at floor level, seated without any jointing and skew-nailed into the floor material with 50 or 75 mm oval nails, punched under the surface. The degree of rigidity achieved by this is minimal – and the newel post's stability depends mainly on the jointed connection to the string and lower step(s); therefore, the gluing, pinning and screwing of these parts (see Figure 10.6) should not be skimped.

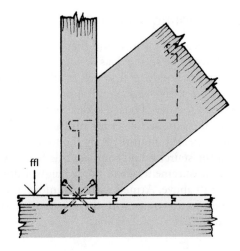

Figure 10.5 (c) Newel post housed into floor

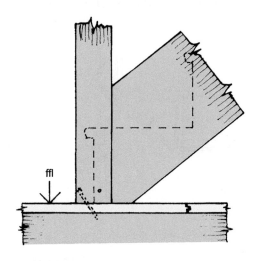

Figure 10.5 (d) Newel post skew-nailed into floor

Bolted to Joist or Nogging

Figure 10.5(e): Finally, on suspended wooden-joisted floors, although more time-consuming, tedious and rarely done in practice, the newel post achieves a far greater degree of rigidity if it is taken through the floor in its full sectional size and coach-bolted to a joist or – more likely in practice – to a solid nogging. According to the precise position of the newel post, the nogging would be trimmed between nearby joists. If not accessible below, pieces of flooring would have to be left out to facilitate the insertion of the bolt.

Note that the type of staircase indicated in Figures 10.5(b)–(e) is nowadays quite common and has the face of its first riser board central to the newel, without any protruding step. Although aesthetically less attractive, it involves less expense.

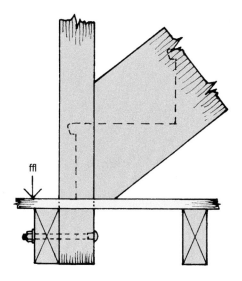

Figure 10.5 (e) Newel post bolted to joist or nogging

10.2.9 Fixing the Newel

The bottom newel post, which will have been morticed and fitted during manufacture, is now ready to be permanently fixed to the outer string. This can be done with the staircase lying on its side or resting up against the landing above and propped up on saw stools, or similar, at the bottom. The mortice and tenon joint should already be drilled to receive 12 mm diameter wooden dowels (pins). The holes should be slightly offset to enable the tapered pins to effect a wedging action when driven in, so drawing up the shoulders of the oblique (uncrampable by normal means) tenons to a good fit against the post.

Procedure

Figure 10.6: If dowel-pins are not supplied, cut off pieces of 12 mm diameter dowel rod, about 50 mm longer than the newel thickness and chisel the ends to

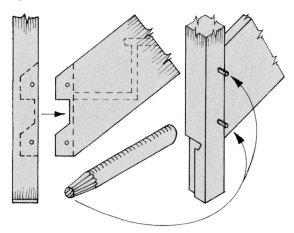

Figure 10.6 Gluing and pinning bottom newel

a shallow taper of about 25 mm length. After trying the newel post into position, coat the joint with PVA (polyvinyl acetate) glue and reposition the newel. This is best done using a claw hammer onto a spare block of wood held against the lower face of the newel. When a reasonable fit has been achieved, a touch of glue is placed into the draw-bore holes and the tapered pins are driven in until no part of the taper remains within the newel – bearing in mind, however, that the lower dowel usually clashes with the step on the other side. Clean off excess glue with a damp rag or paper and then cut off the surplus dowel ends with a fine saw, near the newel's surface. Clean off the remainder with a block plane or smoothing plane.

10.2.10 Fitting a Protruding Step

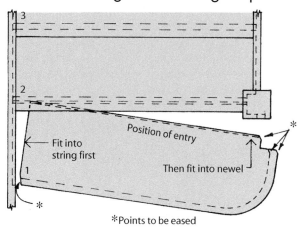

Figure 10.7 Fitting protruding bottom step

Figure 10.7: If the staircase has a bullnose or splay-ended bottom step (as illustrated), which protrudes beyond the newel post, this is the next to be fixed. It should be realized that such steps cannot be attached during manufacture without the newel being permanently in position. The step may have to be fitted and, as shown, this usually involves slight easings to the front end of the tread entering the string housing, and the rear end of the tread and face-edge of the bullnosed riser entering the newel post housings. After a successful dry fit, glue the step into position and drive in the glued string wedges, screw the lower face of the second riser to the back-edge of the tread and, finally, screw the ends of the bottom two risers into the housings of the newel post.

10.2.11 Positioning the Staircase

Figure 10.8(a): Set the staircase back into its ultimate position, ready for the next operation of fitting and fixing the handrail and top newel post. As with the

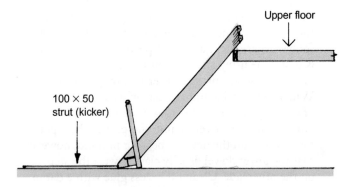

Upper floor

100 × 50
strut (kicker)

Figure 10.8 (a) Positioning staircase to facilitate fixing of top newel and handrail

fixing of the bottom newel and step, the ideal position for the staircase is on its side, but available space rarely permits this, so methods of working *in situ* have to be devised. One method, as illustrated, is to push the staircase forward until enough height has been gained above the landing or upper floor to allow access to complete the work from that level. To make this arrangement safe, a temporary kicker strut or struts of 100 × 50 mm section, should be lodged against the nearest cross-wall and should extend to support the base of the staircase.

10.2.12 Notching the Top Newel

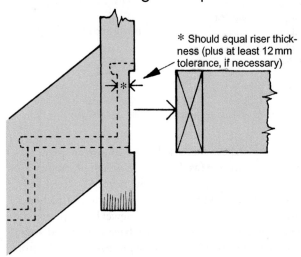

* Should equal riser thickness (plus at least 12 mm tolerance, if necessary)

Figure 10.8 (b) Newel notched to fit trimmer

Figure 10.8(b): Next, the newel may need to be notched-out (housed) to fit over the face of the landing trimmer. If so, this has a certain advantage of providing a good anchorage of the newel – and thereby that side of the staircase – to the top landing. However, whether this needs to be done or not depends on the newel's thickness, the thickness of the riser and whether a tolerance gap is to be allowed between the trimmer and the riser board (Figure 10.8(b)). The main reason for a tolerance gap is to overcome possible problems of the landing being out of square with the staircase.

However, this is usually discovered in the early stages of offering up the staircase and may be proved to be unnecessary.

10.2.13 Fixing the Top Newel and Handrail

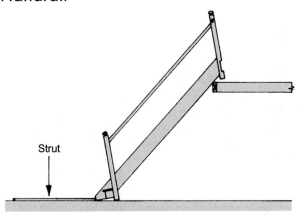

Strut

Figure 10.8 (c) Fixing top newel and handrail

Figure 10.8(c): After checking the dry assembly of the newel post and handrail in relation to both newels – bearing in mind that all three would have been fitted together previously by the manufacturer and draw-bored – glue may be applied to the joints, the handrail located in the lower-newel mortice and held suspended while the top newel is fitted to the string and handrail tenons. Moving speedily, as with all gluing operations, the joints are knocked up and the glue-coated draw-bore pins are driven in to complete the assembly of the skeleton balustrade. Finally, remove the surplus dowels and clean up as described previously.

10.2.14 Preparing Top Riser and Nosing

Figure 10.9: Before the staircase can be set back into position, the top riser and the landing-nosing have to

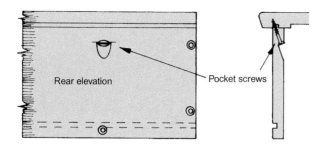

Figure 10.9 Preparing top riser and landing-nosing

be fitted and fixed to each other, to the string housings, the newel-post housings and to the adjacent tread. This operation is often skimped, resulting in a loose top riser and a squeaky top step. To avoid this, attend to all the following points.

Detailed Procedure

Check that the rebated side of the nosing is equal to the flooring thickness and, if thicker, ease with a rebate or shoulder plane. Then measure between the housings and cut the nosing and riser to the correct length. Now, because this particular step cannot have glue blocks set into its inner angle like the other steps (as they would clash with the trimmer), the best way to strengthen the joint is by *pocket screwing*. This is achieved by gouging or drilling shallow niches into the upper back-face of the riser board and by drilling oblique shank holes through these to create at least three fixings to the nosing piece. The riser is also drilled to receive two screws at each end and three or four along the bottom edge for the adjacent tread fixing.

Gluing Up

Next, glue the tongue-and-groove joint between the riser and nosing piece and insert the pocket screws.

Set the step-shaped riser/nosing into the glued housings of the newel and string, up against the glued back-edge of the adjacent tread and insert the two screws at each end, followed by the three or four screws along the bottom edge. Clean off any excess glue on the face side.

10.2.15 Final Positioning and Fixing

Figure 10.10: The staircase is now ready for fixing. Remove the struts and lower carefully into position. Re-check the newel posts for plumb and check that the staircase is seated properly at top and bottom levels. Fix the bottom newel to the floor; fix the top newel to the trimmer by skew-nailing through the side with two 75 mm or 100 mm oval nails or – better still – by pocket-screwing (Figure 10.10(b)); nail the nosing to the landing trimmer with 50 mm or 56 mm floor brads or lost-head nails; then, finally, as described earlier, fix the inner string to the wall.

10.2.16 String-to-skirting Abutment

Figure 10.11: Where wet plastering techniques are being used (as they still are to some extent) and as opposed to dry-lining methods, timber skirting grounds should be fixed at least to the wall-string wall beyond the two extremes of the wall string. This ensures a flush abutment of the skirting where it meets the string. In practice, it is wise to set the grounds back 2 mm more than the given skirting thickness from the face of the string. This combats the effect of the timber ground swelling after gaining an excess of moisture from the rendering/floating coat of plaster. Note that the ground remains swollen, but the subsequent setting coat of finishing plaster, which is usually applied whilst the timber is swollen, remains

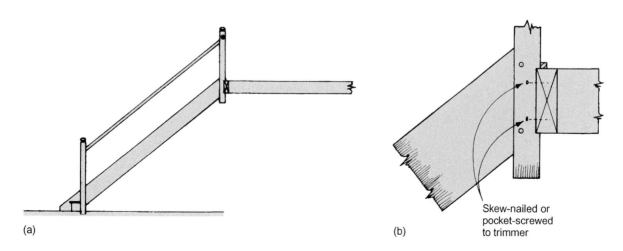

(a)

(b)

Skew-nailed or pocket-screwed to trimmer

Figure 10.10 (a) Staircase finally in position; (b) Skew-nailed or pocket-screwed to trimmer

about 2 mm proud when the ground eventually loses moisture and shrinks back to near normal.

10.2.17 String Mouldings

Figure 10.12: Finally, on the subject of wall strings, it must be mentioned that they might be moulded on their top edge to match the moulded edge – if any – of the skirting member. This entails extra work in the manufacture and/or on site, according to whether the shaped edge is a *stuck* moulding (machined out of the solid timber of the string and skirting), or a *planted* moulding (a separate moulding fixed by nails or pins to the plain, square edges of the string and skirting). As illustrated, only the moulding is bisected at the angle when planted, whereas with the stuck-moulded string and skirting, it will be easier and acceptable to let the bisected angle form a complete cut across the timber.

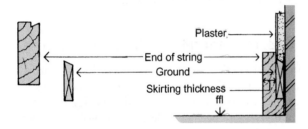

Figure 10.11 String-to-skirting abutment

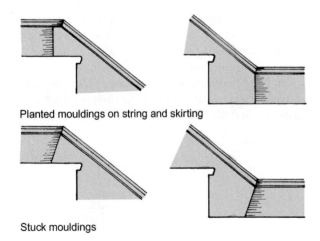

Planted mouldings on string and skirting

Stuck mouldings

Figure 10.12 Planted or 'stuck' mouldings

10.2.18 String-Easings

Figure 10.13: As well as being moulded and forming obtuse and reflex angles, these string/skirting junctions might be required to be swept into a concave shape at the bottom and a convex shape at the top. This shaping is known as *easing* and, according to the

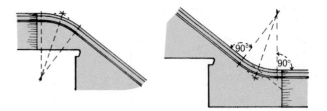

Figure 10.13 String-easings

moulding being stuck or planted, is either formed during manufacture, or fixed on site. It must be said that such work is uncommon nowadays because of the cost and the disinterest generally in moulded work, but could be met on repair, maintenance or refurbishment work.

10.2.19 Protection of Handrails, Newels and Steps

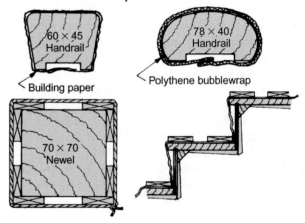

Figure 10.14 Protection of handrails, newels and steps

Figure 10.14: The remaining work on the staircase is best left until the second fixing stage, after the plasterer or dry-liner and other trades are finished. In the meantime, it is good practice to protect the handrail, newels and steps just after a staircase is installed, by wrapping building paper, heavy-gauge polythene or polythene bubblewrap around the handrails and newel posts and securing it with lashings of strong adhesive tape. Newels, which are usually more vulnerable, can be protected with additional wooden corner strips and tied or taped. If the handrail – or the staircase – is of hardwood, it should be sealed with diluted varnish and allowed to dry before being covered. The treads and risers should also be protected for as long as possible, by being covered with building paper or heavy-gauge polythene sheet, held into the shape of the steps by lightly nailed tread boards.

10.3 FIXING TAPERED STEPS

Traditionally, tapered steps – as they are now called – were referred to as *winders* or *winding steps* and they were incorporated into a variety of stair designs, used to change the direction of flight either at the bottom, halfway up, or at the top of the staircase. If the change in direction was 180°, there would be six winding steps, known as a half-space (half-turn) of winders. If turning through 90° – which was more common – there would be three winding steps (the square winder, the kite winder and the skew winder), known as a quarter-space (quarter-turn) of winders.

This terminology equates to landings, identified as quarter and half-space landings. Tapered steps usually replace landings to improve the headroom and when the 'going' of the staircase is greatly restricted. In certain cramped positions, there is often no alternative to them being used at both the top and bottom of the flight. However, tapered steps at middle and high levels of a flight, although not against the current regulations, are generally considered to be potentially dangerous and are usually avoided, if possible, by designers.

10.3.1 Tapered Steps at Bottom of Flight

Figure 10.15: Tapered steps in non-geometrical staircases (those without wreathed strings and wreathed

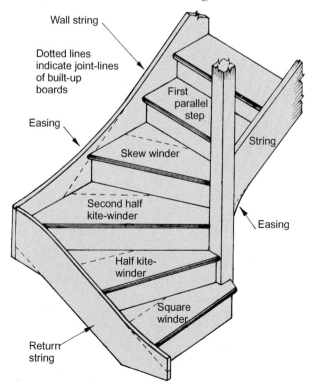

Figure 10.15 Tapered steps at bottom of flight

handrails, but with newel posts) are now mainly used at the bottom of the flight, in the form of four steps, as illustrated, to create a quarter-turn. Although it is possible for some tapered-step arrangements to be completely formed and assembled in the shop, it is more common that they be formed and only partly assembled, then delivered to the site for fitting and fixing. The reasons for this, as with straight flights, are for easier transportation and manoeuvrability through doorways, etc., and for practical issues involved in the fitting and fixing operation. Such a flight would arrive on site separated from the newel posts and balustrade, the top riser board, the landing nosings and apron linings, the return string, the tapered treads and their corresponding riser boards, etc. The operations involved in fitting and fixing this type of staircase are generally the same as already described for straight flights, with certain obvious additions, as follows.

10.3.2 Fitting the Main Flight

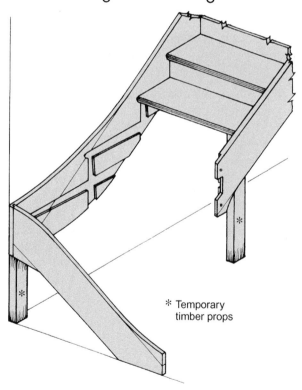

Figure 10.16 Fitting and checking the staircase

Figure 10.16: First, the main flight is offered up and fitted to the landing above and checked for level and plumb. To achieve this, built-up packing will be required at the bottom to compensate for the four missing tapered steps. Alternatively, two short timber props can be used, one under the bottom edge of the extended wall string, the other, as illustrated, up against the outer string, propping up the first available tread board. If supported like this, the staircase should remain firm, because, although not shown in the illustration, the end of the extended wall string butts up to the return wall.

10.3.3 Fitting the Return String

The return string, which connects to the main wall string with a tongued housing joint, is fitted and tried into position, the tops of the long tread-housings then being checked for level. The two strings ought to be at right-angles to each other, but this will depend largely on whether the return wall is truly square or not (which is another good reason for assembling and fitting these steps *in situ*).

10.3.4 Fixing the Main Flight

After these initial operations, the staircase will require repositioning to allow for the fitting and fixing of the newels and handrail. As outlined previously, this can be done by pushing the staircase up onto the landing and supporting the bottom end with packing and struts. The fitting of the skeletal balustrade then follows the sequence:

1. fix bottom newel post by gluing and draw-bore dowelling the string tenons;
2. glue handrail tenon and insert into bottom newel;
3. glue and fit top newel to handrail and outer string;
4. quickly complete the draw-bore dowelling of the unpinned joints;
5. fix top riser and landing nosing – after joining these together.

Back into position again, the main flight is re-checked and fixed as previously described. The tongued housing joint of the return wall-string is glued and fitted and the string fixed to the return wall.

10.3.5 Starting on the Tapered Steps

Finally, starting from the bottom, the tapered steps have to be fitted and fixed. This is the most difficult part of the whole operation and requires great care in checking and transferring details of the tread's site-shape and length from the housings of the strings and newel, to the separate treads.

10.3.6 Using a Pinch Rod

Figure 10.17(a): The treads should be already marked and cut to a tapered shape, usually with tolerances of about 25 mm left on in length to offset any problems that may arise if the return wall (and thereby the

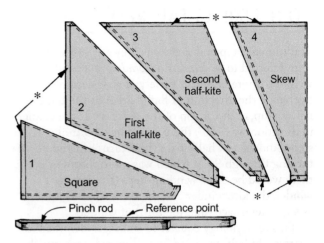

Figure 10.17 (a) Pinch rod and tapered treads. Tolerances in length are indicated by *

return string) should be out of square. A common method for checking lengths in this situation, is to use a pinch rod formed by two overlapping laths (timber of small sectional size). The laths are held together tightly, expanded out to touch the two extremes, then marked across the two laths with a pencil line as a reference point, so that they can be released and put back together when marking the tread and/or riser.

10.3.7 Checking, Cutting, Forming and Fixing

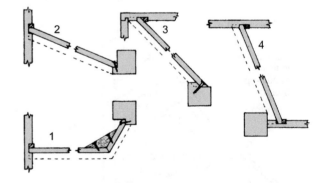

Figure 10.17 (b) Riser-end shapes into string and newel. The dotted lines indicate nosings

Figure 10.17(b): When using such a method, the bottom riser is checked first and cut to length. Then, with the aid of a carpenter's bevel and the pinch rod, the first tapered tread (the *square winder*) is checked, marked and cut to shape. After being tried into position (which often involves easing protruding corners of the tread and/or housings), the tread is fixed to the riser by gluing the joint between the two boards and *glue-rubbing* and panel-pinning the glue blocks on the inside angle. The housings are then glued, the step inserted and glued-wedges are driven into the string housings and screws driven into the newel (as illustrated). On the bottom riser, at least, the wedges cannot be driven-in on the string side and will have to be edge-tapered (planed-off slightly on the underside) and driven-in sideways from the face.

10.3.8 Repetition and Completion

This technique of checking and cutting, forming and fixing, is repeated on the other tapered steps and finalized by the fixing of the last riser to the main flight (riser No. 5). After each step is wedged into position, the bottom of the riser should be screwed to the tread. Sometimes, especially on wider-than-normal flights, 100×50 mm horizontal cross-bearers, on edge, are notched into (or cleated to) the string and newel post at each end, and fixed tight up against and under the back edge of each tapered step.

10.4 FIXING BALUSTRADES

During the second-fixing stage, the uncompleted work on the staircase can be finished. This mainly involves the balusters to the side of the staircase and the balustrade around the edge of the stairwell on the upper landing. The following steps outline the sequence of the operation.

10.4.1 Additional Newel Posts

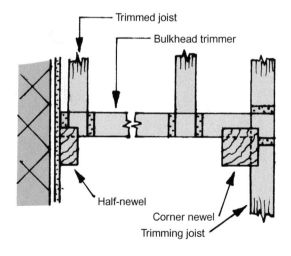

Figure 10.18 (a) Plan view of additional newels

Figure 10.18(a): The number of newel posts needed to form the balustrade to the stairwell depends on the design, but normally only one square-section and one half-section are required in addition to the one already at the head of the stairs. The half-section newel is always used on good class work, because it finishes the

balustrade off properly against the wall adjacent to the bulkhead trimmer and supports the return handrail on a through mortice and tenon joint. On cheap work, the handrail is housed in the wall, minus the half newel. The half newel runs down past and up against the side of the trimmer and is fixed to the finished wall surface with three counter-bored screws. The full-section newel sits in the corner of the stairwell, again running down past the ceiling and up against the trimmer on one side and the trimming joist on the other. Although its rigidity will be gained from the mortice-and-tenon connections of the handrails that will joint into it at 90° to each other, it should still be counterbored and screwed to the trimmer and trimming joists. The preformed handrail mortices in the newels should be used as a datum point to ensure that handrails will be level and/or parallel to the floor.

10.4.2 Level Handrails

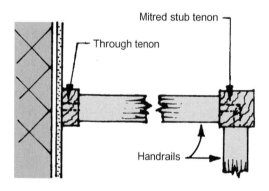

Figure 10.18 (b) Plan view of level handrails, one newel and one half-newel

Figure 10.18(b): In the example used here, two handrails – additional to the first raking handrail used – will be required, one long, one short. If supplied from a joinery works, they may be already tenoned at one end, but not usually at both, as tolerance must be allowed for site variations. Tenons – their size taken from the newel mortices – should be cut and shouldered carefully with a fine saw. The handrail can be laid against the newels, with the shoulder butted against one, while the second shoulder is marked against the other newel. This mark can be squared around the shaped handrail by wrapping a piece of straight-edged cardboard or glasspaper tightly round the handrail until its edge overlaps precisely, then by sliding along to the mark and marking around the edge. Now mark and rip the tenon's length, then shoulder it. Where the tenons intersect in the corner newel, they will require mitring.

10.4.3 Take Care on the Return

Take care when marking the short return-handrail length that the measurement between newels may differ between the top and at floor level, depending on the plumbness of the half newel against the wall surface. All of the tenons – except the one into the half-section newel – should be draw-bored, glued and pinned with 9 mm diameter dowelling. When this is done, the handrails can be installed and, to facilitate this, the fixings of the newels will require to be slackened or – more likely – removed.

10.4.4 Apron Linings and Nosings

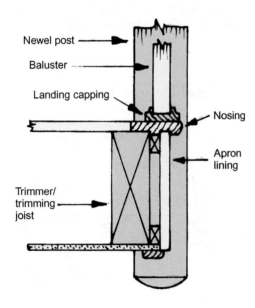

Figure 10.18 (c) Vertical section through apron lining and nosing

Figure 10.18(c): Next in sequence are the apron linings which add a finish to the rough face of the trimmer/trimming joists within the open stairwell. Traditionally, they were made from solid timber of about 21 mm thickness, but nowadays they are more common in plywood or MDF board. Their lateral position in relation to the newel's thickness is critical, because they support the landing nosing above, which in turn supports the balusters that must be dead centre of the newel in order to relate to the underside groove of the handrail. Therefore, they need to be very near the centre of the newel, as seen in the illustration, which usually means packing them out from the joists. This is done with prepared or sawn timber grounds of whatever thickness, but say 18 mm ($\times$ 50 mm), fixed horizontally on the joists to take MDF or plywood, but vertically as so-called *soldier-pieces* at 400–600 mm centres if the apron lining is of solid timber. This is to combat the undesirable effect of *cupping* which may occur to tangentially sawn boards.

10.4.5 Nosing Pieces

Having carefully cut the apron linings in length to fit neatly and tight between newels, they are fixed at top and bottom edges with 56 mm oval or lost-head nails (punched under the surface) at maximum 600 mm centres. Next, the nosing pieces, which are to be attached to the top edge of the apron lining and the joist, being like stair nosings, are usually equal in thickness to the stair treads. Therefore, they will probably be rebated to meet the lesser thickness of the flooring and this should be checked accordingly before fixing. This done, cut carefully to length and nail into position to the joist and apron-lining edges. If the apron lining is of MDF board, it will be better to glue this connection, not nail it, as MDF does not take or hold nails very well in its edges. Use PVA glue or panel adhesive.

10.4.6 String-capping and Landing-capping

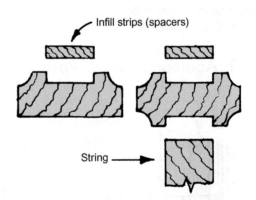

Figure 10.18 (d) Sections through string and landing capping

Figure 10.18(d): Capping is the next item to be fixed and it can be either plain-edged or moulded and grooved on one or both faces. One groove fits the top edge of the string, the other houses the balusters. If there is only one groove, then this houses the balusters and the plain (ungrooved) face is fixed to the string. Similarly, it can also be used on the landing, as illustrated in Figure 10.18(c), fixed to the nosing with 50 mm oval nails. This has the advantage of providing a raised edge for carpet to butt up against. On the string, the capping needs to be cut to an angle to fit against the newels. This can be found with a sliding bevel, set to the angle formed by the junction of the bottom (or top) newel and the string. Set the chop saw (or a mitre box) up with this bevel, cut carefully to length and fix with 38 mm oval nails at about 500 mm centres.

10.4.7 Balusters

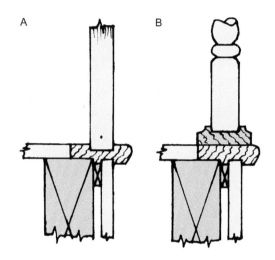

Figure 10.18 (e) Baluster stick A and spindle B

Figure 10.18(e): *Balusters* are also referred to nowadays as *spindles*. Both names refer to spindle-lathe-turned ornamental posts that infill the balustrade at the side of the stairs and stairwell. *Baluster sticks* only differ by being square posts with no ornamental lathe work. Whichever is used, the Building Regulations require that they be set up with a controlled gap between them. This gap should not allow a 100 mm diameter sphere to pass through. In effect, this means that the spacing between the balusters should not be more than 99 mm. The same bevel set up to cut the capping can now be used to cut the balusters to fit the raking balustrade. First, cut one only, bevelled at each end to the precise height taken from the inner face of the newel. Try it in position in a few places and check for plumb. If acceptable, cut the estimated number to this pattern, ready for fixing.

10.4.8 Support from Above

If possible, support the centre of the handrail from above with a timber strut at right-angles to the pitch, to restrict the cumulative effect of the individually fixed balusters pushing up the handrail. Alternatively, fix a single baluster midway between newels and use a sash cramp at this point, square across the balustrade, tightened from the top of the handrail to the underside of the string.

10.4.9 Infill Strips

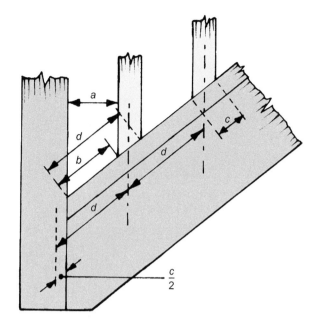

Figure 10.18 (f) Working out spindle spacings

Figure 10.18(f): The correct spacing – and easy fixing – of the balusters is achieved by using short lengths of timber infill-strip between them at top and bottom. This thin strip, machined to fit the shallow groove (of say 6 × 40 mm) in the handrail and the capping (as in Figure 10.18(d)) is cut up into equal angled-lengths, worked out to achieve the required spacing gap. One way of doing this, is to set up a pre-cut baluster in the grooves of the capping and handrail near, say, the bottom newel post and slide/adjust its position until a reading of 99 mm is obtained on a right-angled measurement, as at *a*. The geometrically longer measurement of the angled infill strip will now be obtainable as *b*, between the newel and baluster faces.

10.4.10 Creating Equal Gaps Throughout

Figure 10.18(f): To test the overall spacing of the balusters, the *raking measurement b* just obtained should have the *raking thickness* of the baluster *c* added to it, giving *d*. This value – equalling the baluster centres – should then be divided into the total measurement of the capping between the newels plus the raking thickness of half a baluster at each end, i.e. $c \div 2 \times 2 = c +$ capping length. The odds are that it will not divide equally and therefore adjustments will need to be made by lessening the divisor *d* and dividing again – and again, if necessary – until it works out without a remainder. The difference between the first and final divisor must then be deducted from the first-obtained measurement of the angled infill-strip b.

10.4.11 Example

Assume that the total length of capping between newels = 3.188 m, actual baluster thickness = 40 mm, desired gap a = 99 mm, initially obtained raking-measurement of infill strip b = 129 mm, raking thickness of baluster c = 52 mm. Therefore,

$$b + c = 129 + 152 = 181 \text{ mm baluster centres } d$$

The divisible length of capping between newels plus two times the half-baluster raking-thickness is

$$3.188 \text{ m} + 2 \times 26 \text{ mm} = 3.240 \text{ m}$$

Therefore

$$3240 \div 181 = 17.90$$
$$= 17 \text{ spacings with 90 mm remainder}$$

Trying again,

$$3240 \div 180 = 18 \text{ spacings exactly}$$

The difference between divisors equals 1 mm, therefore the final size for the angled infill-strips b is

$$129 - 1 = 128 \text{ mm}$$

In this example, the resultant gap between balusters will be slightly more than 98 mm. The working out also tells us that, as there are 18 spacings, there must be 17 balusters and 36 angled infill-strips.

10.4.12 Fixing

The infill strips should now be cut and the first two, one in the handrail groove, one in the capping groove, fixed up against the bottom newel post with two, staggered 18 or 25 mm panel pins punched-in. Then the first baluster is fixed, skew-nailed at top and bottom with two 38 mm lost-head oval nails (punched-in) at each end. Then two more angled infill-strips are fixed, and another baluster – and so the process is repeated until the top newel is reached. On the landing, a similar technique of working out and fixing can be used for the stairwell balusters. The only variation is when a grooved capping is not used. In these situations, as was traditional, the balusters are housed into the nosing to a depth of about 6 mm (Figure 10.18(e)) and skew-nailed once from each side.

10.4.13 Newel Caps

Figure 10.18(g): The tops of newels are usually finished in three basic ways:

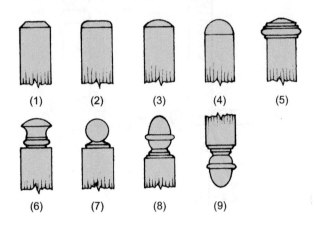

Figure 10.18 (g) Newel caps

1. finished in themselves, shaped in a variety of simple designs such as (1) chamfered-edge, (2) rounded-(quadrant) edge, (3) cross-segmental top and (4) cross semi-circular top.
2. separate recessed, square caps, with projecting moulded edges (5). The caps are glued and pinned or nailed to the tops of the newels.
3. separate spherical ornate caps turned on the lathe with projecting spigots, ready to be glued and inserted into predrilled holes in the ends of the newel posts. Three standard shapes predominate in this range, known as (6) *mushroom cap*, (7) *ball cap* and (8) *acorn cap*.

Usually, on newel posts above ground floor level, the newel is allowed to project down below the ceiling (by minimal amounts nowadays) and should receive an identical cap (9) to that used at the top. Traditionally, these below-ceiling projections were known as *newel-drops* or *pendants*.

11

Stair Regulations Guide to Design and Construction

11.1 THE BUILDING REGULATIONS 2010

11.1.1 *Approved Document (AD) K1, 2013 Amended Edition*

This approved document, as amended in 2013, refers to three categories of stairs:

1. ***private stair***: for dwellings; note that for external tapered steps and stairs that are part of the building, the *going* of each step should be a minimum of 280 mm;
2. ***utility stair***: for escape, access for maintenance, or purposes other than as the usual route for moving between levels on a day-to-day basis;
3. ***general access stair***: in all buildings other than dwellings; note that for school buildings, AD K1's preferred *going* is 280 mm minimum and the *rise* is 150 mm.

In producing a modified version of the approved document's K1 section here, covering most of the points concerning stairs and balustrades only, an attempt has been made to present a clearer picture of stair regulations as a guide, but not as a substitute.

11.1.2 Definitions

The following meanings are given to terms used in the document and a few other definitions have been added for clarity.

Alternating tread stair

Figure 11.1: A stair constructed of paddle-shaped treads with the wide portion alternating from one side to the other on consecutive treads.

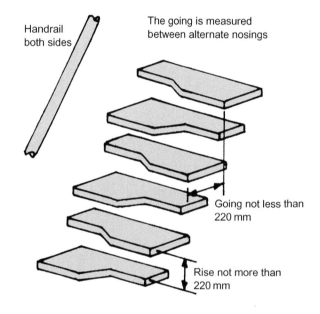

Handrail both sides

The going is measured between alternate nosings

Going not less than 220 mm

Rise not more than 220 mm

Figure 11.1 Alternating tread stair

Balustrade

Figure 11.2: A protective barrier comprising newel posts, handrail and balusters (spindles), which may also be a wall, parapet, screen or railing.

Deemed Length

Figure 11.3: If consecutive tapered treads are of different lengths, as illustrated, each tread can be deemed to have a length equal to the shortest length of such treads. Although not now referred to in AD K1, the deemed length (DL) needs to be established on certain stairs (see Figure 11.21(a)) for the purpose of defining the extremities to which the pitch line(s) will apply.

Flight

The part of a stair or ramp between landings that has a continuous series of steps or a continuous slope.

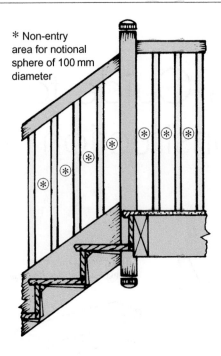

Figure 11.2 Balustrade (see Figure 11.9 re. non-entry area of a *notional sphere*)

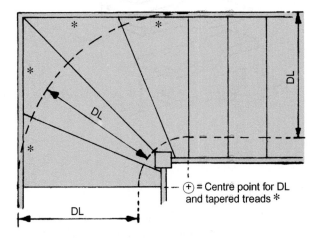

Figure 11.3 Deemed length (DL)

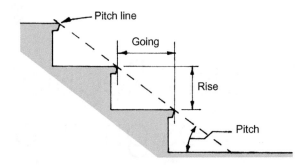

Figure 11.4 Going, rise and pitch

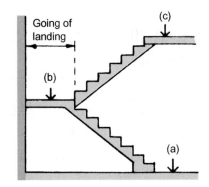

Figure 11.5 Landings = (a)(b)(c)

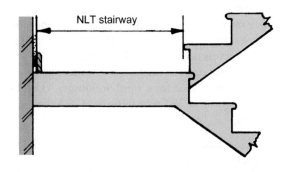

Figure 11.6 Going of landings not less than (NLT) stairway

Going

Figure 11.4: The horizontal dimension from the nosing edge of one tread to the nosing edge of the next consecutive tread above it, as illustrated.

Going of a Landing

Figures 11.5 and 11.6: The horizontal dimension determining the width of a landing, measured at right-angles to the top or bottom step, from the nosing's edge to the wall surface or balustrade.

Helical Stair

A stair that describes a helix round a central void, traditionally known as a geometrical stair.

Nosing

Figure 11.7: The projecting front edge of a tread board past the face of the riser, not to be more than the tread's thickness, as an established joinery rule, and not less than 16 mm overlap with the back edge of the tread below on open-riser steps, as shown in diagram 1.1 in AD K1.

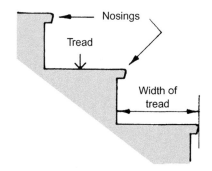

Figure 11.7 Nosings

Pitch

Figure 11.4: This refers to the degree of incline from the horizontal to the inclined pitch angle of the stair.

Pitch Line

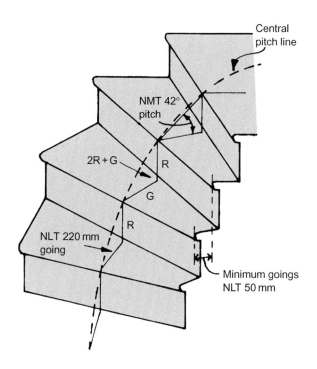

Figure 11.8 Pitch line

Figures 11.4 and 11.8: This is a notional line used for reference to the various rules, which connects the nosings of all the treads in a flight and also serves as a line of reference for measuring the 2R + G rule on tapered-tread steps (explained further on).

Rise

Figure 11.4: The vertical dimension of one unit of the total vertical division of a flight (total rise), as illustrated.

Spiral Stair

A stair that describes a helix round a central column.

Stair

A succession of steps and landings that makes it possible to pass on foot to other levels.

Tapered Tread

Figure 11.3: A step in which the nosing is not parallel to the nosing of the step or landing above it.

11.1.3 General Requirements for Stairs

Steepness of Stairs

In a flight, the steps should all have the same rise and the same going to the dimensions given later for each category of stair in relation to the 2R + G formula.

Construction of Steps

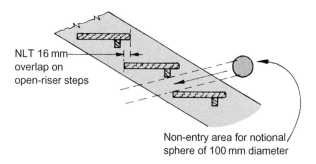

Figure 11.9 Step construction

Figure 11.9: Steps in dwellings should have level treads and may have open risers, but treads should then overlap each other by at least 16 mm. Steps in buildings other than dwellings should use risers that are *not* open. Additional guidance can be found in *Approved Document M, Section 1* (for buildings other than dwellings) and *Section 6* (for dwellings), when external stepped access also forms part of the *principal entrances* and alternative *accessible entrances*, and when they form part of the access route to the building from the boundary of the site and car parking.

Open-riser Stair

Figure 11.9: Steps in dwellings (under *Private stairs*) may have open risers that overlap treads by 16 mm and should be constructed so that a 100 mm diameter sphere cannot pass through the open risers.

Headroom

Figures 11.10 and 11.11: Clear headroom of not less than 2 m, measured vertically from the pitch line or landing, is adequate on the access between levels, as illustrated. For loft conversions where there is not enough space to achieve this height, the headroom will be satisfactory if the height measured at the centre of the stair width is 1.9 m, reducing to 1.8 m at the side of the stair.

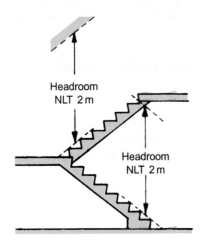

Figure 11.10 Headroom

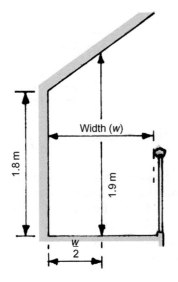

Figure 11.11 Reduced headroom for loft conversions

Width of Flights

Figure 11.12: Contrary to previous regulations quite some years ago (which gave 800 mm as the minimum unobstructed width for the main stair in a private dwelling), no recommendations for minimum stair widths are now given. However, designers should bear in mind the requirements for stairs which:

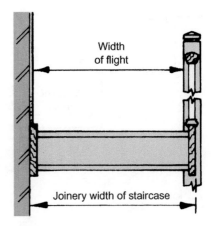

Figure 11.12 Width

- form part of means of escape (reference should be made to *Approved Document B: Fire safety*);
- provide access for disadvantaged people (reference should be made to *Approved Document M*, as amended in 2004).

Dividing Flights

Figure 11.13: A stair in a public building which is wider than 1800 mm should be divided into flights which are not wider than 1800 mm, as illustrated.

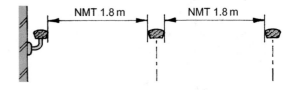

Figure 11.13 Division of flights if over 1.8 m wide

Length of Flights of Stairs for all Buildings

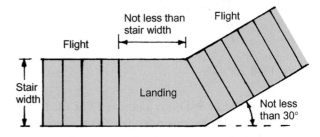

Figure 11.14 Change of direction in flights

Figures 11.14 and 11.16: Stairs having more than 36 risers in consecutive flights, should have at least one change of direction between flights of at least 30° in plan, as illustrated. The number of risers in a flight should be limited to 16 for *Private stairs* and *Utility*

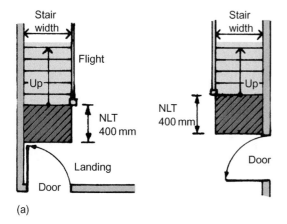

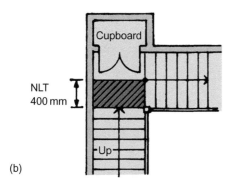

Figure 11.15 (a) Landings next to doors; (b) Cupboard on landing

stairs – and 12 for *General access stairs* (with an exception of no more than 16 in small premises with a restricted plan area). Note that other criteria in paragraph 1.42 of AD K1: *Access for maintenance for buildings other than dwellings*, may also apply.

Landings

Figures 11.5, 11.6 and 11.15: A landing should be provided at the top and bottom of every flight. The width and length of every landing should be at least as long as the smallest width of the flight and may include part of the floor of the building. To afford safe passage, landings should be clear of permanent obstruction. A door may swing across a landing at the bottom of a flight, but only if it will leave a clear space of at least 400 mm across the full width of the flight. Doors to cupboards and ducts may open in a similar way over a landing at the top of a flight (Figure 11.15(b)). For means-of-escape requirements, reference should be made to *Approved Document B: Fire safety*. Landings should be level unless they are formed by the ground at the top or bottom of a flight. The maximum slope of this type of landing may be 1 in 20, provided that the ground is paved or otherwise made firm.

11.1.4 Rise-and-going Limits for Each Category of Stair

Private Stair (Category 1)

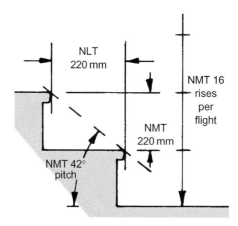

Figure 11.16 Category 1 stair

Figure 11.16: Any rise between 150 mm and 220 mm can be used with any going between 220 mm and 300 mm, subject to the traditionally established design formula of 2R + G, covered here further on. The maximum stair-pitch is 42°. Note that for means of access for *disadvantaged* people, reference should be made to *Approved Document M, Section 6*, as amended in 2004.

Utility Stair (Category 2)

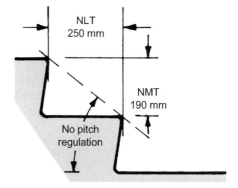

Figure 11.17 Category 2 stair

Figure 11.17: Any rise between 150 mm and 190 mm can be used with any going between 250 mm and 400 mm, subject to the design formula of 2R + G, covered here further on. No stair-pitch regulation is given. Note that other criteria in paragraph 1.42 of AD K1: *Access for maintenance for buildings other than dwellings*, may also apply.

General Access Stair (Category 3)

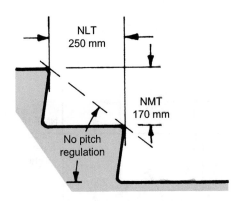

Figure 11.18 Category 3 stair

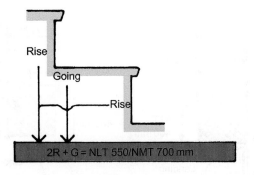

Figure 11.19 Design formula

Figure 11.18: Any rise between 150 mm and 170 mm can be used with any going between 250 mm and 400 mm, subject to the design formula of 2R + G, covered here further on. No stair-pitch regulation is given. Note that for means of access for *disadvantaged* people, reference should be made to *Approved Document M*, as amended in 2004.

Pitch

Figure 11.16: As illustrated, the maximum pitch for a private stair is 42°. Note that recommended pitch angles for the other two categories of stair are not given. However, if the self-limiting criteria of maximum rise and minimum going were used in these two categories, the maximum possible pitch for a category 2 stair would be approx. 34° and for a category 3 stair would be approx. 38°.

The 2R + G Design Formula

Figure 11.19: In all three categories, the sum of the going plus twice the rise of a step (traditionally established as 2R + G) should be not less than 550 mm nor more than 700 mm (subject to the criteria laid down for tapered treads).

Viability Graphs for the Three Categories of Stair

These graphs can be used as a quick test for the viability and regulatory compliance of the various permutations in *rise* and *going* for the three categories of stair Regulations. *Figures 11.20(a)(b)(c):* Any step-sizes that fall into Area A, meet the permitted rise and going, but violate the maximum pitch regulation; Area B meets all the various regulations; Area C violates the

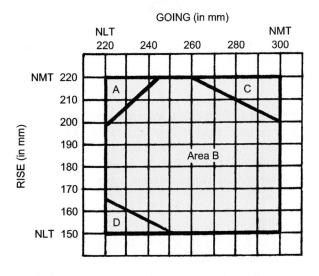

Figure 11.20 (a) Category 1 Private Stair graph

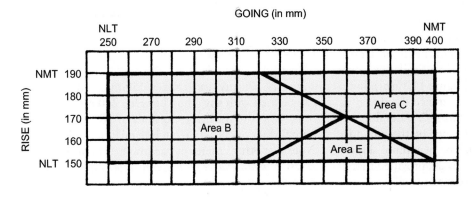

Figure 11.20 (b) Category 2 Utility Stair graph

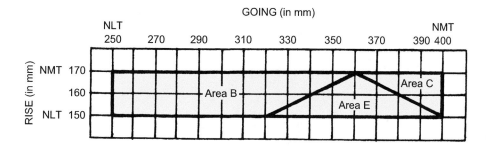

Figure 11.20 (c) Category 3 General Access Stair graph

maximum of 2R + G (700 mm); Area D violates the minimum of 2R + G (550 mm); and Area E meets the various regulations, but violates the non-regulatory, but traditionally established minimum pitch of 25°.

11.1.5 Special Stairs

Tapered Treads

Figures 11.8 and 11.21: For a stair with tapered treads, the going and 2R + G should be measured as follows.

1. If the width of the stair is less than 1 m, it should be measured in the middle, tangentially where the curved pitch line touches the nosings (Figure 11.8).
2. If the width of the stair is 1 m or more, it should be measured at 270 mm from each side, tangentially where each curved pitch line touches the nosings (Figure 11.21(c)). The minimum going of tapered treads should not be less than 50 mm, measured at right angles to a nosing in relation to the nosing

above. Where consecutive tapered treads are used, a uniform going should be maintained. Where a stair consists of straight *and* tapered treads, the going of the tapered treads should not be less than the going of the straight flight.

Alternating Tread Stairs

Figure 11.1: This type of stair is designed to save space and has alternate handed steps with part of the tread cut away; the user relies on familiarity from regular use for reasonable safety. Alternating tread stairs should only be installed in one or more straight flights for a loft conversion and then only when there is not enough space to accommodate a stair which satisfies the criteria already covered for *Private stairs*. An alternating tread stair should only be used for access to one habitable room, together with, if desired, a bathroom and/or a WC. This WC must not be the only one in the dwelling. Steps should be uniform with parallel nosings. The stair should have handrails on both sides and the treads should

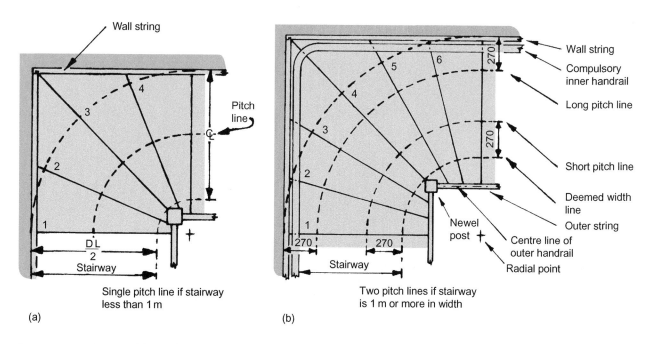

Figure 11.21 (a) Single pitch line; (b) two pitch lines if stairway is 1 m or more in width

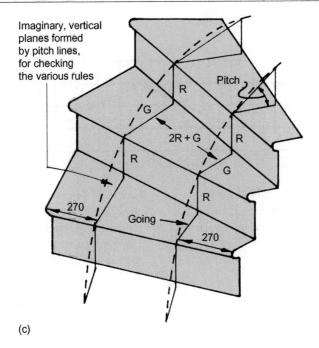

(c)

Figure 11.21 (c) Imaginary vertical planes

have slip-resistant surfaces. The tread sizes over the wider part of the step should have a maximum rise of 220 mm and a minimum going of 220 mm and should be constructed so that a 100 mm diameter sphere cannot pass through the open risers.

Fixed Ladders

A fixed ladder should have fixed handrails on both sides and should only be installed for access in a loft conversion – and then only when there is not enough space without alteration to the existing space to accommodate a stair which satisfies the criteria already covered for *Private stairs*. It should be used for access to only one habitable room. Retractable ladders are not acceptable for means of escape. For reference to this, see *Approved Document B: Fire safety*.

Handrails for Stairs

Figure 11.22 (a)(b): Stairs should have a handrail on at least one side if they are less than 1 m wide and should have a handrail on both sides if they are wider. Handrails should be provided beside the two bottom steps in public buildings and where stairs are intended to be used by people with disadvantages. In other places, handrails need not be provided beside the two bottom steps, as in Figure 11.22(b).

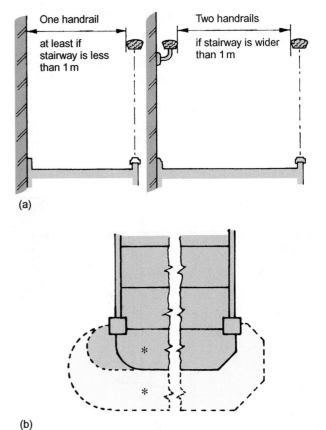

Figure 11.22 (a) Handrails for stairs; (b) Bottom steps*

Handrail Heights

Figure 11.23: In all buildings, handrail heights should be between 900 mm and 1000 mm, measured vertically to the top of the handrail from the pitch line or floor. Handrails can form the top of a guarding, if the heights can be matched.

Guarding of Stairs

Figure 11.2 and 11.22(b): As illustrated, flights and landings should be guarded at the sides:

- in dwellings when there is a drop of more than 600 mm;
- in other buildings when there are two or more risers.

The guarding to a flight should prevent children being held fast by the guarding, except on stairs in a building which is not likely to be used by children under 5 years. In the first case, the construction should be such that

- a 100 mm diameter sphere cannot pass through any openings in the guarding;
- children will not readily be able to climb the guarding (this in effect means that horizontal ranch-style balustrades used in recent years should not now be fitted).

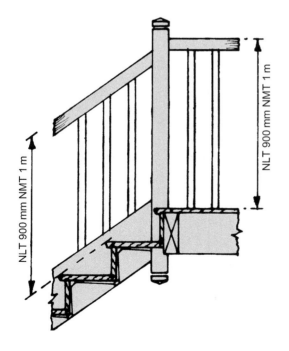

Figure 11.23 Handrail heights

The height of the guarding is given in Figure 11.23 for single-family dwellings in category 1. External balconies should have a guarding of 1100 mm. Although

the stair regulations are not completely covered here, I will end this with a reference to 'buildings other than dwellings', where AD K1/paragraph 1.36(f)(g) infers that the ends of handrails (if not built into newel posts, etc.) should be finished in a way that reduces the risk of people catching their *clothing* on the end projection(s). This is a welcome return of a previous, lapsed regulation in the 1976 Building Regulations, which stated that such open-ended handrails should 'be terminated by a scroll or other suitable means'. I always believed that the original sensible regulation was to stop people catching their *bodies* on the end projection(s).

11.1.6 Additional (Précised) AD K1 Amended Requirements

1.1: For additional guidance when external steps form part of the main entrances and site-access routes, etc., see *Approved Document M, Section 1 (for buildings other than dwellings) and Section 6 (for dwellings)*.

For Buildings other than Dwellings

1.6: Use risers that are *not* open, because: **(a)** The front of an ascending foot or a walking-aid might get caught up beneath a tread without a riser – increasing the risk of a fall; and **(b)** It removes any feelings of insecurity that ascending people may get when looking through the open risers.

1.7: Regarding steps, apply both of the following guidelines: **(a)** Make step nosings noticeable by using a material that will contrast visually, to a minimum 55 mm width on both the treads and the risers. **(b)** Avoid protruding tread-nosings, if possible; if not, ensure that they protrude by no more than 25 mm.

1.8: If a stair soffit is less than 2 m above floor level, protect the area below with either: **(a)** Guarding and low-level cane detection; or **(b)** A barrier giving the same degree of protection.

For Dwellings

1.9: Steps may have open risers (as per the detailed text related herein under Figure 11.9).

For Common Access Areas in Buildings that Contain Flats

1.10: Provide a stair with steps that comply to the following guidelines: **(a)** Make step nosings noticeable, by using a material that will contrast visually, with a 50 mm to 65 mm width on the treads and a 30 mm to 55 mm depth on the risers. **(b)** Use a suitable nosing profile, as shown in K1's Diagram 1.2 (note that this diagram gives three examples: 1) Vertical risers and horizontal treads; 2) A bevelled-edge tread projection, with a maximum 25 mm overlap and a widening-strip bonded to the riser, directly below the nosing projection, and: 3) Two staggered, vertical riser-faces, joined up in mid-riser height by a 60° minimum angled-face to a maximum 25 mm overlap/projection. **(c)** Do not use open risers.

For Buildings other than Dwellings and For Common Access Areas in Buildings that Contain Flats

1.12: Provide all means of escape routes with a minimum clear headroom of 2 m, except in doorways.

Width of Stairs For Buildings other than Dwellings

1.14: For stairs that form part of a means of escape, refer also to *Approved Document B: Fire safety, Volume 2 – Buildings other than dwelling houses.*

1.15: For stair-flights, provide all of the following: **(a)** A minimum stair width between enclosing walls, strings or upstands of 1200 mm. **(b)** A minimum width between handrails of 1000 mm. **(c)** If the flight is more than 2 m wide, divide it into two flights, each with a minimum width of 1000 mm between a central raised-handrail/balustrade arrangement. **(d)** For maintenance access, separate details and referrals are given in AD K1, in paragraph **1.42**.

For Dwellings

1.16: Exceptionally, where severely sloping plots are involved, a stepped change of level within the entrance-storey may be unavoidable. In such instances, ensure that such stairs have flights with a minimum stair-width of 900 mm.

For Buildings other than Dwellings

1.18: Comply with the following: **(a)** Do not have single steps. **(b)** For flights between landings, the number of risers should be: **(i)** Utility Stairs = not more than (NMT) 16 risers. **(ii)** General Access Stairs = 12 risers, but exceptionally NMT 16 in small premises where the plan-area is restricted. **(iii)** Stairs for Maintenance-Access = see details in AD K1, para 1.42.

Landings For Stairs – For all Buildings

1.19: For *means of escape* criteria, refer to *Approved Document B: Volume 1 – Dwelling houses* and *Volume 2 – Buildings other than dwelling houses.*

1.20, 1.21, 1.23 and 1.24: (See Figures 11.5, 11.6 and 11.15(a)(b) herein.)

1.22: Landings, with the following exception, should be level: A landing at the top or bottom of a flight that is formed by the ground, may have a gradient, if: **(a)** the maximum gradient in the direction of travel is 1:60, and **(b)** the surface is paved or made permanently firm.

Handrails For Stairs – For all Buildings

1.34: Provision of handrails (as per the detailed text related herein under Figures 11.22(a) and 11.23).

For Buildings other than Dwellings and For Common Access Areas in Buildings that Contain Flats and Do Not Have Passenger Lifts

1.35: Provide suitable, continuous, raking- and level-handrails on each side of the flights and the landings – and such handrails should precede or continue (in the horizontal plane) ahead of, or past the actual stair-flight at its start and finish by at least 300 mm. The height of the handrails should be at 1 m above the finished floor levels (ffls) and landings; and at 900 mm (measured vertically) above the pitch line to the top of the handrails.

For buildings other than dwellings, the details above still apply, but the height of the handrails should be between 900 mm to 1000 mm (1 m) above the pitch line to the top of the handrails – and between 900 mm to 1100 mm (1.1 m) above ffls and landings.

The illustrated handrail profiles, attached to metal handrail brackets, as shown in AD K1's *Diagram 1.13*, are to be circular-shaped, of 32 mm to 50 mm diameter or elliptical-shaped with a 50 mm major axis and a 39 mm minor axis. The stipulated space between the handrail and the finished wall-surface (controlled by the bracket-design) is 50 mm to 75 mm. And the finger-space below the base of the handrails to the top of the horizontal arm of the brackets is stipulated as being not-less-than 50 mm.

12

Making and Fixing Formwork for *In Situ* Concrete Stairs

12.1 INTRODUCTION

In purpose-built blocks of flats, office blocks, public buildings and the like, flights of stairs and landings are usually formed with steel-reinforced concrete, as opposed to purpose-made timber staircases used in dwellings for one-family occupation. This is because reinforced concrete stairs, firstly, give greater fire-resistance required for buildings with multiple occupancy and/or usage; secondly, are more durable and load-bearing for greater and heavier usage; and, thirdly, cost far less compared with similar stairs and steps of yesteryear that were made of quarried stone. Concrete stairs may also be finished in their first-formed material, or may be designed to have various other material finishes added to their treads and risers. In the latter case, the formwork builder has to take this into account – especially when relating his work to the various *finished floor levels* (ffls).

Although Chapters 10 and 11 dealt with *fitting and fixing timber stairs* and *stair regulations*, the essential criteria for *stair-design and setting out* was not dealt with, as it was not needed; however, this *is* covered here to enable the stair-formwork carpenter/builder to understand and carry out the critical task of forming accurate mould-box structures that produce the client's pre-designed *cast-in-situ* concrete staircase(s) and landing(s). *Cast-in-situ* simply means concrete items (such as staircases or beams, etc.) that are cast in temporary formwork structures, built-up in the item's actual, permanent position; and *precast* refers to concrete items (such as floor-beams or weathered-copings and sills, etc.) that are cast in mould-boxes that are not set up in the component's actual, permanent position.

12.2 DETAILS OF DESIGN

Figures 12.1(a)(b): The first thing to understand in the basic theory of stair design is that one movement of a person's foot *going* forward is referred to as 'the

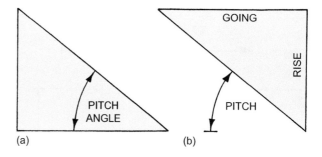

Figures 12.1 **(a)** and **(b)** The concept of a basic triangle transposed into a step

going' of the stair's step and the other foot *rising* up (to another level) is referred to as 'the rise' of the step. Therefore, from a design point of view, all of the step-movements going forward are referred to as 'the total going' (TG) and all of the step-movements rising up are referred to as 'the total rise' (TR). These two terms (TG and TR) are important references to the simple maths and geometrical division required to design flights of stairs that meet the Building Regulations. Mathematically, a step can be related to a right-angled triangle, whereby its base-line represents *the going* of a single step, the adjacent side represents *the rise* – and the ratio of each to the other, forming a hypotenuse, determines the legally important pitch angle of a step, as illustrated at Figures 12.1(a) and (b).

Figures 12.1(c)(d): The above concept of relating a step to a triangle can also be applied to a multiple of steps, as illustrated at Figure 12.1(c), whereby the 'total going' of all the steps (equally divided horizontally between the face of the top step and the face of the bottom step) and – likewise – the 'total rise' of all the steps (equally divided between the lower- and the upper-floor levels), forms a pitch angle according to the ratio of the total going (TG) related to the total rise (TR).

In considering an acceptable pitch-angle, the total rise is of course a constant, but as shown by comparing Figures 12.1(c) and 12.1(d), an adjustment to

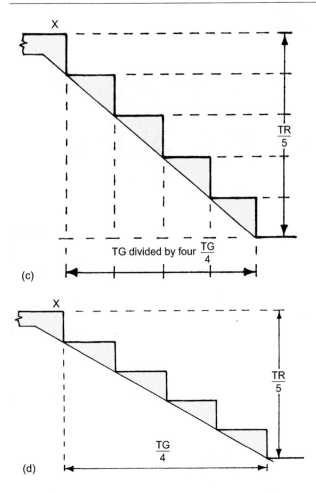

Figure 12.1 (c) A multiple of steps related to the division of the total going and total rise; **(d)** Although the scaled total-rise and the number of divisions at 12.1(c) is equal to the illustration at 12.1(d), it can be seen how an increased going of each step alters the pitch angle. Figure 12.1(c) has a 40° pitch; 12.1(d) has 29°.

the total going alters the angle of pitch. This is an important factor in stair design, especially when trying to fit a staircase into a small space where there is a limited TG.

As seen in Figures 12.1(c) and (d), TR (total rise) is divided by five, unlike TG (total going), which is divided by four. This is because the top steps – marked X in each Figure – are in fact part of the landing and therefore are not included in the horizontal division. As the top surface of a step is called a 'tread' and the vertical face is called a 'riser', as a hard-and-fast rule it can be said that 'in any single flight of stairs, *there must always be one less tread than risers*'. Thereby, when initially designing a staircase and calculating how many steps can be used to meet the various regulations, this simple and basic rule ensures a correct trial-and-error division of TR related to TG.

Although straight flights of stairs are quite common, they are not usually regarded as being very

attractive – and it is not always possible to fit them in when the *going* is restricted, i.e. when there is a limited stairwell length. For this reason, changes in the direction of the stair are often made.

12.2.1 Directional Changes

If you think of directional changes – in any stair design – as they would relate to a spiral staircase having a complete turn of 360°, then it can readily be seen that a *quarter-turn stair* changes direction by 90° and a *half-turn stair* changes direction by 180°. Apart from straight flights, therefore, quarter-turn and half-turn are the two types of turning stairs to be dealt with – and they are turned by introducing either a quarter- or a half-turn landing (an intermediary platform, traditionally referred to as *quarter- or half-space landings*), or tapered (winding) steps. With *tapered* steps (an untraditional term used in the current Building Regulations), because they are considered to be potentially more dangerous than intermediary landings, it is safer to keep them at or near the bottom of a staircase, if at all possible. Also, it must be mentioned that tapered steps create a twisted spandrel-shaped soffit to the underside of that part of the stair, which adds more complexity to the formation of any staircase requiring a continuous, un-stepped soffit.

Figures 12.2(a)(b)(c)(d): These plan-view and elevation diagrams outline the four most common stair arrangements in use. Although variations to these basic designs can be made, they serve to compare the effect on available floor space – and show where the 'headroom' (HR) regulation and 'pitch-line' applies.

12.2.2 Additional Change

Figure 12.3: In the diagrams in Figures 12.2(a), (b) and (d), another change of direction could be achieved by making allowance for a quarter-turn landing at the top of each flight. This would reduce the total going by one step – if required – as indicated in Figure 12.3 and compared with Figure 12.2(b). Although, in my opinion, a single step positioned at the head of a stair (separated from the main flight) can be dangerous, as it is not always noticed.

12.3 DESIGNING STAIRS

The logical detailed-design approach is to first divide TR (the *total rise*) by an intelligent trial-and-error number – say between 12 and 16 – until an acceptable and regulatory-permissible rise has been found. For a Category 3 stair in a building *other* than in Categories

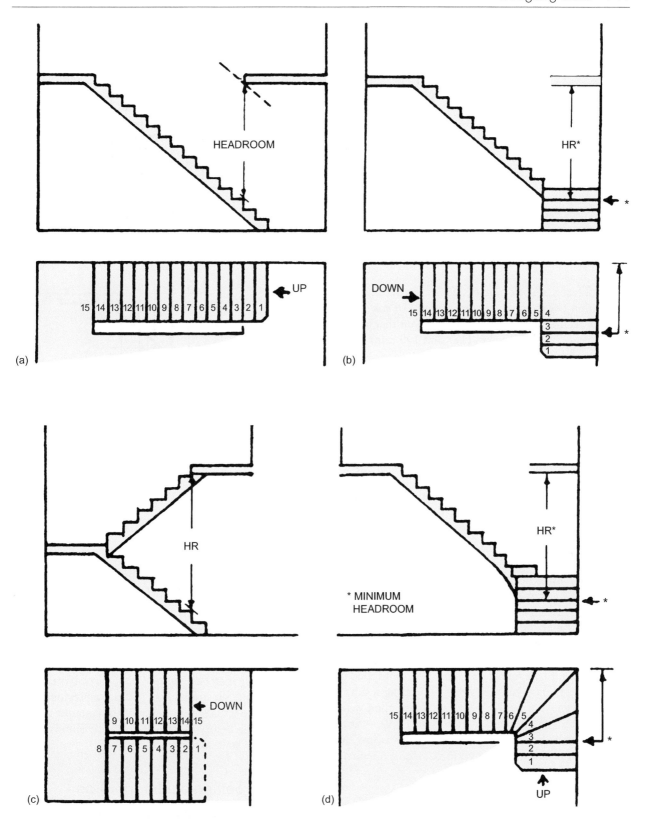

Figure 12.2 (a) Straight-flight stair; (b) Quarter-turn stair via a quarter-turn landing; (c) Half-turn stair via a half-turn landing; (d) Quarter-turn stair via four tapered steps. Note that, at the scaled-design stage, the minimum regulatory headroom (HR) was worked out to be above the third riser (*)

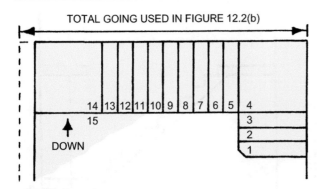

TOTAL GOING USED IN FIGURE 12.2(b)

Figure 12.3 A quarter-turn landing added at the top, reduces the total going by one step and changes the designation to a half-turn stair

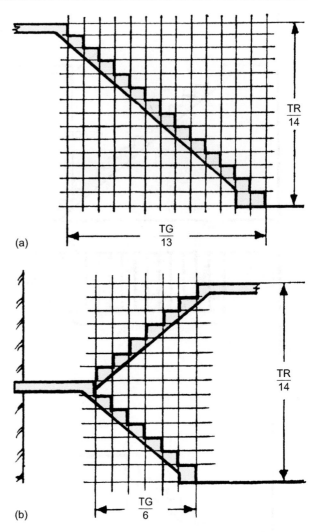

(a)

(b)

Figure 12.4 (a) 14 risers produce 13 goings (because the top landing is excluded); **(b)** This design of a half-turn stair with 14 risers must have one less going than *half* the number of risers

1 and 2 (as in a block of flats), we know that the regulation rise should be *not more than 170 mm* – even though technical research will inform you that a rise of between 7 and 7½ inches (178 and 190 mm) was traditionally considered to be ideal.

Next, we must consider TG (the *total going*). As this dimension might include the going of a landing (the landing's width), this is deducted from TG until the horizontal dimension between the faces of the bottom- and top-risers remains. This resultant dimension is then divided by the number of goings, which will be (remember the rule) *one less than the number of risers* – and this will determine the step-size of the *going*. Once the rise and the going are established, the degree of pitch can be checked quite easily, speedily and accurately by drawing a precise, half scale (1:2) right-angle to the rise-and-going dimensions; then by forming a hypotenuse to the extremities of these lines, thus enabling the *pitch* to be measured with a protractor (preferably the adjustable set-square type). With a 1:2 scale, this can be done on A4-size paper.

Of course, if you have a preference for trigonometry and have a scientific calculator, the pitch angle equals the *opposite* (say, a rise of 192 mm) divided by the *adjacent* (say, a going of 237 mm), all multiplied by the shift tangent. This method of calculating results in a pitch angle of 39.01°. My preferred method with a scaled triangle produced a pitch angle of 39° – close enough to the trigonometry and acceptably below the regulatory 42° maximum pitch for Category 1 stairs.

12.3.1 Graphic Illustration of Basic Designs

Figures 12.4(a)(b): These cross-lined elevational grids serve to illustrate graphically the simple mathematical relationship that exists between the division of *total rise* (TR) and *total going* (TG), of (a) a straight-flight stair and (b) a half-turn stair. Note that, as illustrated

in this example of a two-flight, half-turn stair, instead of *one less going than the number of risers*, there must be *one less going than **half** the number of risers*. Note also that the stairs illustrated here (as in the previous illustrations) are visually representative of concrete stairs – although the theory of TR- and TG-division is the same for all stair-materials.

12.3.2 Setting-out Information

For concrete stairs and landings, a *formwork* (or *shuttering*) carpenter/stair builder usually takes all of his information from the designer's scaled drawings and contract specifications. And from this he proceeds to calculate the precise step sizes – the rise and the going – to enable him to make the required step-template (pitch board).

As a visual example, Figure 12.4(a) proves the rule of there being *one less going than the number of risers*; so, with this visual information, the formwork builder simply needs to divide the TG measurement by 13 and the TR measurement by 14. Therefore, for example, if a total rise of 2.688 m and a total going of 3.081 m was indicated on the drawing of a Category 1 stair, (for private use), the sum would simply be 2688 ÷ 14 = 192 mm rise; and 3082 ÷ 13 = 237 mm going.

12.3.3 Practical Dividing Method with a *Storey Rod*

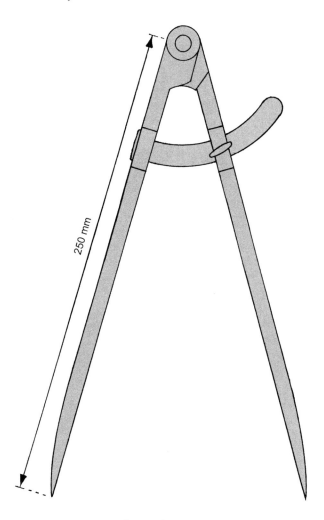

250 mm

Figure 12.5 A pair of wrought-iron dividers with wing-nut adjustment, used traditionally by carpenters and joiners for setting out. Note that the usually dull points of the dividers were easily filed with a mill file, or saw file, until sharp points were produced, as indicated

Figure 12.5: Traditionally, the all-important individual rise of each step was arrived at more tediously by using a prepared timber batten, of ex. 50 or 75 × 25 mm section × 3 m long, as a so-called *storey*

rod, upon which the overall total rise was carefully marked, from 'finished floor level' (ffl) at the bottom to ffl near the top. And between these two extremes, the approximate rise was stepped out with a large pair of carpenter's dividers (Figure 12.5) and a trial-and-error method of repeated mini-adjustments, until the exact division of risers was worked out – and the individual rise was thus determined. This practical method was quite accurate, but relatively time-consuming. However, it must be mentioned that the division of a total rise of 2.688 m (the example used here) was in fact approx. 8 ft 9¾ in, in pre-1970s' imperial measurements. Even though the feet could be easily converted into inches, most storey-divisions required the inches to be converted into fractions. For example, the rise of 192 mm (used above) would be approx. *seven inches and thirty-five sixty-fourths of an inch*. Knowing anecdotally how people generally feel about fractions might explain why a practical method of division evolved in yesteryear before metrication and calculators simplified the task!

12.3.4 Making Step Templates (Pitch-boards) for Marking out the Stair-strings

Figures 12.6(a)(b)(c): Once the rise and the going of the steps have been worked out, the so-called *stair-string* boards need to be marked out carefully with the step-positions. This used to be (and still can be) done with a 610 × 450 mm steel *roofing square* (now called a *metric rafter square*) and a pair of adjustable *stair-gauge fittings*, which were attached to the square's *tongue* and the *blade* – see Figure 12.6(a). Then so-called *pitch-boards* evolved for marking out the step-positions onto the wooden stair-string boards. At first, these were made of thin, solid timber, with separate T-shaped margin-templates (Figure 12.6(b)). Then the margin-template eventually became part of the pitch-board, which was then made from 3.5 mm plywood, hardboard or MDF board, which was sandwiched between two fence-guide battens, as per Figure 12.6(c).

12.4 FORMWORK FOR A HALF-TURN STAIR

Figure 12.7: This elevational-section view of a half-turn concrete staircase, via a half-turn landing, is to be the target referred to in the following notes and illustrations from this point on, in a detailed reference to setting out and forming shuttering for a typical reinforced-concrete staircase. Although not detectable in

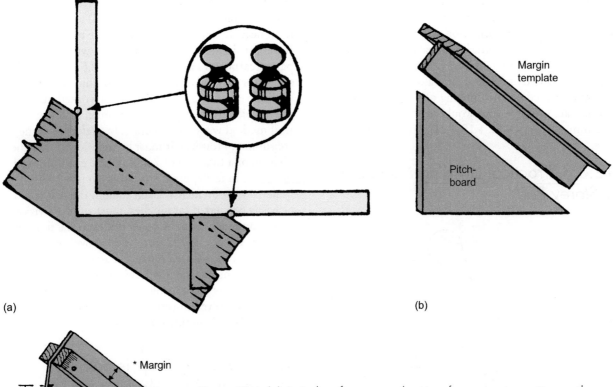

(a)

(b)

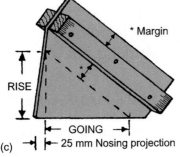

* Margin

RISE

GOING

25 mm Nosing projection

(c)

Figure 12.6 (a) A steel roofing square/metric rafter square in position on the wall string, with stair-gauge fittings attached and step-positions marked out in relation to the margin/pitch line; **(b)** A traditional pitch-board with a separate margin template; **(c)** The combined pitch-board and margin template; note that the extended rise- and going-edges allow the pencil to run on – and once the pitch-board is moved to the next consecutive step-position – the crossed lines create a better intersection of the critical extremes of the hypotenuse on the margin line

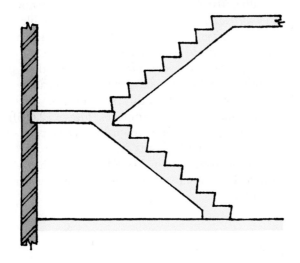

Figure 12.7 A typical half-turn concrete staircase, via a half-turn landing

this elevational view, there would likely be a *small* gap (of, say, at least 50 mm) between the flights (if seen in *plan* view), which would technically determine this design as being a *half-turn stair with an open stairwell.* Such stairwells are usually a regulatory necessity

nowadays to facilitate a continuous handrail around the half-turn landing.

12.4.1 Setting out the Stair-and-formwork's Position on Site

Figures 12.8(a)(b): *Stages (1) to (8)*: The elevational figure (at 12.8(a)) of the essential part of the tar-get-stairwell illustrates a sectional view through one of the stairwell's structural walls (seen on the left), with a preformed landing-bearing recess in its stair-well-face; and, adjacent to that, an elevational view of one of the right-angled return walls of the stairwell, showing the landing-bearing recess in its elevational appearance – and the upper-*landing* or *floor*-bearing recess, in its elevational appearance at the head of the stairs. From stages (1) to (8), indicated numerically in position in Figure 12.8, the setting-out procedure is as follows:

1. Establish the *storey height* (total rise) measurement of the staircase and pencil-mark this horizon-tal point on the wall – bearing in mind that the

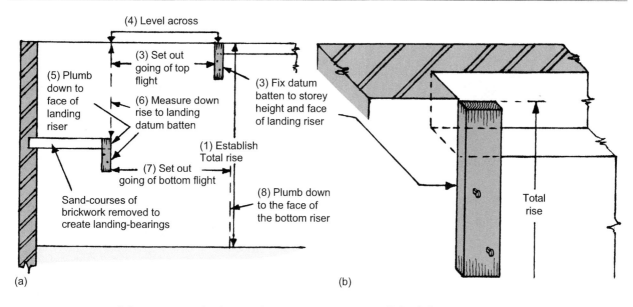

Figures 12.8 (a) and **(b)** Setting out the formwork's position in the stairwell, by following the text-stages (1) to (8)

finished floor levels (ffls) of any other materials to be added later to the basic concrete-treads or floors must be carefully taken into account and allowances made, as necessary.

2. Fix (with cut clasp nails into mortar joints or self-tapping frame screws into bricks or blockwork) a temporary datum-batten (a short length of, say, 75 × 25 mm timber), as illustrated, to the wall, with its top relating to the storey height and its left-hand edge relating to the face of the top landing-riser. Note that the face of the riser's position is determined by measuring laterally the *total going* of the top flight, plus the *going* of the half-landing from the face of the left-hand structural wall. Again, take into account any allowances that may be required for additional, finishing-materials to be added by others to the concrete riser-faces.

3. From the left-hand edge of the first-fixed datum-batten, measure out laterally the top flight's *total going* on the wall, and mark a short, vertical pencil-line.

4. Rest a straightedge on the top edge of the datum-batten, with a spirit level on top, and transfer the storey-height with a short pencil-mark that intersects the short, vertical pencil-line marked on the wall at stage (3).

5. Again, with a straightedge and spirit level, in a vertical position this time, plumb down the vertical intersection created in stage (4) and pencil-mark the wall in the area of the half-landing (just above and below the preformed landing-recess), to represent the face of the bottom landing-riser – and the lateral position for fixing another datum-batten.

6. Measure down the total rise of the top flight to the top of the next, second datum-batten and fix

it to the wall as described before at (2) above; but this time, the *right*-hand face-edge of the batten must be positioned to represent the face of the first landing-riser.

7. On the opposite stairwell-wall, against which the lower stair-flight will be abutted, transfer the level- and lateral-positions of the second datum-batten for the half-landing and fix a third (and final) datum-batten.

8. From the right-hand edge of the half-landing's third datum-batten, measure out laterally the bottom flight's *total going* on the wall and mark a short, vertical pencil-line. Plumb down from this pencil-line mark (with a spirit level) and pencil-mark the face of the lower flight's bottom riser on the wall.

12.4.2 Setting out and Forming the Inner (Wall) Strings

Figures 12.9(a)(b)(c)(d): Once the step's rise and going have been worked out, the required pitch-board can be made, as detailed at Figure 12.9(a), for setting out the exact position of the steps onto the stair-string boards. Note that the pitch-board has to be altered for marking out the so-called *outer string* boards – *after* it has been made to mark out the so-called *inner (wall) string* boards. And for this reason, the pitch-board material needs to be wider on its top, hypotenuse-edge, above the double, wooden fence-battens, as illustrated. The two pieces of fence-material of, say, ex. 50 × 25 mm prepared softwood, must be screwed through to each other (to allow for being repositioned after marking out the wall-string boards) and must be accurately

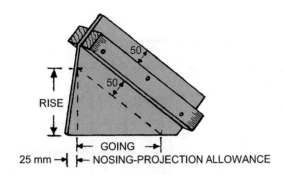

(a)

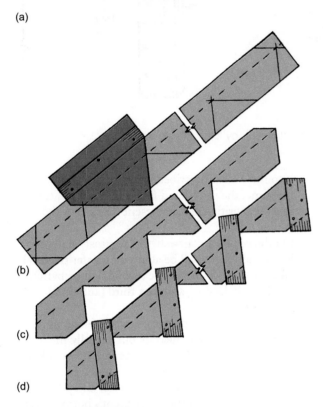

(b)

(c)

(d)

Figure 12.9 (a) A pictorial view of the pitch-board; **(b)** Setting out the wall-strings with the pitch-board; **(c)** The sawn step-shapes and end-waste material removed; **(d)** The 100 × 25 mm sawn timber cleats fixed to the strings

positioned in relation to each other, on each side of the pitch-board.

Next, the inner (wall) string boards – one for the top flight and one for the lower flight – can be made. My scaled examples show 150 × 25 mm sawn boards used as wall-strings and, as illustrated at (b), (c) and (d), one-third of this width (50 mm) is marked/gauged from the top edges with continuous pencil-lines (even though they are shown here (for illustration reasons) as broken lines); these lines are referred to as *margin lines* and they must relate to the hypotenuse (pitch line) on the pitch-board. After carefully marking out the treads-and-risers' positions, as illustrated at (b), the step-shapes are cut out and

the end-waste areas removed, as illustrated at (c); and sawn-timber *cleats* (of, say, 100 × 25 mm section) are nailed or screwed into position, as illustrated at (d). Note that the use of cleats is to strengthen the cut-string's *short grain* in the areas where the riser boards will be fixed. And, because of this, the end-faces of the riser-boards should be nailed or screwed into the cleats – not the short grain of the cut strings.

12.4.3 Alternative Inner (Wall) Strings – and Concrete Waist-thickness

Figures 12.10(a)(b)(c): Two alternative inner (wall) strings (with riser boards in position) are illustrated here to indicate that formwork design is flexible and follows no hard-and-fast rules. Sometimes it is designed to economize on material, as at (a), where the triangular step-offcuts from a 175 × 25 mm cut-string have been used as cleats (note the grain-direction); or, designed to save on labour, as at (b), where steps are *not cut* into the string board. In this example, as illustrated, the top edge of the string is used as the pitch line to mark the hypotenuse of the step-triangle (*), and a pitch-board (made *without a margin*) is marked along the string to give the face-edge positions of the 100 × 25 mm cleats.

At illustration (c), a sectional-view of a few concrete steps, related to their sloping soffit below, is shown to highlight an important reference to a concrete stair's *waist-thickness*. This is determined by a right-angled measurement from the concrete-soffit to the inner edges of the triangular steps, as illustrated. Knowing the size of the waist-thickness, usually shown on a designer's sectional stair-drawings (or scaled from such drawings), will be critical to setting up the undercarriage formwork.

12.4.4 Positioning and Fixing the Inner Strings onto the Walls

Figures 12.11(a)(b): If the setting-out of the stairs' and the landing's position, in relation to the three datum-battens, covered here at the outset, is carried out carefully, positioning and fixing the two inner strings onto the walls, one for the lower flight, one for the upper (the first operation) should be straightforward. As illustrated at (a) and (b), the top edges of the strings (also being the top edges of the risers) simply relate to the tops of the datum-battens – and the bottom edges of the strings should relate to the landing- and floor-levels, whilst also relating to the vertical pencil-lines for the *total going* of each flight, i.e. the bottom riser-face of each flight. If no major errors exist, the strings can be fixed to the

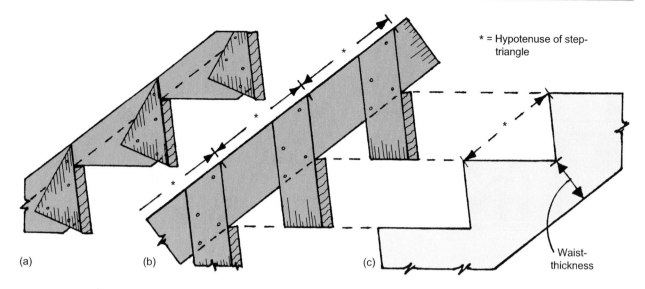

Figures 12.10 (a) and (b) Two alternative inner (wall) strings; (c) The important, structural reference to a concrete-stair's *waist-thickness*

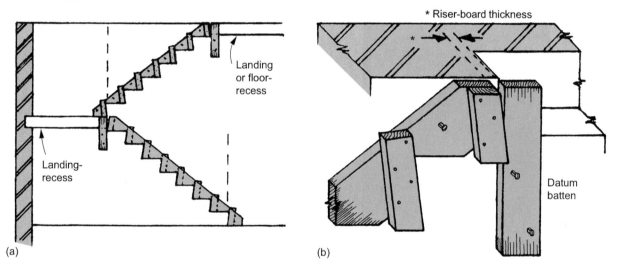

Figures 12.11 (a) and (b) The two inner (wall) strings fixed in position on each opposite wall, related to the setting-out marks and the datum-battens; when the strings have been fixed securely to the walls, the datum-battens will need to be removed

stairwell's walls. Traditionally, if the walls were of relatively 'green' (recently laid) brickwork or blockwork, 75 mm or 100 mm cut-and-clasp nails were used, driven into the fairly receptive bed-joints, or into the faces of receptive blockwork. However, if the stairwell walls are of denser blockwork, then self-tapping frame screws may need to be used to retain the strings in their temporary positions.

12.4.5 Setting up the Undercarriage Formwork and Props

Figures 12.12(a)(b)(c): With the inner strings now fixed, position (or fix temporarily) two short offcuts

of riser boards at two extremes of each string – to enable the waist-thickness (say, 100 mm) to be measured down as illustrated at (b), at right-angles to the pitch, and marked on the wall to determine the sloping soffit-line; then measure down the landing thickness – again, as at (b) – and extend this level line until it intersects with the sloping-soffit line. Then mark parallel pencil lines below these, to represent the 18 mm thick WBP (exterior grade) plywood decking. Next, construct the landings, using 100 × 50mm joists at approx. 400 mm centres, resting on, say, 150 × 75 mm runners. The joists and runners for the upper landing (or floor) can be supported by adjustable, telescopic props, as illustrated; and the joists and runners for the lower, half-turn landing (and the

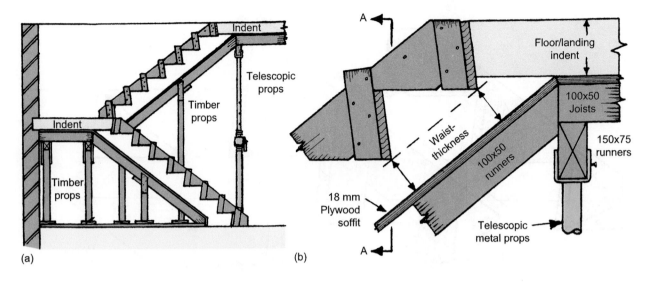

(a)

(b)

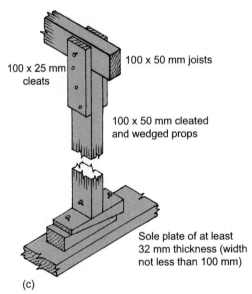

(c)

Figures 12.12 (a)(b)(c) A view of the undercarriage formwork (plywood soffits and decking), the sloping runners, the landing-joists on supporting runners and telescopic props – and diminishing timber-props on timber sole plates and pairs of *folding* wedges. Note that the importance of the non-pointed wedges is to allow the props to be removed more easily when 'striking' (removing) the formwork – lessening the risk of adding stress to the 'green' (newly formed) reinforced-concrete structure

mid-positions of the 100×75 mm sloping runners) can be supported with traditional 75×75 mm or 75×50 mm timber props, cleated at the top and folding-wedged onto a sole plate at the foot – as illustrated at (a) and (c).

12.4.6 Altering the Pitch-board and Setting out and Forming the Outer Strings

Figures 12.13(a)(b)(c)(d)(e): As previously mentioned – and as illustrated here at (a) and

(b) – the pitch-board now needs to be altered to set out the outer strings. This is a simple task of unscrewing the fence-guide battens and re-fixing them to the outer edges to provide a 100 mm (assumed) waist-thickness.

Next, produce two 300 mm widths of 18 mm thick WBP plywood to form the outer strings. Pencil-gauge the waist-thickness from the underside-edges and mark out the steps and landing-abutments, mostly from the underside edges, as shown at (c). Cut and remove the waste-material from both ends of the two strings, as shown at (d) and (e), and fix the cleats.

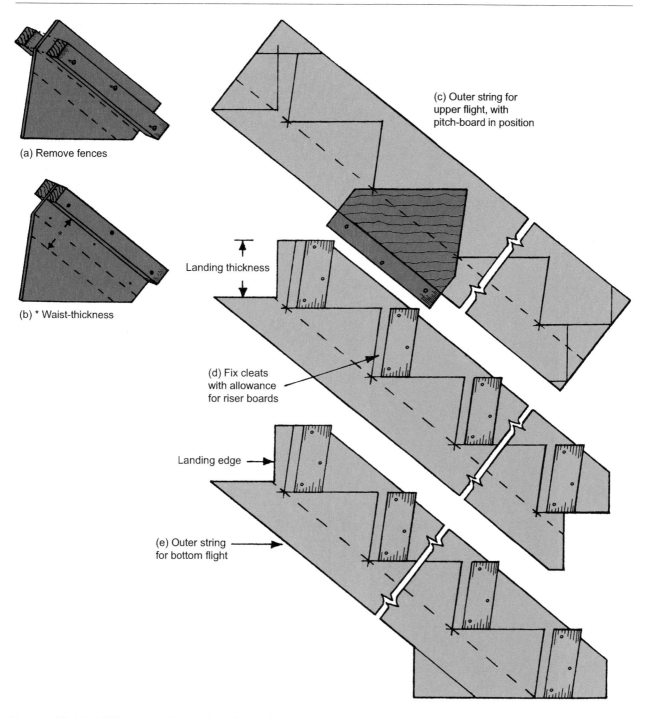

(a) Remove fences

(b) * Waist-thickness

(c) Outer string for upper flight, with pitch-board in position

Landing thickness

(d) Fix cleats with allowance for riser boards

Landing edge

(e) Outer string for bottom flight

Figures 12.13 **(a)(b)** Remove fences from the dual-purpose pitch-board and re-fix to the outer edges to provide a 100 mm (assumed) waist-thickness; **(c)** Set out the outer strings by using the pitch-board from the bottom-edges, in relation to the waist thickness; **(d)(e)** Cut away the waste-material from the four ends and fix the cleats

12.4.7 Fixing the Outer Strings and the Riser Boards

Figure 12.14: Position the outer strings on the sloping soffit-boards, initially with three riser-boards fixed temporarily in position on each flight (in the mid-area and the extremities, to ensure string-verticality and correct lateral-positioning). Then fit and fix the landing-edge fascia boards, where required (at the stairwell-gap of the half-landing, but mostly to the edge of the top floor or landing). Then fit and fix the longitudinal walings and struts, shown here near the top of one of the outer strings – and fix so-called 'kickers' at the base of the strings; fix the bottom edges

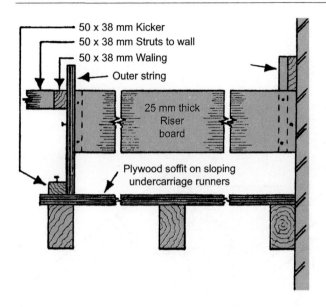

- 50 x 38 mm Kicker
- 50 x 38 mm Struts to wall
- 50 x 38 mm Waling
- Outer string

25 mm thick Riser board

Plywood soffit on sloping undercarriage runners

Figure 12.14 Sectional view A-A through the formwork detailed in Figure 12.12 (b)

of the strings to the kickers and/or to the sloping soffits.

Note that the riser boards, in their entirety, are usually finally fixed after the designed reinforcing-steel has been placed in position and raised-up on small, plastic clipped-spacers across the landing(s) and down the sloping soffits. This raising-up of the steel in the tensile areas of the concrete is to protect it from the effects of fire and corrosion.

Also note that to prevent the riser boards from bulging under pressure from the fluid concrete, support is often added in the central area of the risers, by fixing a cleated, hanging-string, similar to the uncut string illustrated in Figure 12.10(b).

13

Constructing Traditional and Modern Roofs

13.1 INTRODUCTION

Roofing in its entirety is an enormous subject, but the practical issues that involve the first-fixing carpenter on the most common types of dwelling-house roof – dealt with here – are less formidable.

13.1.1 Types of Roof

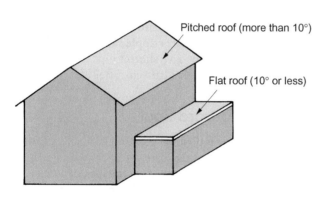

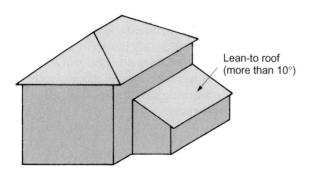

Figure 13.1 Types of roof

Figure 13.1: The three types of roof to contend with are *flat, lean-to* and *pitched.* Timber is used for the skeleton structure to form a carcase and this is covered with various impervious materials. Flat roofs are usually covered with sheet-boarding and (for economy) bituminous felt in built-up layers. Pitched and lean-to roofs are covered with quarried slates (which are very expensive) or non-asbestos fibrous slates, blue/black in colour, a cheaper lookalike, clay tiles (also expensive) or, more commonly, concrete tiles on battens and roofing (sarking) felt.

13.1.2 Knowledge Required

Roofing carpenters nowadays, ideally, require a knowledge of both modern *and* traditional roof construction – and a sound knowledge of at least one of the various methods used for finding lengths and bevels of roofing members. The various methods dealt with later include geometry (to develop an understanding of finding bevels and lengths, but not meant to be used in practice), steel roofing square or metric rafter square, the Roofing Ready Reckoner and last, but not least, the Roofmaster, a revolutionary device invented by Kevin Hodger, formerly a carpentry lecturer at Hastings College in East Sussex.

13.1.3 Traditional Roofing

This consists basically of rafters pitched up from wall plates to a ridge board, such rafters being supported by purlins and struts which transfer the main load to either straining pieces or binders and ceiling joists to an internal load-bearing wall.

13.1.4 Modern Roofing

Modern roofing, using trussed-rafter assemblies, often uses smaller sectional-sized timbers and normally only requires to be supported at the ends. This frees the designer from the need to provide intermediate load-bearing walls and dispenses with purlins and ridge boards. Other differences are that traditional roofing is usually an entire site operation, whereas trussed-rafter roofing involves prefabrication under factory conditions and delivery of assembled units to the site for a much-reduced site-fixing operation.

In figure labels:

Pitched roof (more than 10°)

Flat roof (10° or less)

Lean-to roof (more than 10°)

13.2 BASIC ROOF DESIGNS

13.2.1 Introduction

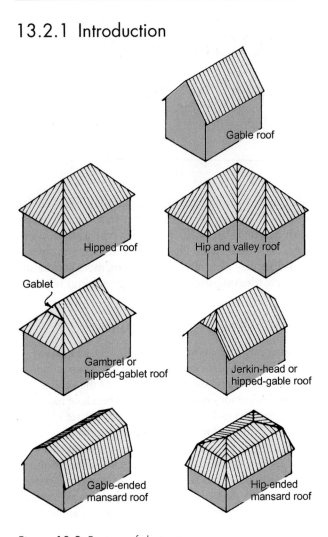

Figure 13.2 Basic roof designs

Figure 13.2: Traditional pitched roofs, by virtue of having been in existence for centuries, predominantly outnumber modern roofs and therefore can be easily spotted for reference and comparison. Their design was often used to advantage in complementing and enhancing the beauty of a dwelling and variations on basic designs can be seen to be infinitely variable. Still occasionally required on individual, one-off dwellings being built, the main features of traditional roofs are described below.

13.2.2 Gable Roof

This design is now widely used in modern roofing because of its simplicity and therefore relatively lower cost. As illustrated, triangular ends of the roof are formed by the outer walls, known as gable ends. Traditionally, purlins took their end-bearings from these walls.

13.2.3 Hipped (or Hip-ended) Roof

This design is also used in modern roofing, but to a lesser extent. The hipped ends are normally the same pitch as the main roof and therefore each hip is 45° in plan. In traditional roofing, the purlins are continuous around the roof and all the old roofing skills are needed. In modern roofing, the hip-ends are either constructed traditionally (by cut and pitch methods) or by using hip trusses supplied with the main trusses; either way, though, a method for finding lengths and bevels will be required at this stage.

13.2.4 Hip and Valley Roof

As illustrated in Figure 13.2, valleys occur when roofs change direction to cover offshoot buildings.

13.2.5 Gambrel Roof and Jerkin-Head Roof

As illustrated, these roofs include small design innovations to the basic hipped and gable roofs. Gambrel roofs can be built traditionally as normal hipped roofs with the full-length hips running through, under the gablets, and the ridge board protruding each end to accommodate the short cripple rafters – and jerkin-head roofs simply have shorter hips. Both types can be built with modern trussed-rafter assemblies.

13.2.6 Mansard Roofs

Apart from their individual appearance being a reason for using such a roof, the lower, steeper roof-slopes, which were vertically studded on the inside, acted as walls and accommodated habitable rooms in the roof space. The upper, shallower roof slopes had horizontal ceiling joists acting as ties, giving triangular support to the otherwise weak structure. Technically (and traditionally), these roofs usually incorporated king-post trusses superimposed on queen-post trusses. This design of roof nowadays is built using modern techniques, especially involving steel beams, taking their bearings from gable ends.

13.3 ROOF COMPONENTS AND TERMINOLOGY

13.3.1 Wall Plates

Figure 13.3(a): 100 × 75 mm or 100 × 50 mm sawn timber bearing plates, are laid flat and bedded on mortar to a level position, flush to the inside of the inner

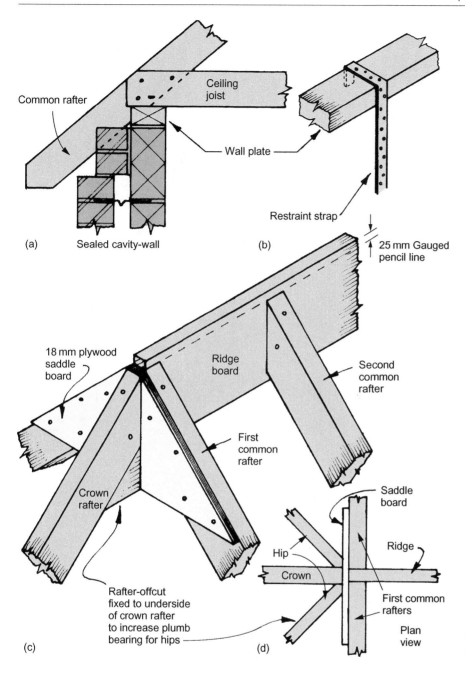

(a)   Sealed cavity-wall

(b)

25 mm Gauged pencil line

Common rafter

Ceiling joist

Wall plate

Restraint strap

18 mm plywood saddle board

Ridge board

Second common rafter

First common rafter

Crown rafter

Rafter-offcut fixed to underside of crown rafter to increase plumb bearing for hips

(c)

Saddle board

Hip

Crown

Ridge

First common rafters

Plan view

(d)

Figure 13.3 (a)–(d) Principal roof components

wall and running along the wall to carry the feet of all the rafters and the ends of the ceiling joists. Nowadays, these wall plates must be anchored down with metal restraint straps.

13.3.2 Restraint Straps

Figure 13.3(b): Vertical, galvanized steel straps, 2.5 mm thick, 30 mm wide, with 6 mm holes at 15 mm offset centres and lengths up to 1.5 m, are fixed over the wall plates and down the inside face of the inner skin of blockwork at maximum 2 m intervals. Additional straps should be used to reinforce any half-lap wall plate joint. Horizontal straps of 5 mm thickness are used across the ceiling joists and

rafters, to anchor the gable-end walls. They should bridge across at least three of these structural timbers and have noggings and end packing fixed between the bridged spaces. These noggings should be at least 38 mm in thickness by half the depth of joist or rafter.

13.3.3 Ceiling Joists

Figure 13.3(a): Like floor joists, these should span the shortest distance, rest on and be fixed to the wall plates, as well as to the foot of the rafter on each side – thereby acting as an important tie and also providing a skeleton structure for the underside-boarding of the ceiling. Usually, 100 × 50 mm sawn timbers are used.

13.3.4 Common Rafters

Figures 13.3(a) and (c): Again, 100 × 50 mm sawn timber is normally used for these load-bearing ribs that pitch up from the wall plate on each side of the roof span, to rest opposite each other and be fixed to the wall plate at the bottom and against the ridge board at the top.

13.3.5 Ridge Board

Figure 13.3(c): This is the spine of the structure at the apex, running horizontally on edge in the form of a sawn board of about 175 × 32 mm section (deeper on steep roofs, depending upon the depth of the common-rafter splay cut +25 mm allowance) against which the rafters are fixed.

13.3.6 Saddle Board

Figures 13.3(c) and (d): The saddle board is a purpose-made triangular board (usually of 18 mm WBP exterior plywood), like a gusset plate, fixed at the end of the ridge board and to the face of the first pair of common rafters. It supports the hips and crown rafter of a hipped end.

13.3.7 Crown or Pin Rafter

Figures 13.3(c) and (d): This is the central rafter of a hipped end.

13.3.8 Alternative Hip-arrangement

This is shown in plan and isometric views in Figure 13.4.

13.3.9 Terminology

Figure 13.4(b): This isometric view of a single-line roof carcase shows the majority of components in relation to each other and defines such terminology as *eaves, verge, gable end*, etc.

13.3.10 Angle Ties

Figure 13.5(a): Sometimes a piece of 75 × 50 mm or 50 × 50 mm timber, acting as a corner tie across the wall plates, replaces the traditional and elaborate drag-on-tie beam used to counteract the thrust of the hip rafter. Alternatively, a simplified, modern, metal drag-on-tie (Figure 13.5(b)) may be used. Either one of these restraints to the hip is recommended, especially on larger-than-normal roofs.

13.3.11 Hip Rafters

Figure 13.5(a): Similar to ridge boards, hip rafters pitch up from the wall-plate corners of a hipped end

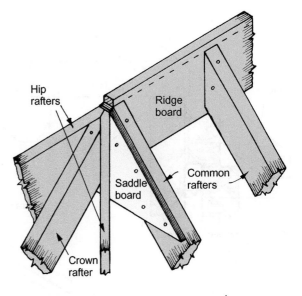

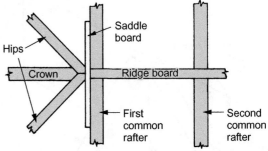

Figure 13.4 (a) Alternative hip-arrangement

to the saddle board at the end of the ridge. They act as a spine for the location and fixing of the jack rafter heads.

13.3.12 Jack Rafters

Figure 13.5(a): These are rafters with a double (compound) splay-cut at the head, fixed in diminishing pairs on each side of the hip rafters.

13.3.13 Valley Rafters

Figure 13.5(c): Valley rafters are like hip rafters, but form an inverted angle in the roof-formation and act as a spine for the location and fixing of the so-called cripple rafters. Sectional sizes are similar to ridge board and hip rafters.

13.3.14 Cripple Rafters

Figure 13.5(c): These are pairs of rafters, diminishing like jack rafters, spanning from ridge boards to valley rafter and gaining their crude name historically by being cut off at the foot.

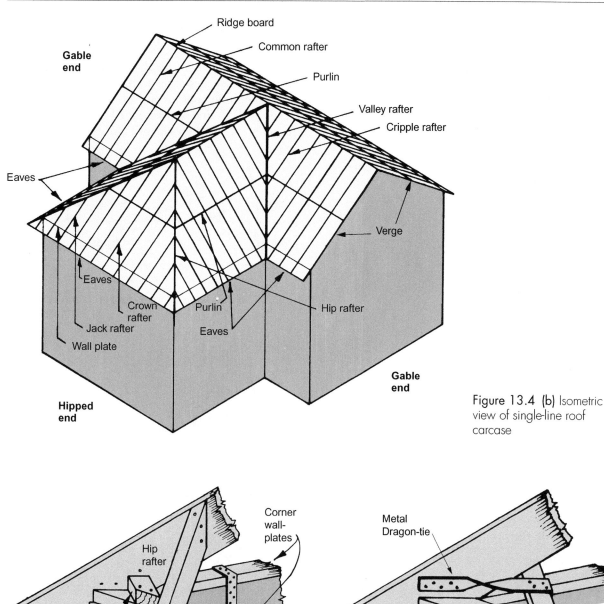

Figure 13.4 (b) Isometric view of single-line roof carcase

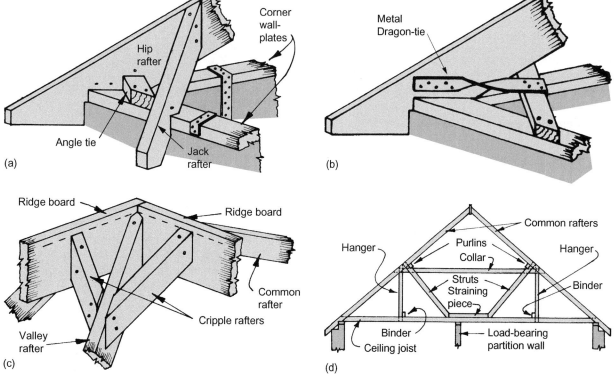

Figure 13.5 (a)–(d) Other roof components

13.3.15 Purlins

Figure 13.5(d): Purlins are horizontal beams, of about 100–150 × 75 mm sawn timbers, that usually support the rafters midway between the ridge and the wall plate, when the rafters exceed 2.5 m in length.

13.3.16 Struts

Figure 13.5(d): These are 100 × 50 mm sawn timbers that support the purlins at about every fourth or fifth pair of rafters. This arrangement transfers the roof load to the ceiling joists and, therefore, requires a load-bearing wall or partition at right-angles to the joists and somewhere near the mid-span below.

13.3.17 Straining Pieces

Figure 13.5(d): These are basically sole plates of 100 × 50 mm or 100 × 75 mm section, fixed to the ceiling joists between the base of the struts. They support and balance the roof thrust and allow the struts to be set at 90° to the roof slope.

13.3.18 Collars

Figure 13.5(d): Collars are 100 × 50 mm sawn ties, fixed to each rafter, sometimes used at purlin level to give extra resistance to the roof-spread and help support the purlins.

13.3.19 Binders

Figure 13.5(d): These are 100 × 50 mm timbers fixed on edge to the ceiling joists with skew-nails or modern framing anchors. They are set at right-angles to the ceiling joists, to give support and counteract deflection of the joists if the span exceeds 2.5 m.

13.3.20 Hangers

Figures 13.5(d) and 13.6(a): These are 100 × 50 mm ties that hang vertically from a rafter side-fixing position near the purlin, to a side-fixing position on the ceiling joist and the binder, usually close to the struts.

13.3.21 Roof Trap or Hatch

Figure 13.6(a): A roof trap is a trimmed and lined opening in the ceiling joists, with a hinged or loose trap door. If hinged, the door either opens up or down. The latter are always latched, the former are often unhinged and loosely fitted. They provide access to the roof void for maintenance of a storage tank (if one exists), pipes, etc., and – if a decked area is provided – storage.

13.3.22 Cat Walk

Figure 13.6(a): This consists of one or two 100–150 × 25 mm sawn boards fixed across the ceiling joists from the edges of the trap, to provide an access path through the roof (usually to the storage tank or decked area). A line of chipboard T&G floor-panels are commonly used nowadays as an alternative.

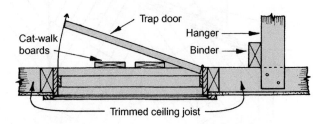

Figure 13.6 (a) Section through roof trap (or hatch)

13.3.23 Lay Boards

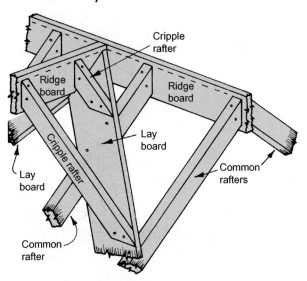

Figure 13.6 (b) Lay boards

Figure 13.6(b): These are valley-located sole-plates of about 175 × 25 mm sawn section, laid flat and diagonally on the ribbed roof structure, to receive the splay-cut feet of cripple rafters forming a valley. This is a popular alternative to using valley rafters, as it is stronger and involves less work.

13.3.24 Eaves

Figure 13.6(c): This is the lowest edge of the sloping roof, which usually overhangs the structure from as little as the fascia-board thickness up to about 450 mm. This is measured horizontally and is known as the *eaves' projection*, which may be open (showing

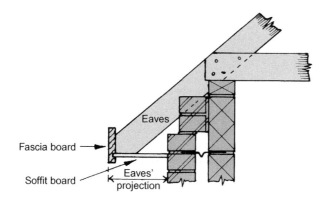

Figure 13.6 (c) *Fascia, soffit and eaves*

the ends of the rafters on the underside, as a feature) or closed (by the addition of a soffit board).

13.3.25 Fascia Board

Figure 13.6(c): The fascia is a prepared board of about ex. 175 × 25 mm, fixed with 75 or 100 mm cut, clasp or oval nails to the plumb cuts of the rafters at the eaves. It provides a visual finish, part of the closing-in at the eaves, and a fixing board for the guttering.

13.3.26 Soffit Board

Figure 13.6(c): This can be fixed to cradling brackets on the underside of closed eaves, as illustrated separately in Figure 13.7(a), between a groove in the fascia board and the wall. This board can be of 9–12 mm WBP exterior plywood or non-asbestos fibreboard. If the skeletal roof is overlaid with a traditional non-breathable roofing felt, the soffit of the eaves should provide cross-ventilation, equivalent in area to a continuous gap of 10 mm width along each side of the roof. (See Figures 13.42(a) to (f).)

13.3.27 Cradling

Figure 13.7(a): Cradling is a traditional way of providing wall-side fixing points for the soffit board. Purpose-made, L-shaped brackets made up from 50 × 25 mm sawn battens, with simple half-lap corner joints clench-nailed together (or screwed), are fixed to the sides of the rafters with 63 mm round-head wire nails (or screws).

13.3.28 Sprocket Pieces

Figures 13.7(a) and (b): These are long wedge-shaped pieces of ex. rafter material, fixed on top of each rafter at the eaves to create a bell-shape appearance or upward tilt to the roof slope. As shown in Figure 13.7(b), this is also achieved by fixing offcuts

of rafter to the rafter sides. Apart from aesthetic reasons, this is done to reduce a steep roof slope, in order to ease the flow of rainwater into the guttering.

13.3.29 Tilting Fillets

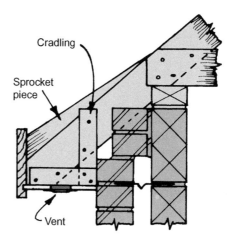

Figure 13.7 (a) *Cradling*

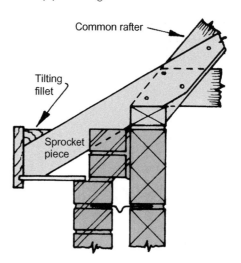

Figure 13.7 (b) *Sprocket piece and tilting fillet*

Figures 13.7(b) and (c): Tilting fillets are timber battens of triangular-shaped cross-section fixed behind the top of a raised fascia board to give it support. The fascia board has been raised primarily to tilt and close the double-layered edge of the tiles or slates, but the increased depth of board also helps the plumber create the necessary *fall* (gradient) in the guttering. Tilting fillets are also used in other places, such as on the top edges of valley boards (Figure 13.7(c)) and back gutters behind chimney stacks, etc.

13.3.30 Valley Boards

Figure 13.7(c): Traditionally, 25 mm sawn boards were used to form a gutter in the valley recess, each board

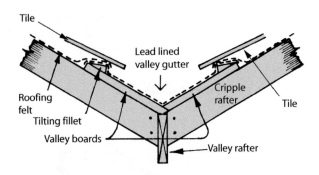

Figure 13.7 (c) *Valley gutter*

being about 225 mm wide. A tilting fillet was fixed at the top edge of each board, ready to be included in the lining of the valley with sheet lead.

13.3.31 Glass-reinforced Plastic (GRP) Valleys

Modern valley linings are now manufactured to a preformed shape, using glass-reinforced plastic (fibreglass). The moulded shape includes a double-roll on each edge to simulate the weathering-upstand of a traditional tilting fillet. These edges should rest on continuous tiling battens, fixed to the cripple rafters; valley boards are not needed.

13.3.32 Verge

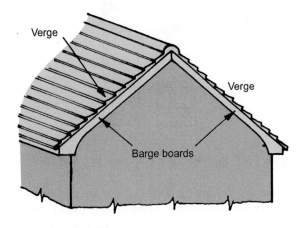

Figure 13.7 (d) *Verge and barge boards*

Figure 13.7(d): This is the edge of a roof on a gable end. It may have a minimum tile-projection and no barge boards, or a greater projection (about 150–200 mm) with barge boards and soffit boards to the sloping underside.

13.3.33 Barge Boards

Figure 13.7(d): These are really fascia boards inclined like a pair of rafters and fixed to a small projection of

roof at the gable-end verges. When projecting from the wall, a soffit board will be required, involving a boxed-shape at the eaves on each side.

13.3.34 Tile or Slate Battens

These are usually 38 × 18 mm, but can be 38 × 25 or 50 × 25 mm sawn, Tanalized (pressurized preservative treatment) battens fixed at gauged spacings on top of the lapped roofing felt. The fixing of these battens is normally done by the slater and tiler, not the carpenter.

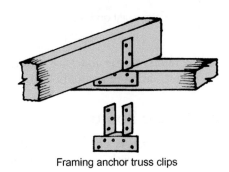

Framing anchor truss clips

Figure 13.7 (e) *Framing anchors*

13.3.35 Framing Anchors

Figure 13.7(e): Galvanized steel framing anchors of various designs for various situations may be used to replace traditionally nailed fixings in certain places. The one shown here replaces skew-nailing of the ceiling joist to the wall plate. The recommended fixings to be used (in *every* hole of the framing anchor) are 3 mm in diameter by 30 mm long sherardized clout nails.

13.4 BASIC SETTING-OUT TERMS

13.4.1 Introduction

All the bevels and lengths in roofing can be worked out by various methods – all of which are based on the principles of geometry. Although the actual method of using drawing-board geometry is not practical in a site situation, it is given in Figure 13.9 as an introduction to understanding how the various bevels and lengths are found.

First, as illustrated in Figure 13.8, the following basic setting-out terms must be appreciated in relation to the sectional view through the pitched roof.

13.4.2 Span

This is a critical distance measured in the directional run of the ceiling joists at wall-plate level, from the

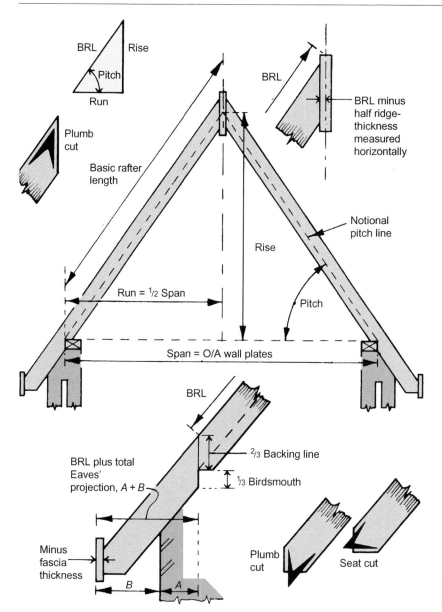

Figure 13.8 Basic setting-out terms (BRL = basic rafter length)

outside of one wall plate to the outside of the other, i.e. overall (O/A) wall plates.

13.4.3 Run

For the purpose of reducing the isosceles roof-shape to a right-angled triangle with a measureable baseline, the span measurement is divided by two to produce what is known as the *run*.

13.4.4 Rise

This represents the perpendicular of the triangle, measured from wall-plate level up to the apex of the imaginary hypotenuse or notional pitch lines running at two-thirds rafter-depth, through the sides of the

rafters, from the outside arris of the wall plates (as illustrated).

13.4.5 Pitch

This is the degree angle of the roof slope. The known rise and the run gives the pitch angle and the basic rafter length – or the known pitch angle and the run gives the rise and the basic rafter length.

13.4.6 Backing Line

This is an important plumb line marked at the base of the setting-out rafter (pattern rafter), marked down two-thirds of its depth to the top of the birdsmouth cut, acting as a datum or reference point for the rafter's length on one side and the total eaves' projection on the other.

13.4.7 Birdsmouth

This is a notch cut out of the rafter to form a seating on the outside edge of the wall plate.

13.5 GEOMETRICAL SETTING-OUT OF A HIPPED ROOF

13.5.1 Bevels and Length of Common Rafter

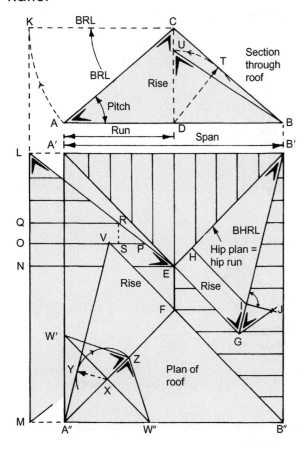

Figure 13.9 Geometry for hipped roof

Figure 13.9: Draw triangle ABC (section through roof) where: AB = span, AD = run (half span), CD = rise, angle DAC = seat cut, angle ACD = plumb cut, line AC or CB = basic rafter length (BRL).

13.5.2 Bevels and Length of Hip Rafter

Figure 13.9: Draw plan of roof, showing two hipped ends (these being always drawn at angles of 45° in equal pitched roofs), denoted as A′E, B′E and A″F, B″F. Now, at right angles to B′E, draw line EG,

equal in length to the rise at CD. Join G to B′: angle EB′G = seat cut, angle B′GE = plumb cut, line B′G = basic hip rafter length (BHRL).

13.5.3 Hip Edge Cut

Figure 13.9: Draw line HI, parallel to EG at any distance from E; then draw line IJ, equal to distance EH, at right-angles to B′I Draw line JG: angle JGI is the edge cut.

13.5.4 Jack (and Cripple) Rafter Bevels and Lengths

Figure 13.9: With radius CA, equal to basic rafter length, describe an arc from A to K. Project K down vertically to form line LM. Join L to E to give an elevated, true shape of roof side. Draw single lines to represent rafters at 400 mm (scaled) centres: angle LEN = jack/cripple edge cut. (The jack/cripple side cut is the same as the common rafter plumb cut.) Line OP = the basic length of the first jack rafter; line QR = the second diminishing jack rafter.

13.5.5 Diminish of Jack (and Cripple) Rafter Lengths

Figure 13.9: The perpendicular line SP, of the right-angled triangle RSP, is equal to the constant diminish of the jack (or cripple) rafters.

13.5.6 Purlin Bevels

Figure 13.9: Angle QLR, on the elevated plan view = purlin edge cut. The purlin side cut is developed within the sectional view through the roof. With compass at D and CB as a tangent at T, describe an arc to cut CD at U, and join U to B: angle DUB = purlin side cut.

13.5.7 Dihedral Angle or Backing Bevel

Figure 13.9: This is for the top edge of hip boards and might only be used nowadays if the roof – on a less cost-conscious, quality job – were to be boarded or sheathed with 12 mm plywood, prior to felting, battening and slating or tiling. On the plan view, establish triangle A″VF, as at B′GE. Draw a 45° line at any point, marked W′W″. With compass at X and A″V as a tangent at Y, describe an arc to cut A″F at Z. Join Z to W′ and W″: angle W′ZX or XZW″ is the required backing bevel.

13.6 ROOFING READY RECKONER

13.6.1 Introduction

The first practical method to be considered for finding the bevels and lengths in roofing, is by reference to a small limp-covered booklet entitled *Roofing Ready Reckoner* by Ralph Goss, published by Blackwell Scientific Ltd, ISBN 0632021969. The tables are given separately in metric and imperial dimensions and are quite easy to follow, once a few basic principles have been grasped.

13.6.2 Choice of Tables

Figure 13.10: The tables cover a variety of roof pitches up to 75°, giving the various bevels required and the diminish for jack rafters. Basic rafter lengths (BRL) and basic hip rafter lengths (BHRL) must be worked out from the tables – which show a given measurement for the hypotenuse of the inclined rafter, in relation to base measurements of metres, decimetres and millimetres (or, in separate tables given, feet, inches and eighths of an inch) contained in the run of the common rafter.

13.6.3 Determining the Run and Pitch

To use this method, first the span of the roof must be measured from the bedded wall-plates and halved to give the run. Also, the pitch must be known – and if not specified, should be taken, with the aid of a simple protractor, from the elevational drawings.

13.6.4 Example Workings Out

As an example, take a hipped roof of 36° pitch, with a span of 7.460 m. Halve this to give a run of 3.730 m. Then, referring to the tables taken from the booklet and illustrated here in Figure 13.10, work out the lengths of the common rafters and the hip rafters.

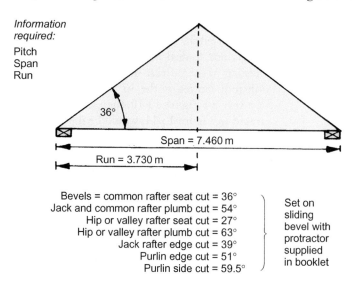

Information required:
Pitch
Span
Run

36°

Span = 7.460 m

Run = 3.730 m

Bevels = common rafter seat cut = 36°
Jack and common rafter plumb cut = 54°
Hip or valley rafter seat cut = 27°
Hip or valley rafter plumb cut = 63°
Jack rafter edge cut = 39°
Purlin edge cut = 51°
Purlin side cut = 59.5°

Set on sliding bevel with protractor supplied in booklet

Table for 36° pitch

Run of rafter	0.1	0.2	0.3	0.4	0.5	0.6	0.7	0.8	0.9	1.0
Rafter length	0.124	0.247	0.371	0.494	0.618	0.742	0.865	0.989	1.112	1.236
Hip length	0.159	0.318	0.477	0.636	0.795	0.954	1.113	1.272	1.431	1.590

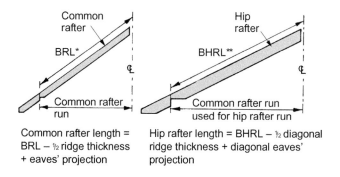

Common rafter
BRL*
Common rafter run

Hip rafter
BHRL**
Common rafter run used for hip rafter run

Common rafter length = BRL – ½ ridge thickness + eaves' projection

Hip rafter length = BHRL – ½ diagonal ridge thickness + diagonal eaves' projection

Figure 13.10 Roofing Ready Reckoner

Common Rafter

Length of common rafter for 1 m of run = 1.236 m
Length of common rafter for 3 m of run = 3.708 m
Length of common rafter for 0.7 m of run = 0.865 m
Length of common rafter for 0.03 m of run = 0.0371 m
Length of common rafter for 3.730 m of run = 4.6101 m

Therefore the basic rafter length (BRL*) = 4.610 m

Speedier Working Out

It must be mentioned that a speedier mathematical method would be to multiply the total common rafter run by the 1 m run-of-rafter figure given in the booklet's tables, i.e.:

3.730 × 1.236 = 4.610 m = BRL

Hip Rafter

Length of hip rafter for 1 m of run = 1.590 m
Length of hip rafter for 3 m of run = 4.770 m
Length of hip rafter for 0.7 m of run = 1.113 m
Length of hip rafter for 0.03 m of run = 0.0477 m
Length of hip rafter for 3.730 m of run = 5.9307 m

Therefore, the basic hip rafter length (BHRL**) = 5.931 m

Speedier Working Out

This involves multiplying the total common rafter run by the 1 m run-of-rafter figure given directly under that column for the length of hip, in the booklet's tables, i.e.

3.730 × 1.590 = 5.9307 m (call this 5.931 m BHRL)

Information required: Rafter centres (spacings)

Jack rafters 400 mm centres decrease 494 mm
Jack rafters 500 mm centres decrease 618 mm
Jack rafters 600 mm centres decrease 742 mm

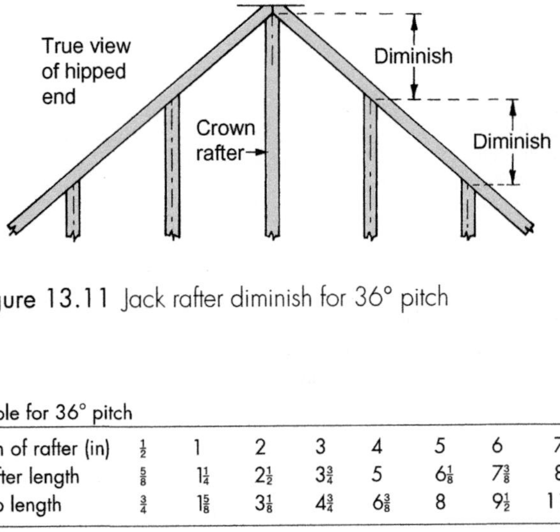

Figure 13.11 Jack rafter diminish for 36° pitch

13.6.5 Jack Rafter Diminish

Figure 13.11: Jack and cripple rafters should diminish in length by a constant amount in relation to the length of the main rafter (crown or common). The *Roofing Ready Reckoner* gives a set of figures for each different pitch, shown on the same page as the bevels and tables, to deal with the decrease according to the spacing of the rafters. Assuming rafters to be spaced at 400 mm centres on the 36° pitched roof described in the text for Figure 13.10, then the diminish, as illustrated in Figure 13.11, would be 494 mm.

13.6.6 Imperial-dimensioned Tables

Figure 13.12: As a comparison – and perhaps as an alternative for any diehards in the industry, still using feet and inches – the following example is given, using the imperial tables in the *Ready Reckoner*, expressed in feet, inches and eighths of an inch. Again, take a hipped roof of 36° pitch, but with a span of 24 ft 5in (24 feet, 5 inches). Halve this to give a run of 12 ft $2\frac{1}{2}$in. Then, referring to the tables taken from the

Table for 36° pitch

Run of rafter (in)	$\frac{1}{2}$	1	2	3	4	5	6	7	8	9	10	11
Rafter length	$\frac{5}{8}$	$1\frac{1}{4}$	$2\frac{1}{2}$	$3\frac{3}{4}$	5	$6\frac{1}{8}$	$7\frac{3}{8}$	$8\frac{5}{8}$	$9\frac{7}{8}$	$11\frac{1}{8}$	$12\frac{3}{8}$	$13\frac{5}{8}$
Hip length	$\frac{3}{4}$	$1\frac{5}{8}$	$3\frac{1}{8}$	$4\frac{3}{4}$	$6\frac{3}{8}$	8	$9\frac{1}{2}$	$11\frac{1}{8}$	$12\frac{3}{4}$	$14\frac{1}{4}$	$15\frac{7}{8}$	$17\frac{1}{2}$

Run of rafter (ft)	1	2	3	4	5	6	7	8	9	10
Rafter length	1-$2\frac{7}{8}$	2-$5\frac{5}{8}$	3-$8\frac{1}{2}$	4-$11\frac{3}{8}$	6-$2\frac{1}{8}$	7-5	8-$7\frac{7}{8}$	9-$10\frac{5}{8}$	11-$1\frac{1}{2}$	12-$4\frac{3}{8}$
Hip length	1-$7\frac{1}{8}$	3-$2\frac{1}{8}$	4-$9\frac{1}{4}$	6-$4\frac{1}{4}$	7-$11\frac{3}{8}$	9-$6\frac{1}{2}$	11-$1\frac{1}{2}$	12-$8\frac{5}{8}$	14-$3\frac{5}{8}$	15-$10\frac{3}{4}$

Jack rafters 16 in centres decrease $19\frac{3}{4}$ in
Jack rafters 18 in centres decrease $22\frac{1}{4}$ in
Jack rafters 24 in centres decrease $29\frac{5}{8}$ in

Figure 13.12 Imperial tables

booklet and illustrated here, work out the lengths of the common rafters and hip rafters as follows:

Length of common rafter for 10 ft of run	$= 12$ ft $4\frac{3}{8}$ in
Length of common rafter for 2 ft of run	$= 2$ ft $5\frac{5}{8}$ in
Length of common rafter for 2 in of run	$= 2\frac{1}{2}$ in
Length of common rafter for $\frac{1}{2}$ in of run	$= \frac{5}{8}$ in
Length of common rafter for 12 ft $2\frac{1}{2}$ in of run	$= 15$ ft $1\frac{1}{8}$ in
Length of hip rafter for 10 ft of run	$= 15$ ft $10\frac{3}{4}$ in
Length of hip rafter for 2 ft of run	$= 3$ ft $2\frac{1}{8}$ in
Length of hip rafter for 2 in of run	$= 3\frac{1}{8}$ in
Length of hip rafter for $\frac{1}{2}$ in of run	$= \frac{3}{4}$ in
Length of hip rafter for 12 ft $2\frac{1}{2}$ in run	$= 19$ ft $4\frac{3}{4}$ in

13.6.7 Conclusion

It must be realized that, because of the involvement with fractions of an inch, the imperial method of working out lengths of rafters and hips, unlike the metric method, does not offer a speedier working out. However, it can be done by using inches as the basic units, providing the fractions of an inch are changed to decimals, whereby $\frac{1}{2}$ in $= 0.5$, $\frac{1}{4}$ in $= 0.25$, $\frac{1}{8}$ in $= 0.125$, $\frac{1}{16}$ in $= 0.0625$ and $\frac{1}{32}$ in $= 0.03125$.

Therefore, on the last working out, the pitch was 36°, the common rafter run was 12 ft $2\frac{1}{2}$ in. Converted to inches and decimals, this would become 146.5 in. This figure is then multiplied by 1.236 (the 1 m run-of-rafter figure for 36°), therefore:

$$146.5 \times 1.236 = 181.074 \text{ in}$$
$$= 15 \text{ ft } 1.074 \text{ in}$$

Therefore the length of common rafter for a 12 ft $2\frac{1}{2}$ in run is 15 ft $1\frac{1}{16}$ in. The $\frac{1}{16}$ in less than the previous working out is insignificant in roofing.

13.7 METRIC RAFTER SQUARE

13.7.1 Introduction

Figure 13.13: The next method to be considered involves the use of a traditional instrument known as a *steel roofing square*, now metricated and called the *metric rafter square*. The one referred to here was manufactured by I & D Smallwood Ltd, and came with an explanatory booklet on its use. Other, similar rafter squares on the market, adequate for the job, do not have the protractor facility incorporated in the Smallwood square, which is personally preferred by the author. The booklet, which is well illustrated, clearly explains the elements of roofing and the application of the square.

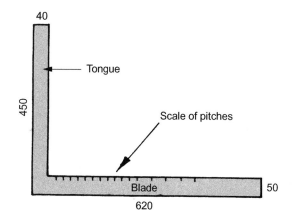

Figure 13.13 Metric rafter square model No. 390 or 390B

13.7.2 Protractor Facility

The various settings for different roof bevels – which are not easy to remember – are usefully given on the faces of the tongue. These are related to traditional settings, based on geometric principles. The main innovation, though, is that the square has a protractor facility in the form of a scale of pitches on the inner edge of the blade, enabling any pitch angle up to 85° to be set up quickly and easily.

13.7.3 Common Rafter Plumb and Seat Cuts

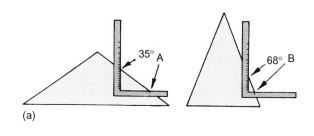

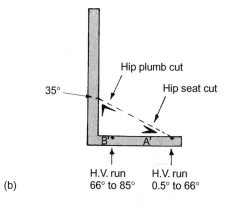

Figure 13.14 (a) Using protractor facility to find common rafter plumb and seat cuts; **(b)** finding hip bevels

Figure 13.14(a): By using the protractor facility to set up the pitch, common-rafter plumb and seat cuts are quickly found. This is achieved by using a common-rafter-run point A, set at 250 mm on the inner edge of the tongue, as illustrated, and by rotating the square until the degree figure for the roof pitch registers on the inner edge of the blade. This works for pitch angles up to 66°. Should greater angles than this be required, then a common-rafter-run point B, set at 50 mm on the inner edge of the tongue, is used in relation to the scale of pitches on the blade. This will give angles from 66° to 85°.

13.7.4 Hip (or Valley) Rafter Plumb and Seat Cuts

Figure 13.14(b): These bevels are also found by using the protractor facility related to the roof pitch. This is achieved by using a hip or valley-rafter-run point, marked 'H.V. run' on the square, set at 354 mm on the inner edge of the tongue and, by rotating the square until the degree figure for the roof pitch registers on the inner edge of the blade. This will give hip or valley plumb and seat cuts for roofs up to 66°. Should the roof be steeper than this, then a hip or valley-rafter-run point, set at 71 mm on the inner edge of the tongue for roofs from 66° to 85°, is used in relation to the scale of pitches on the blade.

13.7.5 Common and Hip (or Valley) Rafter Lengths

As is normal practice, these rafter lengths are worked out in relation to the run. Some figures for this are given on both sides of the blade, against the pitch required, related to a 1 m run. A more definitive set of figures are given in the booklet, varying from 1° up to 89.5° pitch, in increments of 0.5°. Hence, a small and simple calculation will be necessary, whereby, as with the *Roofing Ready Reckoner*, the total common rafter run is multiplied by the 1 m run-of-rafter figure given on the blade or in the booklet, there being, as before, one figure for common-rafter lengths and another for hip (or valley) rafter lengths.

13.7.6 Stair Gauge Fittings (British Pattern)

Figure 13.15 *Stair gauge fittings model No. 385*

Figure 13.15: To increase the square's use from an instrument to a working tool, stair gauge fittings, as illustrated, are an available option for this – or any roofing square. The fittings are thumb-screwed to the square's edges to act as stops, for setting and marking repetitive bevels, thus avoiding the use of a separate carpenter's bevel.

13.8 ALTERNATIVE METHOD FOR THE USE OF THE METRIC RAFTER SQUARE

13.8.1 The Scaled Method

A method used traditionally for its speed and simplicity in finding the main rafter lengths and bevels was known as the scaled method. This lends itself easily to metrication, from its original use in imperial dimensions – whereby inches represented feet and twelfths-of-an-inch represented inches. Figure 13.16 should help in grasping this method, by visualizing the roofing square within the roof, either by imagining the square scaled up to roof-size or the roof scaled down to square-size.

13.8.2 Stair Gauge Fittings (American Pattern)

Figure 13.17: These American pattern stair gauge fittings (also illustrated on the sides of the square in Figure 13.18(b)) are almost an essential requirement – if you can obtain them – for this particular roofing method. This is because they relate accurately to pre-set measurements on the edges of the blade or tongue of the square – and enable very accurate readings of the scaled hypotenuse measurements required.

As the name of these fittings implies, they were originally meant to be used on the square for setting

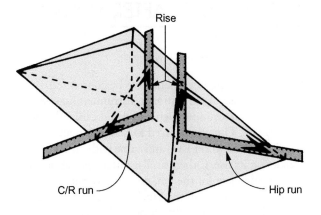

Figure 13.16 Visualizing the square within the roof (C/R = common rafter)

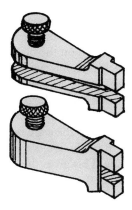

Figure 13.17 American-pattern stair-gauge fittings

out the steps of a staircase or the shuttering for concrete stairs – a use for which, of course, they can still be put, if necessary.

13.8.3 Finding the Rise

Figure 13.18(a): With the scaled method, it will be necessary to determine the rise before proceeding further. The key to this is the protractor facility, which will simplify the task. First, use the square as already described in the text for Figure 13.14, by laying it on a straight piece of rafter or straightedge material, relating to common-rafter-run point A on the inside edge of the tongue and – in this example – 32.5° pitch on the inside edge of the blade. Mark the bottom, outer edge of the tongue with a marking knife, chisel or a sharp pencil. This will give the seat cut and, being greater in length on this shallow pitch than the plumb cut would be on a given width of straightedge material, the plumb-cut mark available against the blade will not be needed.

13.8.4 Common Rafter Length, Plumb and Seat Cuts

Figure 13.18(b): The next step is to attach a stair gauge fitting to the outer edge of the blade, set carefully at, say, 310 mm, representing a scaled run of roof of 3.100 m. Line up the blade to the previously found seat cut bevel marked on the straightedge, as illustrated, then carefully attach the other stair gauge fitting to the outer edge of the tongue against the straightedge. This will automatically produce the scaled rise at 198 mm (representing a true rise of 1.98 m). These settings determine the plumb and seat cuts and provide a scaled measurement of the common rafter length. The scale used is one-tenth full size (1:10) and is easily achieved by moving the decimal point by one place:

3.100 m run divided by 10 = 0.310 mm on the blade
0.198 mm on the tongue times 10 = 1.980 m rise

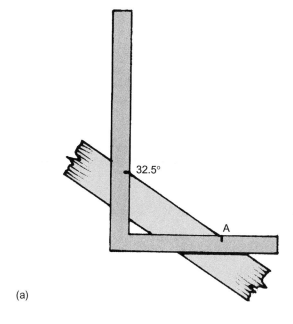

(a)

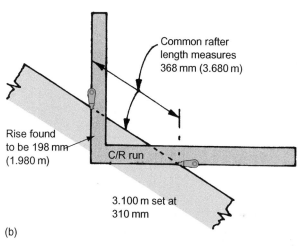

Common rafter length measures 368 mm (3.680 m)

Rise found to be 198 mm (1.980 m)

C/R run

3.100 m set at 310 mm

(b)

Figure 13.18 (a) First step to finding the rise by scaled method; (b) second step to find the rise, bevels and length

The scaled measurement of the rafter length on the hypotenuse – measured between the sharp arrises of the stair gauge fittings – is also brought up to size by multiplying by 10, i.e. 0.368 mm × 10 = 3.680 m.

13.8.5 Finding the Hip-rafter Run

Figure 13.19: To find the hip rafter length, first find the hip rafter run. As this represents – in an equal pitched roof – a 45° diagonal line in plan, contained within the run of the common rafter and the run of the crown rafter, forming a square, it follows that the scaled common-rafter run, set on both the blade and the tongue, as illustrated, gives the scaled measurement of the hip run at 45° on the hypotenuse.

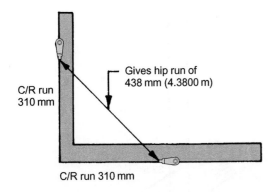

Figure 13.19 Settings to find the hip run

13.8.6 Hip Rafter Length, Plumb and Seat Cuts

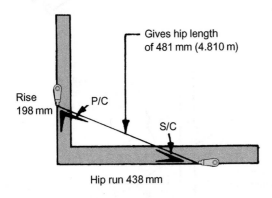

Figure 13.20 Hip rafter length, plumb and seat cuts

Figure 13.20: Now alter the position of the stair gauge fittings and set the scaled hip run on the blade, and the scaled rise on the tongue, as illustrated. This will give the scaled hip rafter length, the plumb cut (P/C) and the seat cut (S/C).

13.9 BEVEL-FORMULAS FOR THE ROOFING SQUARE

13.9.1 Hip Edge Cut

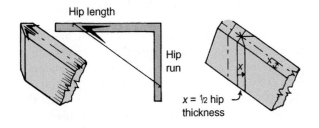

Figure 13.21 Hip edge cut

Figure 13.21: This equals the scaled hip length on the blade and the scaled hip run on the tongue. As illustrated, the bevel is found on the blade. This is applied to the edges of the hip plumb cut and enables the hips to fit into the heads of the crown rafter and the first common rafters, against a saddle board – or, alternatively, to fit against each other and the saddle board. A simpler way of finding the edge bevel, is to measure and mark half the hip's thickness *x* in from the second plumb cut each side. This gives the three points on the top edge for marking the edge bevels. To understand this more fully, see Figure 13.33(a) 'Setting out a hip rafter'.

13.9.2 Hip Backing Bevel (Dihedral Angle)

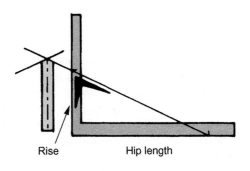

Figure 13.22 Hip backing bevel

Figure 13.22: This equals the scaled hip length on the blade and the scaled rise on the tongue. As illustrated, the bevel is found on the tongue. This was used traditionally when the roof was to be boarded and was applied by planing the top edge from both sides to a top centre line.

13.9.3 Jack or Cripple Side (Plumb) Cut

This is the same formula as for the common rafter plumb cut, with the scaled common rafter run set on the blade, the scaled rise on the tongue and the required bevel found on the tongue.

13.9.4 Jack or Cripple Edge Cut

Figure 13.23: This combines with the above cut to make a compound angle to fit against the hip or valley rafters. The formula equals the scaled common-rafter length on the blade and the common rafter run on the tongue. The bevel, A, is found on the blade.

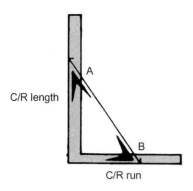

Figure 13.23 Jack edge cut (bevel A) and purlin edge cut (bevel B)

13.9.5 Purlin Edge Cut

Figure 13.23: This bevel is applied to the surface that the underside of the rafters rest on, at the junction of the hips or valleys, to form a mitred edge against the sides and under the centre of the hip or valley rafters. The formula is the same as for the jack or cripple edge cut, except that the required bevel B is found on the tongue.

13.9.6 Purlin Side Cut

Figure 13.24: This combines with the above cut, to complete the mitred faces against and under the hips or valley rafters. The formula equals the scaled common rafter length on the blade and the rise on the tongue. The bevel required is found on the tongue.

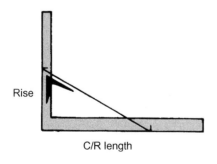

Figure 13.24 Purlin side cut

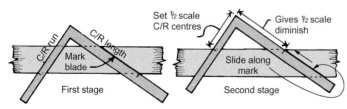

Figure 13.26 Jack rafter diminish

13.9.7 Purlin Lip Cut

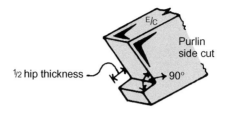

Figure 13.25 Purlin lip cut (E/C = edge cut)

Figure 13.25: As illustrated, this is simply marked at 90° to the purlin side cut in relation to the amount of hip projection below the rafters.

13.9.8 Jack or Cripple Rafter Diminish

Figure 13.26: As illustrated, this is found in two stages. First, by setting the jack edge-cut formula on the square and marking this – against the edge of the blade – on a straightedge. Then by sliding the blade along the mark until the projecting tongue registers the common-rafter centres at a half full-size scale (say 400 mm centres divided by 2 = 200 mm on the tongue). The measurement now showing on the projecting blade will be equal to a half full-size scale of the diminish.

13.9.9 Bevel Formulas

Table 13.1 on the next page gives bevel formulas in quick reference style.

13.10 ROOFMASTER SQUARE

13.10.1 Introduction

Figure 13.27: The final method to be considered as an alternative to the roofing square and the Roofing Ready Reckoner, is the innovative, recently re-marketed tool (or instrument) known as the Roofmaster. It consists of an anodized aluminium blade, resembling a 45° set square, engraved with easy-to-read laser-etched figures and markings on each face side. Attached to the blade is a pivoting and lockable, double-sided fence (or arm),

Table 13.1 Bevel formulas

Component	Required bevel	Blade setting	Tongue setting	Side for marking
Common rafter	plumb cut	C/R run	rise	tongue
Common rafter	seat cut	C/R run	rise	blade
Hip rafter	run of hip	C/R run	C/R run	hypotenuse measurement
Hip rafter	plumb cut	hip run	rise	tongue
Hip rafter	seat cut	hip run	rise	blade
Hip rafter	edge cut	hip length	hip run	blade
Hip rafter	backing bevel	hip length	rise	tongue
Jack rafter	side cut	C/R run	rise	tongue
Jack rafter	edge cut	C/R length	C/R run	blade
Purlin	edge cut	C/R length	C/R run	tongue
Purlin	side cut	C/R length	rise	tongue
Lay board	plumb cut	C/R length	C/R run	blade
Lay board	seat cut	C/R length	C/R run	tongue

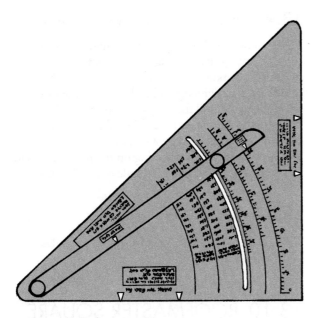

Figure 13.27 Roofmaster square (artistic representation only)

through which the blade is slid when locking the fence to a required setting. The instrument comes with a booklet which is well illustrated and clearly explains the application and use.

13.10.2 Basic Concept

The basic concept of this revolutionary square is that the only knowledge needed to access all the different bevels required in the cutting of a roof is the pitch angle. If, for example, the pitch to be used is 35°, that would be the only reference required to set up the Roofmaster as many times as necessary, and *when* necessary, to determine the different bevels, such as

common-rafter plumb and seat cuts, hip-rafter plumb and seat cuts, purlin side and edge cuts, jack and cripple rafter plumb and edge cuts, etc.

13.10.3 Main Features

Three main features are engraved on each side of the blade as follows.

- A set of separate, graduated segmental arcs, each referenced to a specific roof member and calibrated to give the required bevel, which is obtained simply by setting the adjustable fence to the required pitch angle numbered on the selected arc.
- Information panels on each face side of the blade indicate which edge to mark for the selected angle cut. This valuable facility eliminates yet another area of confusion associated with other roofing squares.
- Tables for length of common, hip and valley rafters per metre of run, radiating around the remaining segmental arcs, are calibrated to line up the multiplier figure for use in obtaining the true or basic length, obtained simply – once again – by setting the adjustable fence to the required pitch angle numbered on the pitch-angle arc.

13.10.4 Handling the Roofmaster

Figure 13.28: As with other roofing squares, the angles required will always be acute angles contained within the right-angled triangle formed with this tool, by the right-angle of the blade and the varying position of the adjustable fence, acting as the triangle's hypotenuse. As illustrated, the setting edge of the fence, which is clearly marked, always butts against the edge of the roof member.

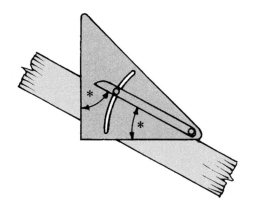

Figure 13.28 Acute angles required*

13.10.5 Reverse and Opposite Marking-positions

Figure 13.29: An important operational fact to realize is that once the fence has been locked into the required position, the Roofmaster can be used on opposite side-edges of the timber to mark the same bevel (Figures 13.29(a) and (b)) or, if necessary, can be reversed (turned over) to give either left- or right-handed cuts (Figure 13.29(c)).

13.10.6 Finding Cutting Angles

To apply the plumb and seat cuts to a common rafter for a roof to be pitched, say, at 35°, first, select the arc designated for common-rafter cuts of 16–45°; second, lock the adjustable arm on number 35; third, hold the tool with the setting edge of the fence firmly against the rafter material; and finally mark the edges indicated by the information panels on the blade, for the required cuts.

Note that this sequence of operations is carried out on each relevant arc to obtain the different cut-angles for all of the roof members.

13.10.7 Rafter Lengths

Figure 13.30: The common rafter and hip or valley rafter lengths per metre of run are engraved on each side of the blade, for roof pitches of 16–45° on one side and 45–75° on the other. To obtain the true or basic length of one of these rafters, the adjustable arm must be set to the correct pitch angle figure on the common-rafter arc. The two sets of table numbers now displayed – and indicated by reference headings – along the setting edge of the arm will be the common rafter length per metre of run and the hip or valley length per metre of run. When the figures shown are multiplied by the common rafter run, the true or basic lengths of rafters will be obtained.

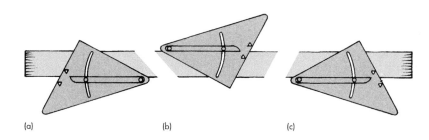

(a)   (b)   (c)

Figure 13.29 Reverse and opposite marking-positions

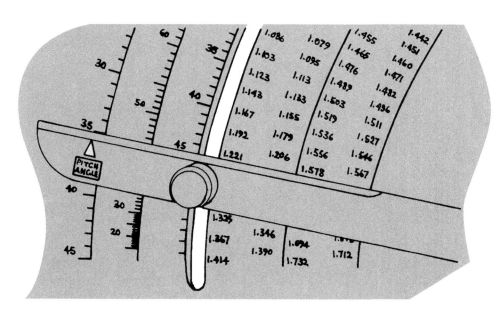

Figure 13.30 Rafter lengths per metre of run

13.10.8 Example Working Out

Figure 13.30: Assume a roof pitch of 35° with a span of 6.486 m.

1. Set the arm to number 35 on the common rafter arc.
2. Record the table numbers displayed under reference headings on the adjustable arm, thus:

   common rafter = 1.221; hip or valley rafter = 1.578
3. Divide the span by 2 to find common rafter run:

   $6.486 \div 2 = 3.243$ m
4. True length of common rafter is

   $1.221 \times 3.243 = 3.959$ m

   True length of hip or valley rafters is

   $1.578 \times 3.243 = 5.117$ m

13.11 SETTING OUT A COMMON (PATTERN) RAFTER

13.11.1 Technique (for any Roofing Tool or Bevel)

Figure 13.31(a): Select a piece of rafter material which is straight and without twists and mark a common-rafter plumb cut A at the top of the rafter's face side, representing the centre of the ridge board. Next, mark another plumb cut B into the body of the rafter, at half the ridge-board thickness measured at right-angles to plumb line A, representing the actual plumb cut against the ridge board. Transfer these lines squarely across the top edge and make a shallow saw cut on the waste side of the edge-line (Figure 13.31(b)) to provide an anchorage for the hook of a tape rule.

13.11.2 Marking the True or Basic Length

Figure 13.31(c): Place the hook of the tape rule into the saw cut on the rafter's top edge, run the rule down to the eaves' area and mark the common rafter length – after having worked this out from one of the methods previously covered. Square this mark across the top edge and, from this point, mark another plumb cut C. This important plumb cut is known as the *backing line*, which is now measured for depth, divided by three and marked down two-thirds to indicate the corner of the birdsmouth D, from which point the common rafter seat-cut is marked.

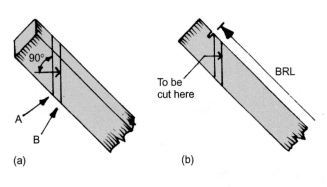

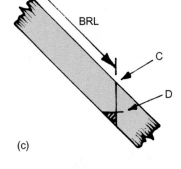

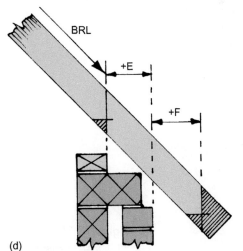

Figure 13.31 (a) First step; (b) second step; (c) third step; (d) fourth step

13.11.3 Working out the Eaves' Projection

Figure 13.31(d): Finally, the rafter must carry on down, from the backing line, across the outer skin of brickwork E, to a distance measured horizontally from the face of the wall and known as the eaves' projection F. As referred to earlier (Figure 13.8, showing basic setting-out terms), the total projection horizontally equals E + F minus the thickness of the fascia board. The concealed projection E is equal to the wall-thickness, minus wall plate. The visible eaves' projection F can be scaled from the elevational drawing showing the roof. For example, assume a wall of 275 mm, a visible eaves' projection of 200 mm, a wall plate 100 mm wide and a fascia board 20 mm thick. The sum would be

$$E = 275 - 100 = 175 \text{ mm}$$
$$F = 200 - 20 = 180 \text{ mm.}$$

Total eaves' projection $= 175 + 180 = 355$ mm.

13.11.4 Adding the Eaves' Projection

With the metric rafter square, this value (355 mm) can be set on the tongue while the blade is lined up to the backing-line plumb mark, the square being slid up or down until the blade lines up precisely and the 355 mm point on the tongue can be marked at the end of the rafter to become the plumb cut for the fascia board.

With the Roofmaster or the *Roofing Ready Reckoner*, the sum of 355 mm can be used as the figure for the run of the total projection, multiplied by the common-rafter table figure for, say 35°, i.e. 0.355 × 1.221 = 0.433 m (433 mm), to be measured down the top edge of the rafter, from the backing line to the top of the fascia-board plumb cut. If a soffit board is to be used, this plumb cut is usually marked down a half to two-thirds its depth and a seat cut established at this point.

13.11.5 Cutting and Checking the First Pair

Figure 13.32: The pattern rafter just marked should be double-checked (a wise trade saying is 'check twice, cut once'), as this first rafter is used for marking out the rest. Next, square the face marks across any edges not yet done. Cut all the bevels carefully, including the birdsmouth and write the word 'PATTERN' boldly on the face of the rafter. Next, only one common rafter should be marked and cut from the pattern, the pair laid on level ground with an offcut of ridge board between the plumb cuts and a tape rule stretched across the birdsmouth-cuts to check the span. If only slightly out, adjust the rafters to the correct span, then check the plumb cuts for a good fit against the ridge-board

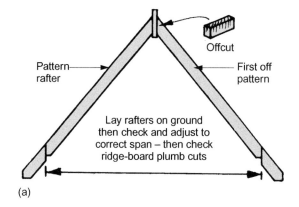

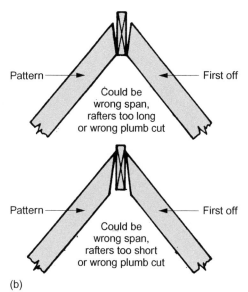

Figure 13.32 (a) Cutting and checking first pair; (b) possible errors

offcut. If satisfactory, the rest of the common rafters may be cut, keeping any cambered or sprung edges on top. If unsatisfactory, go through a checking procedure, as suggested in Figure 13.32(b).

Note that some carpenters leave the marked fascia plumb-cuts to be checked with a string line after all of the rafters have been fixed and then they cut them off *in situ*, for better alignment. However, if a soffit seat-cut is required, this is very awkward to cut *in situ* and should be done at the initial cutting stage.

13.12 SETTING OUT A CROWN (OR PIN) RAFTER

13.12.1 Slight Difference

The procedure for setting out a crown rafter is identical to that for setting out a common rafter, except that instead of deducting half the *ridge-board* thickness from the first plumb cut – as described in the common-rafter text to Figure 13.31(a) – half the *common-rafter* thickness is deducted. Therefore, the first plumb cut A represents the geometrical centre of the hip end and the second plumb cut B, again measured into the body of the rafter, at right-angles to the plumb line, represents the actual plumb cut against the saddle board.

13.13 SETTING OUT A HIP RAFTER

13.13.1 The Technique

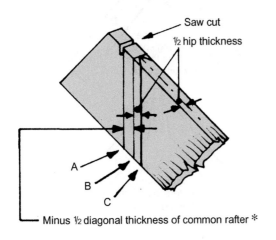

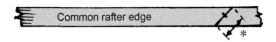

Figure 13.33 (a) Setting out a hip rafter

Figure 13.33(a): This technique is applicable for any roofing tool or bevel. First, mark a hip-rafter plumb cut A at the top of the rafter's face side, representing the geometrical centre-point intersection of the hip end. Next, mark another plumb cut B into the body of the rafter, measured at right-angles to plumb line A and equal to half the common rafter thickness measured diagonally at 45°, as illustrated. This represents the arris or sharp edge of the compound-angled plumb cut. Transfer these lines squarely across the top edge and make a shallow saw cut on the waste side of edge-line A, to provide an anchorage for the hook of a tape rule – and mark the centre of edge-line B to provide the central point for the opposing hip edge cuts.

The simple way of finding this bevel, as explained previously in the text to Figure 13.21 (bevel formulas), is to mark another plumb cut C, again into the body

of the rafter on each side, measured at right angles to line B and equal to half the hip's thickness. Where these plumb lines meet the top edge opposite each other, together with the centre-mark of edge-line B, they present the three points for marking the hip edge cut bevels, as illustrated.

13.13.2 Marking the True or Basic Length

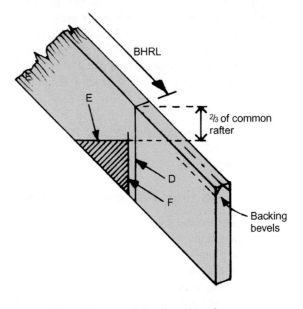

Figure 13.33 (b) Hip with backing bevels

Figure 13.33(b): Place the hook of the tape rule into the saw cut on the rafter's top edge, run the rule down to the eaves' area and mark the hip rafter length – after having worked this out from one of the methods previously covered. Square this mark across the top edge and, from this point, mark a hip rafter plumb-cut D, representing the external corner of the half-lapped wall plate. If the hip rafters were receiving backing bevels (dihedral bevels) on their top edges, plumb line D – the important backing line reference for rafter-length – would also be the line to measure down and mark the same two-thirds measurement worked out for the common rafter (as illustrated) to form the birdsmouth after marking a hip rafter seat cut E at this depth.

13.13.3 Addition for Loss of Corner

Figure 13.33(c): Now, as the external corner of the bedded wall-plate is usually removed, to allow a square abutment of the hip plumb-cut against the wall plate, it follows that the amount removed has to be added to the hip plumb-cut in the birdsmouth. As illustrated, the amount removed diagonally from the corner geometrically equals half the thickness of the hip

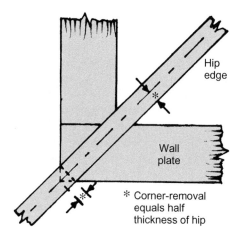

Figure 13.33 (c) Addition for loss of corner

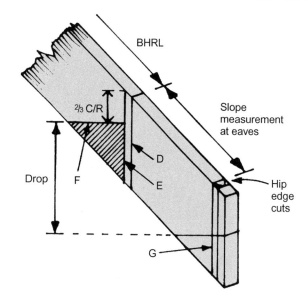

Figure 13.33 (d) Hip without backing bevels

rafter. Therefore, if the hip was 32 mm sawn thickness, 16 mm would be cut off the corner of the wall plate and 16 mm would be added back into the birdsmouth, as indicated by plumb line F in Figure 13.33(b).

13.13.4 Hip Rafters Without Backing Bevels

Figure 13.33(d): This simplified hip rafter, left with a square top-edge, is more in evidence nowadays, since the economic omission of boarding (sarking) which was fixed over the whole roof surface, under the sarking felt, some years ago in a number of individual properties in the UK. Therefore, having marked out the top-end of the hip rafter, as described in the text to Figure 13.33(a), hook in and run the tape rule down to mark the length. As before, square this across the top edge and from this point, mark a hip plumb cut D, representing the external corner of the wall plate. Next, the amount to be removed diagonally from the wall-plate corner (half the thickness of the

hip rafter), say 16 mm, is measured into the body of the rafter, at right-angles to plumb line D, and marked to become hip plumb cut E. This time, it is this *inner plumb line* which, as illustrated, is marked down with the same two-thirds measurement worked out originally for the common rafter. At this depth, a hip seat cut F is marked to form the birdsmouth.

13.13.5 Hip Rafter Eaves' Projection

Figure 13.33(e): The total eaves' projection for the common rafters, as described in the text to Figure 13.31(d), was worked out to be – as an example – 355 mm from the plumb cut backing-line to the plumb cut for the fascia board, measured horizontally (or 433 mm measured down the slope of the rafter, as worked out by the *Roofing Ready Reckoner* or the Roofmaster for a 35° pitch). Using the common-rafter working out of

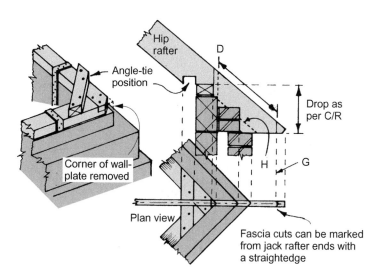

Figure 13.33 (e) Hip eaves' detail

355 mm again for the hip-rafter eaves' projection, bearing in mind that the hip is at 45° in plan and is therefore diagonally longer, you can find the horizontal diagonal projection with the metric rafter square. Do this by setting 355 mm on the blade and 355 mm on the tongue and by measuring the diagonal, which, in this example, produces a figure of 502 mm. A figure, only fractionally more accurate than this, can be found by the other two methods; by using 45° as the pitch and the common-rafter-table figure per metre of run, i.e.

$$0.355 \times 1.414 = 0.50197 \text{ m (say 502 mm)}$$

13.13.6 Down the Hip-Edge

Figures 13.33(d) and (e): Using the *Ready Reckoner* or the Roofmaster to find an alternative measurement for applying down the slope of the rafter, for the same 35° pitch, the sum would be the run of the common-rafter eaves' projection, multiplied by the hip table-figure:

$$0.355 \times 1.578 = 0.560 \text{ m (560 mm)}$$

This would be applied from the top of the backing line D to the top of another hip plumb cut to be marked at G, from which hip edge cuts are marked, as illustrated, against which the mitred fascia boards will be fixed.

13.13.7 Additional Notches

Figure 13.33(e): As illustrated, additional birdsmouth notches might be necessary to clear wall projections which clash with the hip's extra depth. Alternatively, the hip may be reduced in depth on the underside, from a point starting from and equal to the bottom outer-edge of the corner wall-plate, indicated by the dotted line H. Also, if corner angle-ties are to be used, the notches for them should now be marked ready for cutting on the inner edges of the extended birdsmouth seat cuts.

13.13.8 Practical Considerations

Figures 13.33(f) and (g): Finally, it must be mentioned that the length of the hip and its eaves' projection, rather than relying completely on geometrical principles, is often determined by practical methods, especially if the walls at the hipped end are out of square or the wall plates are out of level. One of a few practical methods is to reduce the width of a truly square-ended rafter offcut to equal two-thirds the common-rafter backing-line height (height above the birdsmouth (*)), fix it temporarily as illustrated in Figure 13.33(f), to be flush to the diagonally offcut wall plate and rest a length of hip rafter on it edgeways and on the crown/common rafter intersection at the top, as indicated in Figure 13.33(g).

13.13.9 Use for Marking or Checking

Where the hip rafter rests at the top, in a self-centralizing position, the sides of the actual hip-edge plumb cuts can be marked and where it rests at the bottom, the actual plumb cut for the birdsmouth can be marked. Then the seat cut for the birdsmouth can be marked down (in the case of hips without backing bevels) at two-thirds the common-rafter plumb-cut depth (*). Alternatively, hip rafters that have already been marked out by calculation can still be checked by this technique, prior to being irreversibly cut.

13.13.10 Variation of Method

Figures 13.33(g) and (h): An interesting variation on the above method is to omit the rafter offcut which equalled two-thirds' backing-line height, rest the unmarked length of hip rafter on the actual, diagonally-offcut wall-plate edge instead (Figure 13.33(h)), and again on the crown/common rafter intersection at the top, mark these two extreme resting points on the bottom-face of

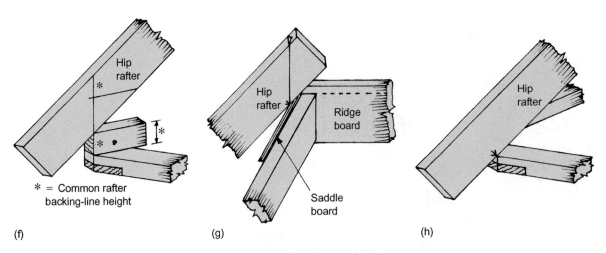

* = Common rafter backing-line height

(f)   (g)   (h)

Figure 13.33 (f)–(h)

the rafter and square them across the edge. This length of hip material can now be used as a rod, to lay diagonally on the face of a hip rafter previously set out by calculation, to check that the rod-marks relate exactly to the marked corner of the birdsmouth at one end while relating to the top edge of the hip edge cuts at the other. Alternatively, hip rafters can be set out this way, without length-calculation, by first marking out the eaves' projection and birdsmouth and then by laying the rod to run up diagonally from the birdsmouth corner, pivoting as necessary at the top of the hip until the mark on the rod relates to the top edge, where it is marked to become the inner plumb cut of the hip edge cuts.

13.13.11 The Eaves' Drop

Figure 13.33(e): The practical way to deal with the hip eaves' projection, is to mark out and cut the soffit seat-cut only, and leave the hip-edge plumb cuts for the fascia board, to be marked and cut *in situ* by string-line or straightedge, once the hips, commons and jack rafters are fixed. The position of the hip rafter seat cut for the soffit board is measured down vertically from the birdsmouth seat cut, at a distance known as *the eaves' drop*. This drop measurement is taken from the pattern common rafter.

13.14 SETTING OUT JACK RAFTERS

13.14.1 The Calculated Diminish

As mentioned earlier in the text to Figure 13.11, jack rafters, being equally spaced, should diminish in length by a constant amount in relation to the length of the common rafter. From information given in the *Roofing Ready Reckoner*, the accompanying booklet to the *metric rafter square* and the booklet that comes with the *Roofmaster*, working out and applying the diminish, to find the length of each pair of diminishing jack rafters, is not difficult to understand, but again there is a practical method that can be used.

13.14.2 The Practical Method

Figure 13.34(a): After the skeletal structure of a hip-ended roof has been pitched (the sequence yet to come), the next operation is to fix the jack rafters in pairs on each side of the hip rafters; start by working away from one of the first pair of common rafters. Ideally, you will need a steel roofing square, but an improvised wooden square will do and can be quickly and easily made out of, say 50 × 25 mm battens, to resemble a miniature builder's-square or, as illustrated

at (a), to resemble a tee square, simply constructed with a lapped, clench-nailed or screwed joint.

13.14.3 The Technique

Figures 13.34(b) and (c): As illustrated, the tee square or the roofing square is held firmly and squarely against the common rafter, with the tongue resting on the hip-rafter edge, while being slid up or down until the rafter-spacing mark or measurement relates exactly to the inner edge of the hip. At this point a mark is made and squared across to the other edge on the crown-rafter side of the hip. Edges A and B of this squaring, establish the highest points of the positions of the first pair of jack rafters.

13.14.4 Centres of Rafters

Figure 13.34(d): Next, mark the spacing-distance of the rafters (usually 400 mm centres on traditional roofs) on top of the wall plate, away from the vertical face of the common rafter. In practical terms, this will be 350 mm, the distance *between* the common and the first jack rafter. The spacing for the first jack rafters on either side of the crown rafter will be minus the thickness of the saddle board, so the distance between these rafters will be 350 − 18 = 332 mm. This dimension can also be found by measuring squarely across from edge mark B (Figure 13.34(c)) to the face-side of the crown, ready to be reproduced on the wall plate. All following pairs of jack rafters will conform to normal rafter spacing.

13.14.5 Marking the Length

Figure 13.34(e): Having gathered a selection of varied offcuts and lengths of rafter material to be used up as jacks, mark out and cut the birdsmouth on each (if not the plumb cut for fascia) and lay aside. Next, take a tape rule or, preferably, a measuring batten (rod), and carefully check the distance by laying the rod in a pitched position, from the edge of the wall plate E (Figure 13.34(d)), up to the hip-edge mark A. Then, mark these points onto the rod and transfer the rod to the prepared jack rafter, to lay in a diagonal position, relating to the corner of the birdsmouth E and being marked at A, on the top edge of the jack rafter. From this point, the jack edge cut and jack side cut are marked, ready for cutting.

13.14.6 Fixing Pair After Pair

Figure 13.34(c): To avoid overloading one side of the hip rafter, jacks should always be fixed in diminishing pairs, exactly opposite each other. Once the first pair are fixed, they can be used for relating to and continuing the squaring technique for the next pair, as indicated at C and D.

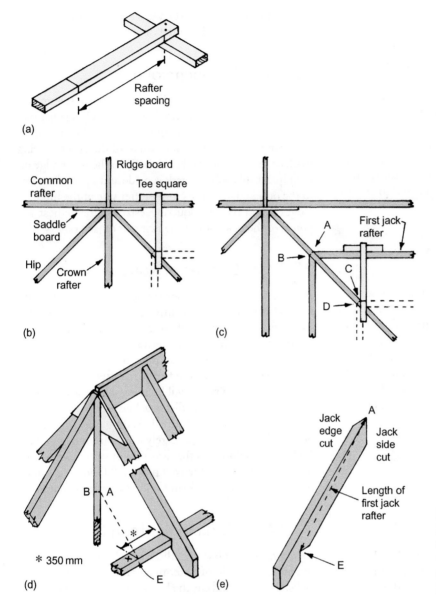

(a)

(b)

(c)

(d)

* 350 mm

(e)

Figure 13.34 (a) Tee square; **(b)** part-plan view. Marking position of first pair of jack rafters; **(c)** part-plan view. Marking position of second pair of jack rafters; **(d)** centres of rafters; **(e)** marking the length

13.14.7 Setting Out Valley Rafters

Figure 13.35(a)–(c): Geometrically, a valley rafter is identical to a hip rafter, but inverted. The plumb cut, the seat cut and the calculated length are the same. However, there are a few variations. First, because the valley rafter usually stems from a ridge-board junction, the reduction at the top from F to G (shown at (a)), would be set in squarely at half the diagonal thickness of the *ridge board* (not the common rafter). Second, the plumb cut of the birdsmouth can be shaped with hip-edge cuts to allow a proper abutment against the internal angle of the wall plate. This is shown here at (b), marked out between plumb lines H and I, equalling half the valley-rafter thickness, measured out squarely. The third and final variation is that when the feet of

the cripple rafters are fixed, unlike jack-edges on hips, their top edges should protrude above the valley-rafter edge by the backing-bevel depth, as indicated in Figure 13.35(c).

13.15 PITCHING DETAILS AND SEQUENCE

13.15.1 Jointing the Wall Plates

Figure 13.36: Wall plates for gable-ended roofs are only required on the two side walls from which the roof pitches. If the walls exceed standard timber lengths of

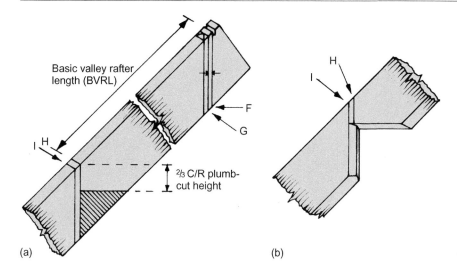

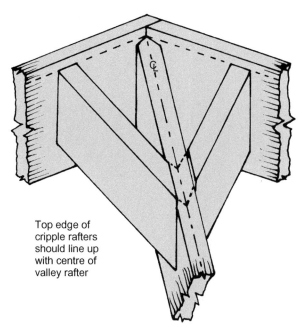

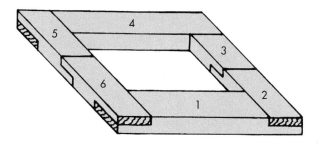

Figure 13.35 **(a)** Setting out of valley rafters; **(b)** valley birdsmouth

Top edge of cripple rafters should line up with centre of valley rafter

Figure 13.35 **(c)** Protruding cripple edges

Figure 13.36 Sequence of lap-jointing for wall plates

as where intermediate lengthening joints may be necessary. As illustrated, a sequence of lap-jointing should be worked out to allow successive pieces of plate to be dropped onto the open-joint and the mortar, rather than be pushed under a wrong-sided lap joint – which tends to trap the mortar in the under-joint.

13.15.2 Bedding the Wall Plates

Bedding the wall plates is best done as a team effort between carpenter and bricklayer. The bricklayer lays and spreads the mortar, beds and levels the plates and, in the case of a hip-ended roof, the carpenter checks that they are square (by using the 3–4–5 method) and parallel to each other across the span. To achieve this, in the case of a gable-ended roof, slight lateral adjustments of the plates may be made – within reason. The 3–4–5 method referred to, is a practical application of Pythagoras' theorem of the square on the hypotenuse of a right-angled triangle being equal to the sum of the squares on the other two sides. When used for squaring the wall plates, using metres as units, mark 3 m along the edge of one plate, 4 m along the edge of the other and, to prove a true right-angle, the diagonal measurement between these points should be 5 m.

13.15.3 Pitching a Gable Roof

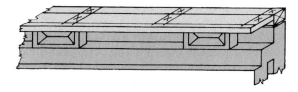

Figure 13.37 **(a)** Marking rafter positions on the plate and ridge board

6.3 m, the wall plates would have to be joined in length with half-lap joints, whereby the lap equals the width of the timber. In the case of a double hip-ended roof, the wall plates would be required on all four sides. These would be half-lap jointed on the four corners, as well

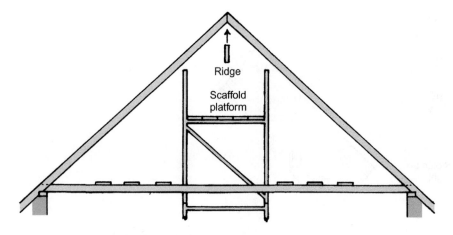

Figure 13.37 **(b)** Inserting the ridge board

Figure 13.37(a): When the bedded plates are set, the joints can then be fixed together with wire nails or screws (screws being less likely to disturb the 'green mortar'). Then the ridge board may be laid out against the wall plate, if possible, and the positions of the common rafters spaced out to the required centres (usually 400 mm) and marked across both members. Next, in any of the clear areas indicated by the marking out, the vertical restraint straps are fixed over the plates, close to each end, one on each side of a lap-joint, and not more than 2 m apart – and then the ceiling joists, with any sprung edges kept uppermost, are fixed adjacent to the rafter marks, either by skew-nailing with 75 or 100 mm round-head wire nails or by being fixed into pre-fixed, shoe-type metal framing anchors. The ceiling joists, which are also fixed to the plates of any internal cross-walls, act as a working platform and should be close-boarded with an area of scaffold boards.

13.15.4 Technique and Sequence

Figure 13.37(b): At each end of the roof, a pair of rafters is pitched and fixed to the wall plates *and* the ceiling joists, their plumb cuts supporting each other at the apex. This is at least a two-man job, with one man at the foot of each rafter. An interlocking scaffold can be erected through the ceiling-joist area and, from this, the ridge board is pushed up between the rafter plumb-cuts and fixed into position. Ideally, a third man would do this. On each fixing, one 75 mm round-head wire nail is driven through the top edge of the rafter and two – one on each side – are skew-nailed through the sides into the ridge board. The nails first driven into the rafters' top edges usually pierce through the ridge board on each side, causing problems unless both opposite rafter-heads are fixed at the same time. To manage this single-handedly, leave the first nail-head protruding until the nail on the other side is partly or fully driven in.

13.15.5 Important Foot-Fixings

The foot of each rafter must be fixed to the wall plate and the ceiling joist (the latter acting as an important structural tie) with at least three 100 mm round-head wire nails – one skew-nailed above the birdsmouth into the plate, which, as well as fixing, also tightens the rafter against the ceiling joist, and the other two driven in squarely through the side of the ceiling joist, into the rafter, as previously indicated in Figure 13.7(a).

13.15.6 Adding Purlins and Struts

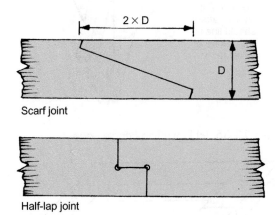

Figure 13.37 **(c)** Purlin half-lap joint and scarf joint

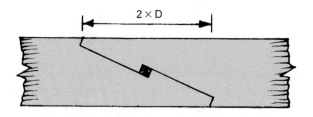

Figure 13.37 **(d)** Ridgeboard scarf joint

Figures 13.37(c) and (d): According to the length of the roof (from gable to gable) and whether the purlins or ridge board need to be extended in length by jointing – as illustrated – a few more pairs of rafters may need to be fixed at strategic positions before the purlins are offered up and fixed by skew-nailing from above, through the available rafters, into the purlin edges. The struts, set out by using a rod and/or rise-and-run principle for lengths and bevels, are then fixed into position. Next – to complete the main structure of the roof – the remainder of the rafters are filled in.

13.15.7 Restraint Straps

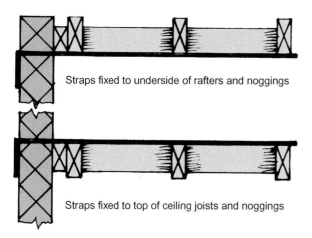

Straps fixed to underside of rafters and noggings

Straps fixed to top of ceiling joists and noggings

Figure 13.37 (e) Restraint straps

Figure 13.37(e): Whether the gable walls are to be built before or after the roof is pitched, it must be remembered that horizontal restraint straps of 5 mm thick galvanized steel must be fixed at maximum 2 m centres across the ceiling joists and rafters, to be built into (to help stabilize) the gable-end walls.

13.16 PITCHING A HIPPED ROOF (DOUBLE-ENDED)

13.16.1 Setting out the Wall Plates

Figure 13.38(a): After the wall plates have been half-lapped (see Figure 13.36) and bedded into position, they must be checked for being level, parallel and square-ended. When the mortar has set and the joints have been fixed, the rafter positions can be set out. This will allow the vertical restraint straps to be fixed in any of the clear areas – as described for the gable-roof wall plates. To set out, check the actual span across the wall

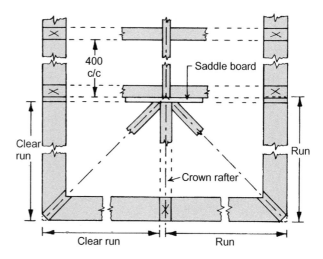

Figure 13.38 (a) Setting out a hipped end

plates and divide by two to find the run. Mark this as a centre line on the wall plate at each hip-end and split the thickness of the crown rafter on each side, to be squared across the plates. The clear run is now measured and should equal the run minus half the crown-rafter thickness. This measurement is now marked in from each side of each hip-end and represents the face of the saddle board. The thickness of the saddle board – usually 18 mm WBP plywood or OSB – is now marked across the plates and this line is equal to the face of the first pair of common rafters at each end. The other rafters are spaced out between these pairs, at specified centres from one end, regardless of an odd spacing at the other end.

13.16.2 Marking and Cutting

As mentioned earlier in this chapter, most – if not all – of the marking and cutting of the various components in the roof should be done on the ground, laid out on pairs of saw stools (or Workmates etc.), then hoisted or man-handled up to the roof. Nowadays, the components may be cut with a compound-angled mitre saw.

13.16.3 Pitching Technique and Sequence

First, the ceiling joists are fixed, but only those in the middle area that attach to common rafters – not those at each end that attach to jack rafters. As before, a boarded area is laid with scaffold in position. The first pair of rafters at each end are pitched and fixed. The marked ridge board is inserted and fixed, after the rafters have been braced diagonally down to the wall plates. Then, the saddle boards are fixed at each end with 63 mm round-head wire nails (at least six in each board), followed by the crown rafters and then the hips – or the hips and then the crown rafters, if using the alternative hip-arrangement illustrated in Figure 13.4(a).

13.16.4 Adding Purlins and Temporary Struts

After fixing a few more pairs of common rafters, strategically placed to relate to any lengthening joints in the purlins or ridge board, it will be found to be advantageous to fix the purlins, after checking that the hips are not bowed and, if necessary, bracing them straight with diagonal battens down to the wall plates. Fixing the purlins at this stage reduces the struggle against a full complement of potentially sagging rafters and provides an intermediate ledge upon which to manoeuvre the rafters on their way up to the ridge. At this stage, the purlins should be supported, if only with temporary struts until the joists at each end are complete and the binders have been added.

13.16.5 Finishing Sequence

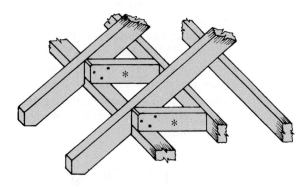

Figure 13.38 **(b)** Return-joists on hipped end*

Figure 13.38(b): Finally, cut and fix the jack rafters, the remaining ceiling joists each end, the binders (skew-nailed from alternate sides into the top of each ceiling joist), the struts and straining pieces, the remaining common rafters, the hangers and, if required, collars to complete the main structure of the roof. On hipped roofs, as illustrated, short return-joists are fixed to the feet of the crown and jack rafters at the hipped ends. These return-joists only provide fixing points for the plasterboard ceiling at the wall-plate edge and, unlike ceiling joists, do not act as structural ties. For this reason, the *New Build Policy Guidance Notes BGN 9B*, recommend that horizontal restraint straps or timber ties (like binders) should be fixed across the tops of a minimum of three ceiling joists and be attached to the side of the crown and every second jack rafter.

13.17 FLAT ROOFS

13.17.1 Introduction

The structure of flat roofs is very similar to that of suspended timber floors, especially when kept level for the provision of a ceiling on the underside. If the ceiling is unimportant, or not required, the joists may be set up out of level to create the necessary *fall* (roof slope). Like floors, the roof joists span across the shortest distance between the load-bearing walls – or the longest distance when intermediate cross-beams are used – at similar centres to floor joists. They are subject to the same rules as floors regarding strutting being required when the span exceeds 2.5 m. Suitable joist sizes can be obtained either by structural design or by reference to the TRADA (Timber Research and Development Association) Span Tables.

13.17.2 Anchoring

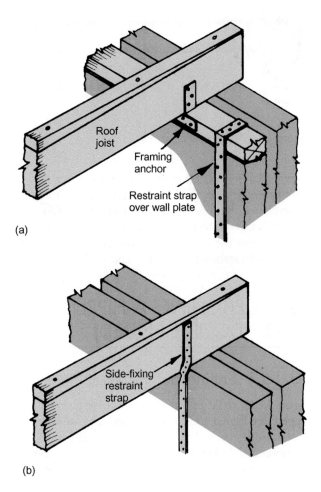

Figure 13.39 Flat roof details: **(a)** anchoring; **(b)** side-fixing restraint strap

Figure 13.39(a) and (b): As illustrated, the roof joists are either fixed to the wall plates by skew-nailing with round-head wire nails or by framing anchors. Galvanized restraint straps are fixed over the plates and to the inside face of the wall, as described for pitched roofs, at maximum 2.0 m centres – or, without wall plates, the joists may be anchored to the top of the walls with twisted side-fixing restraint straps.

13.17.3 Creating a Fall

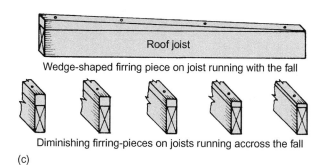

Figure 13.39 (c) Firrings

Figure 13.39(c): Wedge-shaped timber fillets, or diminishing parallel-fillets, known as firring (or *furring*) pieces, are fixed to the upper joist edges to create the necessary *fall* or roof slope. The recommendations for these falls vary between as low as 1 in 80 (25 mm in 2.0 m) and 1 in 40 (25 mm in 1.0 m). The *New Build Policy Guidance Notes BGN 9A* recommend the latter as a minimum fall for flat roofs. An alternative method of creating a fall, available nowadays, is to keep the structural roof truly flat and create the fall *above* the roof deck by using a system of pre-designed tapered roof insulation boards.

13.17.4 Flat Roofs Against Buildings

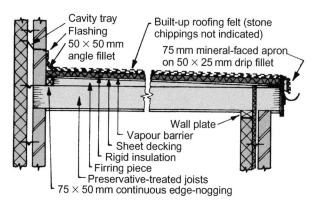

Figure 13.40 (a) Flat roof butted to adjacent walls

Figure 13.40(a): Flat roofs may be independent (as on a detached garage, Figure 13.40(c)), or have one or more

edges butted up to the face of an adjacent building. Sometimes, on single-storey buildings such as bungalows, the abutment is weathered-in with the pitch of the main roof at eaves' level. In the case of abutments to walls, the roofing material is turned up the wall and is covered by a lead flashing chased into the wall at a minimum height of 150 mm above the roof.

13.17.5 Cavity Trays

If the wall is of cavity construction, a cavity tray should be built in or be inserted to lap onto the top edge of the flashing, as illustrated in Figure 13.40(a). However, this creates a problem in conversion work, where a lean-to or flat roof is butted-up to a wall as an afterthought, and involves the bricklayer in cutting away a brick-course of the existing wall in a piece-meal operation to allow for building in a cavity-tray system.

13.17.6 Joists Into or Against a Wall

Figure 13.40(a) shows the joists at right angles to the adjacent wall, built into the outer skin of brickwork. They may also be carried on galvanized, type TW (timber to wall) joist hangers – a popular method in conversion work, as the top flange of the hanger is easily cut into a mortar-bed joint in the brickwork. This can be done with a disc angle-grinder and/or a traditional plugging chisel. When the hangers are inserted, any gaps above the flange should be caulked with bits of slate or strips of sheet lead.

13.17.7 Roof Fabric

The most common covering to flat roofs for some years now, has been bituminous roofing felt, although it has a limited life-span up against more traditional coverings such as mastic asphalt, sheet lead or copper. However, high-performance (torched-on) roofing felts are now available, built-up as before in three layers and high-performance GRP (glass-reinforced plastic – also referred to as *fibreglass*) flat roofs are also being marketed with a 25-year guarantee.

13.17.8 Decking Material

Sheet decking is laid and fitted on the roof joists, firrings or counter-battens, with cross-joints staggered in a similar way to flooring panels, fixed down with 50–56 mm by 10 gauge annular-ring shank nails at recommended 100 mm centres. Types of decking material include WBP exterior grade plywood, Sterling OSB board, moisture-resistant or bitumen-coated chipboard and pre-felted chipboard with roofing felt bonded onto the top surface.

13.17.9 Board Thicknesses

Thickness of decking is related to the joist-spacing – plywood can be 12 or 15 mm for joists at 400 mm centres, Sterling board 15 mm and chipboard 18 mm. When joists are at 600 mm centres, the thickness should increase to 18 mm for plywood, 18 mm for Sterling board and 22 mm for chipboard. Nails should be at least two times the thickness of the board, so only the 22 mm chipboard qualifies for the 56 mm nails mentioned above. Decking should be laid with a 3 mm expansion joint between boards and board-joints should be covered with 100 mm wide bitumen felt strip to protect the edges of the boards while awaiting the arrival of the roofing specialist.

13.17.10 Projecting Eaves and Verges

Figures 13.40(b) and (c): When projecting eaves and/or verges are required for design purposes or to allow for ventilated soffits (as is necessary for *cold roofs*), the joists can be extended in length across the wall and made to extend on the sides. As illustrated three-dimensionally in Figure 13.40(c), this is achieved by the formation of short, projecting return-joists fixed at right-angles to the sides of the outer joists by using type TT (timber to timber) joist hangers, framing anchors or simply by skew-nailing. Even if the fascia boards are to be kept close to the face brickwork, these short return-joists will be needed for

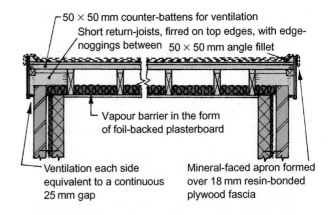

Figure 13.40 (b) Verge details

fascia-fixings and for better continuity of the whole roof structure.

13.17.11 Insulation to Flat Roofs

The Building Regulations' *Approved Document F2* states that excessive condensation in roof voids over insulated ceilings must be prevented, otherwise the thermal insulation will be affected and there will be an increased risk of fungal attack to the roof structure. This applies only to roofs where the insulation material is at ceiling level (cold roofs). Where the insulation is kept out of the roof void, placed on the deck (warm roofs), the risks of excessive condensation developing are not present and these roofs, therefore, are not covered.

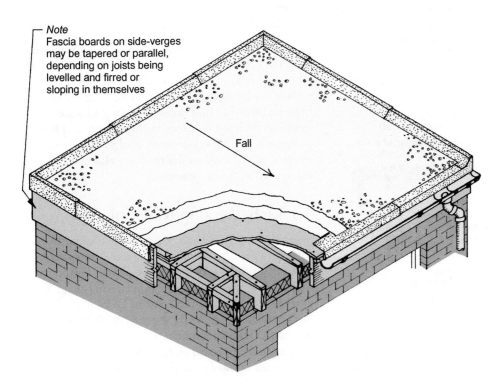

Figure 13.40 (c)
Partly-exposed view of independent flat roof (Note: Counter-battens for ventilation – as shown in Figures 13.40 (b)+(e) – omitted here.)

13.17.12 Categories of Flat-roof Construction

There are three categories. The first, which is preferred nowadays because a better balance between heat-loss and condensation-control can be achieved, is known as a *warm deck flat roof*, which does not require ventilation of the roof void. The second, which *does* require ventilation and is more susceptible to condensation, is known as a *cold deck flat roof*. This type of roof, although covered by the Building Regulations of England and Wales, is not recommended in the Building Standards Regulations of Scotland (in which warm roof constructions are recommended). The third category of roof is known as a *hybrid flat roof*, covering certain roofs which do not fall within the warm or cold category. This is because some structural decks are themselves composed of insulating materials, such as woodwool or, in other cases, insulation is added above the deck in addition to insulation at ceiling level.

13.17.13 Warm Deck Flat Roof

Figure 13.40(d): The main feature of this roof is that the insulation is placed above the structural deck in the form of rigid boards, such as Thermazone polyurethane foam roofboards. These particular boards are 600 × 1200 × 50 or 80 mm thick; their edges are rebated for interlocking to avoid cold bridging and they have bitumen glass fibre facings on each side.

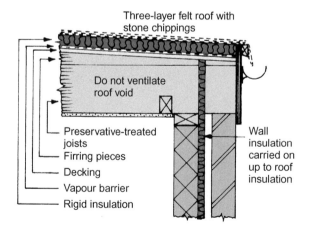

Figure 13.40 (d) Warm deck flat roof (eaves' detail)

13.17.14 Construction

Figure 13.40(d): The rigid insulation is bonded or fixed with mechanical fastenings to a layer of felt vapour barrier, which has been fully bonded to a sheet decking. The decking is fixed to tapered firring pieces which are fixed to the joists. To avoid cold bridging

around the perimeter of the roof, any cavity insulation, as illustrated, should be carried on up to meet the deck insulation. In this type of construction, the roof-void is not to be ventilated. All roof timbers should be preservative-treated. The waterproof covering to the roof is recommended to be of three-layer high performance built-up felt or GRP.

13.17.15 Cold Deck Flat Roof

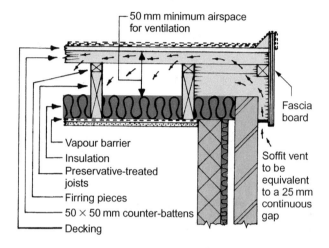

Figure 13.40 (e) Cold deck flat roof (verge detail)

Figure 13.40(e): The main feature of this roof is that the insulation is placed *below* the structural deck, between the joists at ceiling level. To avoid cold bridging, the ceiling insulation should join up to the cavity insulation, as illustrated, with care being taken not to block the perimeter air vents. These vents, positioned in the soffit area, can be continuous or intermittent and must be on opposite sides of the roof for cross-ventilation. Where this is not possible, then this type of roof should not be used. The opening of the vents, incorporating an insect screen mesh, should be equivalent to a continuous 25 mm gap. Additional, intermediate roof vents are recommended for spans over 5 m and for roofs with an irregular plan-shape.

13.17.16 Construction

Figure 13.40(e): *Approved Document F2* also recommends that there should be a free airspace of at least 50 mm between the insulation and the underside of the roof deck. This will not normally cause a problem if the continuous air vents can be placed at right-angles to the joists on opposite sides of the roof, but in situations where the air vents are parallel to the joists, as illustrated, airflow across the joists can be achieved by positioning 50 × 50 mm counter-battens at joist-spacings across the roof joists, fixed on top of the firring pieces and to the short, projecting return-joists. Under the insulation (which can be a rigid or flexible type, but not loose), a vapour barrier should be placed at ceiling level, using minimum 500 g polythene or metallized polyester-backed plasterboard. All roof timbers, as before, should be preservative-treated and the waterproof roof-covering should be as recommended for the warm deck flat roof.

13.18 DORMER WINDOWS AND SKYLIGHTS

13.18.1 Introduction

Roof lights in the form of dormer windows and skylights are usually found in roof spaces used for storage or habitation. Both of these windows involve a trimmed opening in the roof slope and the use of thicker trimming and trimmer rafters, according to the size of opening and the amount of trimmed rafters to be carried. The trimming rafters that carry the trimmers and their load of trimmed rafters, can also be – and usually are nowadays – formed by fixing two common rafters together, as indicated in the illustration.

13.18.2 Dormer Windows

Figure 13.41: These traditional constructions protrude vertically from the eaves or middle area of the roof and have triangular sides known as cheeks, framed up from minimum 100 × 50 mm sawn studs, sheathed

externally with ex. 25 mm diagonal boarding (parallel to the roof slope) or 15 mm WBP plywood or Sterling OSB board. If the cheeks are to receive tile or slate cladding, a breather membrane, such as Tyvec Supro (not a roof underlay), should be fixed to the sheathing behind the preservative-treated tile battens. The windows can be uPVC from the outset, but are usually specified initially (for economy) to be the wooden stormproof casement type, with wide rebates for 14 mm double-glazed sealed units.

13.18.2 Dormer Roof

Figure 13.41: The roof may be flat and the 100 × 50 mm minimum joists firred to slope backwards to the main roof or to fall to a front gutter and corner down-pipe which discharges onto the main roof. Dormer roofs may also be segmental, semi-circular, etc., or pitched with a gable end or hipped end and tiled or slated in keeping with the main roof. Typical construction details of the skeletal dormer are shown in the illustration. All timbers not rated according to BS 5268: Part 5, should be preservative-treated and the waterproof roof-covering should be as recommended for the warm-deck and cold-deck flat roofs.

13.18.3 Ventilation and Condensation Control

Figure 13.42(a): Where there is a well-ventilated space within a pitched roof, above the insulation at ceiling level, a vapour control layer is not normally required;

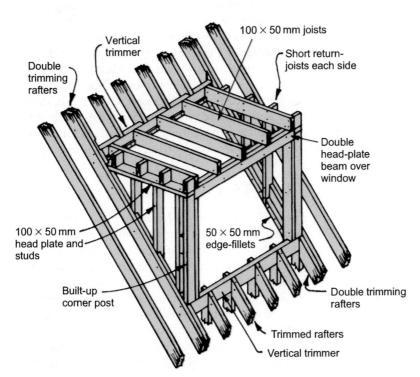

Figure 13.41 Skeleton dormer with window omitted

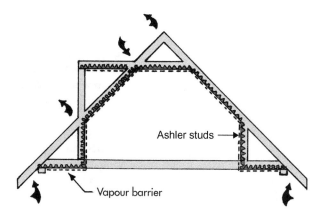

Figure 13.42 (a) Ventilation and continuous insulation

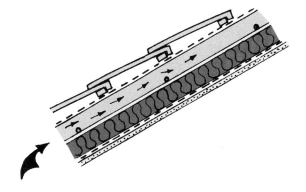

Figure 13.42 (b) Maintaining minimum 50 mm airspace by using push-fit baffle boards

but where there is limited space above the insulation, making it difficult to ventilate effectively, as in the case of a loft room (attic), a vapour control layer *would* be necessary, as and where indicated in the illustration. The whole loft area, including the dormer, should be insulated up against the dwelling-side of the construction, against the vapour barrier, between the timbers of the dormer cheeks, the joisted roof, the rafters where affected, the ceiling joists, the limited areas of floor joists behind the ashlering and between the ashler studs as well.

13.18.3 Detailed Recommendations

Figure 13.42(a): In recommending the above, BS 5250 also recommends that the insulation on inclined and vertical faces must be held firmly in place to prevent slipping and that where there is limited space for the insulation, such as above the ceiling to the sloping rafters, a minimum 50 mm clear airspace between the insulation and the sarking should be provided. To achieve this, it may be necessary to increase the size of the roof and dormer timbers to accommodate the correct thickness of insulation.

13.18.4 Ventilation Requirement

Figures 13.42(a)–(c): To ensure the 50 mm clear airspace is maintained, BS 5250 suggests that consideration be given to installing some form of inert baffle boarding between the roof timbers to restrict the insulation. This recommendation could be met by using strips of 9 mm thick Sundeala board (to BS EN 622-3) or 12 mm thick insulation board (conforming to BS EN 622-2), easily cut on site, or pre-cut by the supplier's mill from the imperial-sized, metricated boards of 2.440 × 1.220 m. The strips could be cut slightly oversize and pushed into a friction fit and/or positioned against small protruding pre-set nails, as indicated in Figure 13.42(b). A separate baffle board,

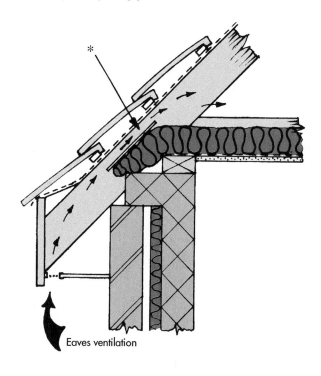

Figure 13.42 (c) Maintaining minimum 25 mm airspace by using baffle boards*

about 200/250 mm wide, is recommended between each rafter at the eaves, above the structural wall, again to restrict the insulation and so ensure that a minimum 25 mm clear airspace is maintained at this point, as indicated in Figure 13.42(c).

13.18.5 Low-level and High-level Ventilation

Figures 13.42(a) and (c): With roofs pitched above 15°, low-level eaves' ventilation should normally be equivalent to a continuous 10 mm gap, but where there are areas of limited space above the insulation, as with an attic or loft room, the opening area of the vents – incorporating an insect screen mesh – should be increased to be equivalent to a continuous 25 mm gap, as indicated in Figure 13.42(c), and high-level ventilation openings at the ridge should be provided, equivalent to a continuous 5 mm gap.

13.18.6 Mid-roof Ventilation

Figure 13.42(a): Additional ventilation openings should be provided when there are obstructions in the ventilation path, such as a roof light or dormer window. These vents should be placed immediately below the obstruction, equivalent to 5 mm × obstruction-length and immediately above the obstruction, equivalent to 10 mm × obstruction-length. Finally, to achieve cross-ventilation within the dormer's cold-deck flat roof, counter-battens, as described earlier, will have to be used.

13.18.7 New Regulation Requirements

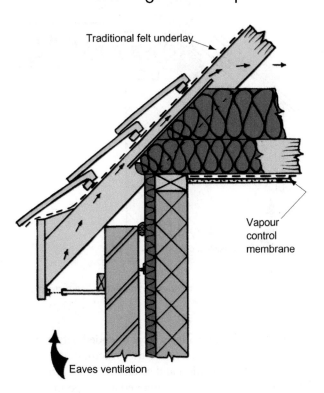

Figure 13.42 (d) Ventilated 'cold roof' with roofing felt underlay, 270 mm ceiling-insulation and wider baffle-board

Figure 13.42(d): As outlined in Chapter 8, under section 8.2.2, regarding floors, radical changes have been made to the regulations concerning the *Conservation of fuel and power*. These initially came into force in April 2002 as Approved Documents (AD) L1 and L2 and have now been amended and expanded into four separate documents known as ADL1A, ADL1B, ADL2A and ADL2B. Regarding the details shown in the roof

section at Figure 13.42(c), the insulation would now need to be upgraded to comply with the new regulations. By using the Elemental Method of achieving tabled U-values in the thermal elements – only allowed in ADL1B and ADL2B – the insulation would need to be increased, as shown in Figure 13.42(d). This could be in the form of mineral wool quilt, such as Crown Wool, of 100 mm thickness laid between the ceiling joists, with an additional 170 mm thickness laid over and across the ceiling joists. To avoid cold-bridging, partial-fill or full-fill insulation in the cavity walls should continue up to connect with the roof insulation.

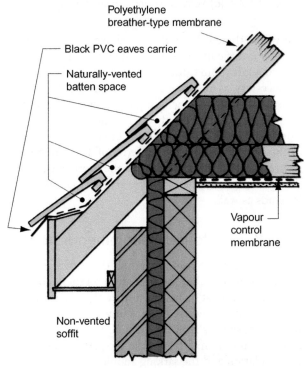

Figure 13.42 (e) Non-ventilated 'cold roof' with breather membrane, 270 mm ceiling-insulation and full cavity-fill

Figure 13.42(e): Research in recent years has established that condensation in roof voids (lofts), that came about with the introduction of loft-insulation four or five decades ago, is caused through there being an impermeable membrane over the rafters in the form of traditional roofing felt under the tile-battens. Although proper ventilation eliminates this problem, it also adds to heat-loss. It was only when a breather-type membrane – such as Tyvec Supro – was introduced to replace roofing-felt, that roof ventilation became unnecessary. Any moisture vapour trapped in the roof void escapes through the breathable membrane into the tile-batten space.

As illustrated in Figure 13.42(e), Tyvec® rigid PVC Eaves Carriers are laid over the edge of the fascia board and are fixed to the splayed tilting fillet with large-headed galvanized clout nails, of 20 or 30 mm length. The carriers, which are 1.3 m long, must overlap each other by 100 mm when joined end-to-end. Tyvec® double-sided butyl tape is then applied to the Eaves Carrier rebate. The breather membrane is draped horizontally over the rafters, so as to form a minimal 10 mm drainage-sag (maximum 15 mm) under the tile battens, and fixed with non-ferrous staples or nails at maximum 300 mm centres. The membrane is lapped onto the carrier by 150 mm and is finally bonded to this by peeling off the backing paper from the previously bonded Tyvec® Butyl Tape. Subsequent layers of breather membrane are fixed progressively up the roof, each with a 150 mm overlap. Note that because the breather membrane 'drapes' slightly in this example of a non-ventilated 'cold roof', it is referred to by Dupont™ Tyvec® as being of 'unsupported application' and the membrane that they recommend for this is Tyvec® Supro.

To upgrade this roof further, a Tyvec® Supro *Plus* membrane could be used. This has the advantage of having an integral adhesive lap tape on its top edge, allowing all horizontal edges to be bonded to produce a more thermally efficient sealed roof system. However, to achieve this, the breather membrane is not draped and needs to be fixed to the rafters in a taut condition with 25×50 mm counter-battens fixed over. These should be fixed with 65×3.35 mm

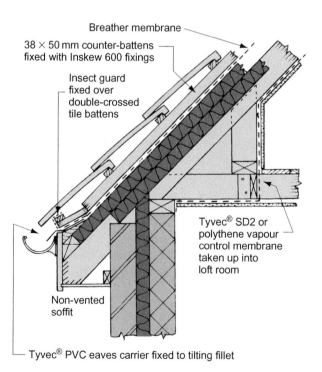

Breather membrane

38 × 50 mm counter-battens fixed with Inskew 600 fixings

Insect guard fixed over double-crossed tile battens

Tyvec® SD2 or polythene vapour control membrane taken up into loft room

Non-vented soffit

Tyvec® PVC eaves carrier fixed to tilting fillet

Figure 13.42 (f) Non-ventilated 'warm roof' with counter-battens, breather membrane and rigid insulation between and over rafters

galvanized or sherardized round-headed wire nails, ready for tile-battening and tiling to complete.

Figure 13.42(f): This detail of a loft room (to be compared with Figure 13.42(a), which shows a loft room in a ventilated 'cold roof'), is an example of a non-ventilated 'warm roof' construction, achieving higher thermal efficiency. Because the breather membrane is close to the rigid insulation fixed over the rafters, it is referred to as being of 'supported application'. For this, Dupont™ Tyvec® recommend the use of either of their above named membranes or a third membrane known as Tyvec® Solid, which is of lighter weight. The sealing of horizontal lapped edges on this membrane is optional, but, if required to achieve a sealed roof system, it is easily done from above with Tyvec® single-sided tape.

The two layers of rigid insulation shown here could be 48 mm Celotex GA2000. One layer is laid over the rafters, ready for fixing – which is done through the counter-battens – and the other is fitted friction-tight between the rafters (usually fitted later from inside the loft space). As illustrated, a timber stop-batten, equal to the insulation thickness, is fixed behind the fascia board. This has two purposes: 1) It stops any downwards slide of the roof-loaded insulation; 2) It gives upper support and a fixing for the fascia board and tilting fillet.

The fixing of the counter-battens is a critical operation, because special stainless steel nails – known as *Inskew 600* (600 meaning 6.00 mm Ø) *fixings* – are used, and they have the important task of holding the counter-battens *and* the rigid insulation to the rafters. These fixings, from Helifix Ltd, are also known as *helical warm roof batten fixings*. They are headless and, as the name suggests, have a spiral shape. Sizes range from 75 mm to 150 mm in 5 mm increments and from 150 mm to 250 mm in 10 mm increments. Fixings up to 125 mm can be driven in by hammer-only, but longer fixings need the assistance of a PDA (power driver attachment) or a so-called Hand Supportal tool. The length of a fixing is determined by the thickness of counter-batten, plus the insulation, plus a minimum 35 mm into the rafter. As minimal 38×50 mm counter-battens are recommended in the example roof shown in Figure 13.42(f), Inskew 600 fixings would need to be not less than $38 + 48 + 35 = 121$ mm long. The nearest available size, therefore, would be 125 mm. Helifix Ltd recommend that the spacing of fixings has to be calculated according to varying criteria, but should not be less than 150 mm, nor more than 400 mm centres. Finally, because it would not be practical – or safe – to fit and lay the rigid insulation over the whole of the roof in an unfixed state, awaiting fixings through the counter-battens, the insulation is laid in manageable amounts and short lengths of counter-batten are fixed progressively to hold it.

13.19 SKYLIGHTS (ROOF WINDOWS)

13.19.1 Introduction

These windows usually lay in the same plane as the roof slope and traditionally consisted of a glazed skylight window, hinged on the underside at the top and fixed to a raised curb or lining. Although metal aprons of lead, zinc, etc. were fixed to the back and side edges of the skylight, these windows were not always weathertight.

13.19.2 Roof Windows

Figure 13.43 Roof window (skylight)

Figure 13.43: Modern skylights, referred to as roof windows by the manufacturers, are a different proposition. These skylights are very sophisticated and reliable. They are made from preservative-impregnated Swedish pine, clad on the exterior with aluminium. The casements are of the horizontal pivot type with patent espagnolette locks, seals and draught excluders, and are double glazed with sealed units. The windows, suitable for roof pitches between 20° and 85°, are easily fixed to the rafters with metal L-shaped ties provided. Metal flashings and full fitting and fixing instructions are also supplied with each unit.

13.19.3 Loft Conversions

If these windows are to be installed in a roof already tiled or slated, as in a loft conversion, depending on the size of the window, it may be necessary to shore up the roof temporarily within the loft, to enable the opening to be made. This shoring-up can be as simple as fixing two 100 × 50 mm timber plates, flat-faced across the rafters, one slightly higher than the proposed opening, the other slightly lower; 100 × 50 mm sole plates are fixed across the ceiling joists and similar-sized struts are fitted between the high and low plates. If possible, the struts should be at right-angles to the roof slope. The affected rafters can now be cut and removed safely to enable a top and bottom trimmer to be fixed in position. The faces of trimmers should also be at right-angles to the roof slope. When the trimmers are fixed to the trimming rafters and to the ends of the newly trimmed rafters, the shoring can be released.

13.20 EYEBROW WINDOWS

13.20.1 Introduction

Figure 13.44 (a) Eyebrow window

Figure 13.44(a): The eyebrow window is another form of roof light, serving roof spaces used for storage or habitation. This type of window is similar in most respects to a dormer window, whereby a trimmed opening in the roof slope will be required with triangular studded cheeks to the sides of the opening. These cheeks, which emanate from the trimming rafters, are not seen externally – unlike the cheeks of a dormer, which are seen. Vertical timbers known as ashler studs are fixed to the underside of the window trimmer, the top edge of which should protrude about 100 mm above the roof slope to form the apron below the window sill. As with the dormer, the window, in a present-day construction, would initially be of the wooden stormproof casement type. To keep the window independent, to allow for future replacement, the old practice of resting the ceiling joists on the head of the window frame should be avoided. Instead, they should span from the upper trimmer in the roof to the double-plated beam at the head of the structural window opening.

13.20.2 Eyebrows and Steep Pitches

Figure 13.44(b): Roofs with tiled eyebrow windows need to be of a steep pitch of about 50–60° to accommodate the shallower pitch of the curved roof to the raised eyebrow, which should not be less than 35°. This is because the only tiles that can be used on roofs with eyebrows are non-interlocking 165 mm × 265 mm ($6\frac{1}{2}$ in × $10\frac{1}{2}$ in) plain tiles, which

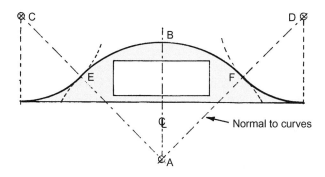

Figure 13.44 (b) Setting out the eyebrow shape

are not recommended for roofs below 35° pitch. As illustrated, the geometry for the serpentine shape of the eyebrow is best established to suit the predetermined window height and width.

13.20.3 Setting out the Eyebrow Shape

Figure 13.44(b): First, in scaled form (usually done by the architect or a draughtsperson), draw the outline of the window frame with a centre line A–B drawn vertically through it. By trial and error, try the compass or trammel point at different positions on the centre line until a suitable segment appears to fit above the frame. Now determine the outer limits of the segment by judging a suitable point E, near the base of the window frame and draw a line through it, radiating from the centre A. Measure the straight distance between E and B and set the same distance from B to mark point F. (Or transfer E from B with a compass, to establish F.) Establish another line radiating from centre A through F. These lines are geometrical normals to the curves required in reverse positions (cyma reversa) and at the predetermined points along these normals, as at C and D, the centres can be established to complete the eyebrow (serpentine) shape.

13.20.4 Method of Construction

Figure 13.44(c): The construction details of traditional eyebrow windows can be found to vary to some extent, thereby creating variations in the methods of construction that are used. However, the following method is recommended. Assuming that the skeletal structure of the main roof is nearing completion and a free space between the double trimming rafters is already established to accommodate the eyebrow window, the first step is to position and fix the 100 × 50 mm floor plate A to the floor joists, ready to carry the ashlering. Then, against the inside face of the double trimming rafters, fix a vertical stud B on each side of the opening, extended in height to run up about 50 mm above the estimated roof curve. Now fix the trimmer C between the vertical

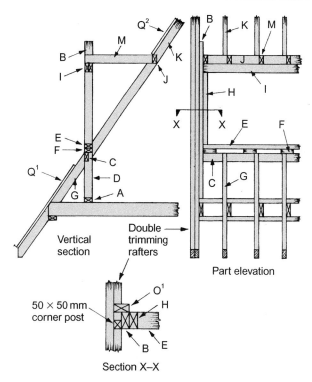

Figure 13.44 (c) Method of construction

studs B, by forming housing joints, nailing or with modern fasteners such as Mafco CL-type framing anchors.

13.20.5 Sub-sill/Apron to Window

Next, cut and fix the ashlering D to the floor plate and trimmer, at centres equal to rafter-spacings. Then cut and fix the sub-sill E between the vertical studs B, with the top edge protruding at least 100 mm above the roof to act as an apron below the window; the sub-sill is also fixed to 100 × 50 mm offcuts F, pre-fixed to the trimmer. Now complete the bottom section by cutting and fixing the trimmed rafters G to the wall plate at the bottom and the trimmer C at the top. Above the sub-sill now cut and fix the vertical sub-frame stud H to the face of stud B on each side of the opening, equal in height to the window frame plus a minimum tolerance of 7 mm for fitting and for the lead apron to be dressed under the sill. Studs H act as bearings for the double head-plate (beam I) over the window, which should now be cut to length, fixed together by stagger-nailing through the face side at about 400 mm centres, then fixed in position at each end with 100 mm round-head wire nails.

13.20.6 Upper Trimmer and Trimmed Rafters

Now start the ceiling structure by levelling across from the top of the head-plate beam I and marking the side of the trimming rafters on each side of the opening. This establishes the underside of the upper trimmer J, which should now be cut and fixed in a similar way to that described for the lower trimmer. Next, with birdsmouth cuts at the foot and plumb cuts at the head, cut and fix the trimmed rafters K from trimmer J to the ridge board.

13.20.7 Ceiling above the Window

The 100 × 50 mm ceiling joists M can now be fitted and fixed. They rest on the window-head beam I and are skew-nailed flush to its front edge, while at the other end they are fixed to the face of the trimmer J in various ways similar to fixing the ends of the trimmers C and J. As illustrated in the elevational view in Figure 13.44(c), the ceiling joists M should be positioned and fixed immediately to the side of an alignment with the trimmed rafters K.

13.20.8 Studded Side Cheeks

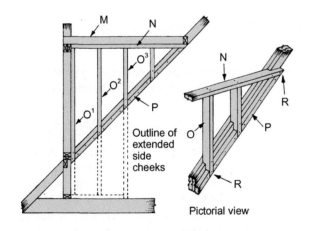

Figure 13.44 (d) Studded side-cheeks

Figure 13.44(d): Sometimes the side cheeks of dormer and eyebrow windows – or a part of them, as illustrated by the broken lines – are carried on down to the floor and the line of ashlering each side of the window takes up more of the available floor space, giving an increased vertical face to the dwarf walls of the room. However, the following notes assume that the cheeks are to be triangular.

13.20.9 Triangular Cheeks

First, notch the bottom of stud O^1, as seen at R in the pictorial illustration, to lap onto the trimming rafter, then cut squarely to length to finish 50 mm below the top of the window-head beam. Fix to the side of stud B and through the half-lap notch to the trimming rafter

with 100 mm wire nails. Next, notch the head-plate N, in a similar way to O^1, to fit against the trimming rafter near trimmer J and cut to length to fit squarely on top of stud O^1 and fix in position. Now, at 400 mm centres, fit and fix the remaining vertical studs O^2, O^3 etc., similar to O^1, fixed at the head and through the half-lap notch. Finally, cut and fix the 50 × 50 mm edge fillets P as shown.

13.20.10 Making the Full-size Template/Sheathing

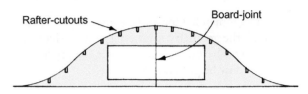

Figure 13.44 (e) *Making the template/sheathing*

Figure 13.44(e): The next important step is to lay out one or two sheets of 12–18 mm WBP plywood or Sterling board and by scaling measurements from the working drawing, set out the eyebrow shape full size by using the geometrical method already described. The idea is that these boards should be set out to include the whole vertical face of the eyebrow window. This will give the dual advantage of the board(s) being fixed against the front of the studded structure already erected, to act as a template for the formation of the eyebrow shape, while also becoming a permanent sheathing.

13.20.11 Marking the Rafter Cut-outs

Once the eyebrow shape is marked, it should be cut out with a jigsaw. Next, the eyebrow-rafter positions – equal to the main roof common-rafter centres – must be marked around the curved edge of the template and set down for depth, as indicated in Figure 13.44(e), giving the appearance of a castellated edge. To simplify this, the one- or two-piece template can be laid out flat on the roof slope, either immediately below the window opening, with the template's base resting against temporary nails driven in near the fascia plumb-cut, as indicated at Q^1 (Figure 13.44(c)), or on the roof slope above the window, with the base resting against the ceiling joists M, as indicated at Q^2. Either way, the template must be in an exact position laterally, i.e. the centre of the template must be equal to the centre of the window-width. Resting against the rafters like this, their positions can be easily marked onto the curved-template edge and lines drawn down at right-angles to the base, equal in depth to the eyebrow-rafter plumb cut, as illustrated in Figure 13.44(f).

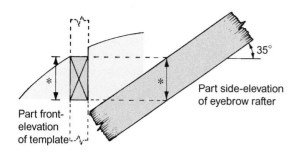

Figure 13.44 (f) Cut-outs equal plumb-cut depth*

13.20.12 Cutting the Window Opening

Before removing the template/sheathing from the roof to make the cut-outs, it should be offered up and adjusted for its true position against the face of the studded structure, then marked around the inside edge to outline the window opening. This, then, can be cut when making the cut-outs for the eyebrow rafters – which should now be done, again with the aid of a jigsaw.

13.20.13 Pitching the Eyebrow Roof

Figure 13.44(g): It must be appreciated that, because the eyebrow rafters are pitched at the same angle, say

35°, around the serpentine profile, the surface shape and its effect on the eaves is geometrically similar to a cylinder resting at 35° to the horizontal plane, whereby its base, representing the eaves, would be at 55° to the horizontal plane. From a side elevation, this would be at right angles (35° + 55° = 90°) to the cylinder/roof slope, as indicated between points A, B and C in the illustration. This can be proved in a practical way by bending a strip of 6 mm plywood, say 265 mm wide, equal to the tile depth, over the eyebrow shape near the edge and noting that it would follow the same path indicated at A–B. If you can imagine this plywood strip being equal to a line of tiles laid side by side, you should then appreciate that the unequal projection of the eaves' edge must be like this for the tiles.

13.20.14 Fixing the Template

The template/sheathing board(s) should now be fixed lightly in position, to be stable enough to work against, but able to be removed easily if necessary. This is because the studwork yet to be built-up behind the template might be awkward to fix properly in certain places with the template in position.

13.20.15 Eyebrow Rafters

The eyebrow rafters, seen in their diminishing lengths in Figure 13.44(g), have an acute-angled cut at one end

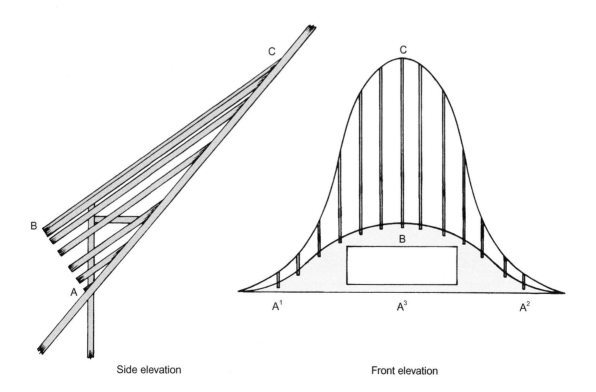

Figure 13.44 (g) Formation of eyebrow rafters: left, side elevation; right, front elevation

for fixing to the top edges of the main roof's rafters, like sprocket pieces, and are either finished with a plumb cut at the other, front end, or with a shallow plumb cut and a seat or soffit cut. Mostly, eyebrow eaves' rafters are left visible, but can be found covered up with a soffit lining. Whether visible or not, a strip of 6 mm WBP plywood will be required, either for fixing on top of the rafters in the area of the projecting eaves, if the rafter-ends are left visible, or for shaping to a diminish and fixing on the underside of the rafters, if it is to be lined with a soffit.

13.20.16 Finding the Angles

Figure 13.44(h): The illustrations used here assume a main roof pitch of 50° and a pitch of 35° for the eyebrow rafters. The main angle required is for the sprocket cut. Thinking of this, as illustrated, in the form of right-angled triangles, the cut for this very acute angle would be

$$a - b = 50° - 35° = 15°$$

Also illustrated is the setting out from a roofing instrument, to provide an alternative way of visualizing the angle required. If a soffit seat-cut is needed, we already know this to be 35° and, therefore, a plumb cut, if required, would be 90° − 35° = 55°.

13.20.17 Finding the Lengths

On this type of roof, a practical approach to finding the diminishing rafter-lengths will be more expedient. One way of doing this, is to establish the longest rafter in the crown position and then use it as a guide to finding the length of its neighbour on each side. First, cut the 15° sprocket angle on one end and lay the rafter in position, resting in the profile cutout and on the main rafter.

Because the sprocket cut is shallow and therefore long, small adjustments up or down until the cut is properly seated will ensure that the rafter is at its required angle of 35°. Now fix this position with a temporary nail. At the eaves' end, judge the approximate amount of projection required, with a margin for error of, say 100 mm, and mark for the initial cut. If considered necessary, check this by squaring down to a point near the apron upstand, as indicated at A–B–C in Figure 13.44(g).

13.20.18 Cutting and Fixing the Eyebrow Rafters

The first rafter is now removed and tried in the cut-outs on either side and, again, a judgement is made regarding the approximate amount of projection required. The side of the rafter is marked accordingly and the marked length is transferred to other timbers to make two more rafters. Rafter number one can now be fixed back in place with the temporary nail through the sprocket-cut again. Newly cut rafters two and three can now be used to determine the diminished length of the next pair, rafters four and five. Then rafters two and three can be fixed like rafter number one with a temporary nail through the sprocket-cut into the main rafters. This sequence is carried on down the eyebrow shape in pairs of rafters, one on each side of the crown position, until the ends are reached at main-roof level. Only the sprocket-cut ends are nailed, the other ends should be unfixed for now, resting in the cutouts.

13.20.19 Marking the Eaves' Edge

The precise eaves' edge can now be determined, as mentioned earlier, by pinning down a strip of 6 mm plywood over the rafter-ends of the eyebrow and marking

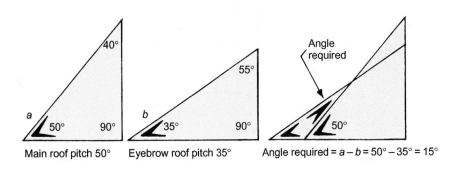

Main roof pitch 50°   Eyebrow roof pitch 35°   Angle required = a − b = 50° − 35° = 15°

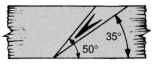

50° − 35° = 15°
Setting out from roofing instrument

Figure 13.44 (h) Finding the angles

the line on each rafter-top. This task can be made easier by using the three points A^1, A^2 and B, referred to in Figure 13.44(g). First, on each extreme of the eyebrow, at the lowest and smallest sprocket, mark points A^1 and A^2 and projecting, say, 75 mm from the vertical face of the sheathed window. Strike a line across the tops of the trimmed rafters from point A^1 on one side to point A^2 on the other and mark the line midway at A^3. Now square up from this middle point A^3 to mark point B on the eyebrow rafter in the crown position. This can be done with a straightedge, with one end resting on mid-point A^3 and the other being adjusted at point B up against a roofing or try-square. These three points, marked on the rafter-tops, greatly assist in positioning the ply and establishing the eaves' edge.

13.20.20 Finishing the Eaves

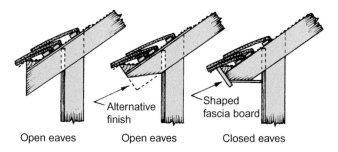

Figure 13.44 (i) Eaves' finish

Figure 13.44(i): Next, remove the temporarily nailed rafters and mark and cut the required eaves' finish, related to the eaves'-edge line just established. Three optional finishes are shown in Figure 13.44(i). The two open-eaves details are most common and seem to be more in keeping with the aesthetics of eyebrow windows. The 6 mm WBP plywood is shown fixed on top of the rafters, to give a better visual finish on the exposed underside. This has to be wide enough to mask the greatest projection at the crown. A tilting fillet is also fixed at the eaves' edge, to give the required uplift to the eaves' tiles. This may need a few strategic half-depth saw cuts to ease it around the serpentine shape, on top for hollows and on the underside when going over the crown.

13.20.21 Closed Eaves

The closed-eaves detail in Figure 13.44(i) requires a narrow fascia board, backed up by a tilting fillet, as shown. The fascia board, of about 75 mm width and normal thickness, must be shaped to follow the eyebrow curvature. The soffit lining, of 6 mm WBP plywood, needs to be fixed to a level seat cut – to be less

problematic geometrically – and is best fixed before the fascia to allow the outer edge to be trimmed against the rafter-ends, if necessary. When the eaves' cuts are known or have been decided, marked out, cut and completed, refix the eyebrow rafters in their previous positions, adding a few more 100 mm wire nails as final fixings.

13.20.22 Fixing the Side-support Studs

Figures 13.44(j) and (k): The final studwork directly behind the sheathing consists of two vertical studs per

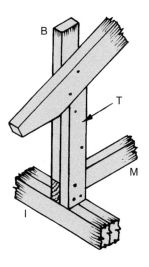

Figure 13.44 (j) Studding on window beam

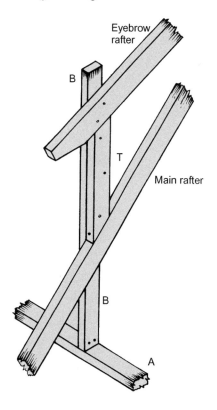

Figure 13.44 (k) Studding on rafter and floor plate

eyebrow rafter and a row of staggered noggings. The first, vertical stud B, is a side-support stud which, as shown, either rests squarely on the ceiling joist, when the studding is on the window beam, or rests squarely on the floor plate which continues through on each side of the window opening. With the sheathing/template still in position, studs B should now be cut and fixed. They are cut with a square-ended allowance, to project at the top, skew-nailed at the base and side-fixed to the eyebrow rafters. Side-fixings should also be made where studs B rest against main-roof rafters (Figure 13.44(k)).

13.20.23 Fixing the Load-bearing Studs

Now that the eyebrow rafters are supported by the side-support studs B, the sheathing/template can be removed, if considered necessary, to facilitate the fixing of studs T, which tend to get more awkward to fix as they diminish in height. These studs are load-bearing studs which, as shown, are cut with an acute angle at the top to fit under the eyebrow rafter and either rest squarely at the bottom, on the window beam, or are cut with another acute angle to rest on the main-roof rafter. With an eyebrow pitch of 35°, the top angle would be $90° - 35° = 55°$; and with a main roof pitch of 50°, the bottom angle would be $90° - 50° = 40°$. Whether removing the sheathing/template or not, studs T, when cut and fitted, should be side-fixed to studs B, skew-nailed to the window beam, where applicable, and an edge-fixing nail should be driven through the sharp point of the acute angles. 100 mm round-head wire nails should be used throughout.

13.20.24 Fixing the Final Noggings and Sheathing

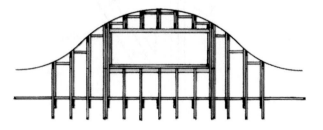

Figure 13.44 (l) Staggered noggings

Figure 13.44(l): To give the studded eyebrow structure more rigidity, 100×50 mm noggings should be cut in and fixed as near to the top as possible, as indicated in the exposed elevational view. Because the noggings should be kept in a horizontal position, they will take on a staggered appearance. Finally, fix the sheathing/template to the whole studded structure with 50 mm

round-head wire nails (or screws) at approximately 200 mm centres and cut off the square-ended projections of studs B, flush to the top of the eyebrow rafters, following the top edge of the sheathing/template.

13.21 LEAN-TO ROOFS

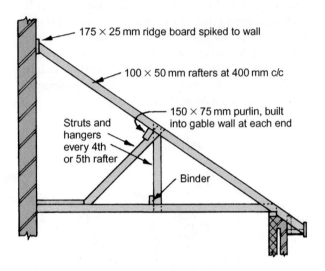

Figure 13.45 Traditional lean-to (double) roof

Figure 13.45: This type of roof, usually found on parts of the building that extend beyond the main structure, comprises mono-pitched rafters *leaning* on the structural wall in various ways. This connection to the wall was usually in the form of a wall plate resting on wrought-iron corbels built into the wall at about 1 m centres, or a wall plate bedded on continuous brick corbelling, projecting from the face of the wall. If the potential thrust of a particular lean-to roof can be discounted, the connection may be simply a ridge board fixed to the wall to take the plumb cuts of the rafters.

13.21.1 The Ceiling-Joists' Wall-connection

Traditionally, ceiling joists were built in or cut into the main wall. Nowadays, TW type joist hangers could be used. Technically, without purlins, this roof would be termed a *single roof* and would be restricted to a span of about 2.4 m.

13.22 CHIMNEY-TRIMMING AND BACK GUTTERS

Figure 13.46: When a chimney stack passes through a roof, the rafters are trimmed around it in a similar

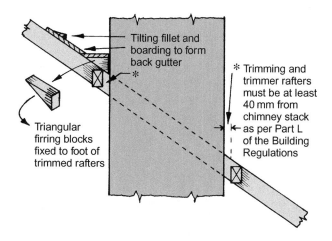

Figure 13.46 Chimney trimming and back gutter

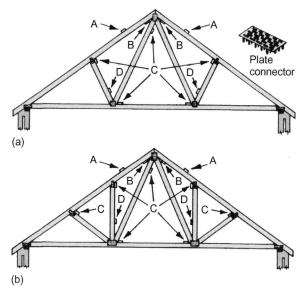

Figure 13.47 (a) The Fink or W truss; (b) the fan truss

way to trimmed openings for dormer or eyebrow windows and skylights. The trimmer rafters can be stophoused, fixed with TT type joist hangers, CL type framing anchors, or simply butt-jointed against the trimming rafters – and may be vertical or leaning to the roof pitch. The former position is preferred, as this allows the trimmed rafters a birdsmouth bearing on the trimmers. Triangular blocks, boarding or sheeting material, as illustrated, form the usual back gutter to the stack.

13.23 TRUSSED RAFTERS

13.23.1 Introduction

As mentioned briefly in the beginning of this chapter, roofing on domestic dwellings now predominantly comprises factory-made units in the form of triangulated frames referred to as trussed rafters. These assemblies are made from stress-graded, prepared timber, to a wide variety of configurations according to requirements. Most shapes have a named reference and the two most common designs used in domestic roofing are the Fink or W truss, and the Fan truss. All joints are butt-jointed and sandwiched within face-fixing plates on each side. These plates are usually of galvanized steel with integral, punched-out spikes for machine-pressing onto the joints. After initial positioning of the trusses, they must be permanently braced.

13.23.2 Bracing

Figures 13.47(a) and (b): The bracing-arrangement details given here and in the illustrations – as well

as the other references to truss rafters – are based on current information given in the technical manuals obtained from Gang-Nail Systems Ltd, a member company of International Truss Plate Association.

- *Braces A*: 75 × 25 mm or 100 × 25 mm temporary longitudinal bracing, used to stabilize the position of the trusses during erection.
- *Braces B*: 97 × 22 mm (minimum size) permanent diagonal bracing, forming 45° angles to the rafters, should run from the highest point on the underside of a truss to overlap and fix to the wall plate, starting at a gable end and running zigzag throughout the length of the roof, fixed to every truss with two 3.35 mm × 65 mm long galvanized wire nails. There should be not less than four braces (two on each slope) of any short-length duopitched roof. All joins of incomplete bracing-lengths should be overlapped by at least two trussed rafters. The angle of bracing given above as ideally 45°, should not be less than 35° or more than 50°.
- *Braces C*: 97 × 22 mm (minimum size) permanent longitudinal bracing, with fixings and overlap allowances as for diagonal bracing, positioned at all *node points* (points on a truss where members intersect), with a 25 mm offset from the underside of the rafters (top chords) to clear the diagonal bracing, extended through the whole length of the roof and butting tight against party or gable walls.
- *Braces D*: 97 × 22 mm (minimum size) permanent diagonal web chevron bracing, each diagonal extended over at least three trusses, required for duo-pitched spans over 8 m and monopitched spans over 5 m.

13.23.3 Advantages

One of the advantages of trussed-rafter roofs is the clear span achieved, without the traditional need for load-bearing partitions or walls in the mid-span area. Another advantage, to the building designer, is that the specialist truss-fabricator will only need basic architectural information to plan the truss layout in detail for the building designer's approval.

13.23.4 Site Storage and Handling

It is important to realize that although the trusses are strong enough to resist the eventual load of the roofing materials, etc., they are not strong enough to resist certain pressures applied by severe lateral bending. These pressures can have a delaminating effect on the metal-plated joints and are most likely to occur during truss delivery, movement across the site, site storage and lifting up into position – especially the motion of see-sawing over the top edges of walls when the truss is laying on its side face.

Recommendations

Storage on site should be planned to be as short a time as possible, preferably not more than 2 weeks. In bad weather, stored trusses should be protected by a waterproof cover, arranged to allow open sides for air ventilation. The trusses should be stored on raised, level bearers to avoid distortion and contact with the ground.

Vertical Storage

This is the preferred method. The trussed rafters are stored in an upright position, stacked close together against a firm, lean-to support at each end, resting on bearers at the position where the wall plates would normally occur, built up to ensure that any eaves' projection clears the ground and any vegetation present.

Horizontal Storage

This alternative method, where trussed rafters are laid flat, stacked up upon each other, requires a greater number of bearers which should be carefully arranged to give level support at close centres and be directly under every truss joint. This is to reduce the risk of joint-damage and long-term deformation of the trusses. If subsequent sets of bearers are used higher up in the stack to take another load of trusses, the bearers must be placed vertically in line with those below.

Inverted Storage

A third method of storage, preferred by some manufacturers and builders, is to invert the trussed rafters and support them on built-up side frames. These frames, resembling braced stud-partitions, are secured with raking struts and are set up parallel to each other and at the same span as the roof's wall plates (the length of the bottom chord or ceiling tie). The height of the two side frames must be more than the rise of the roof, to ensure that the apex of the upside-down trusses clears the ground.

Manhandling

On wide-span trusses – which are more liable to joint-damage from sideways-bending – it may be necessary to use additional labour to provide support at intermediate positions. When carrying trusses across the site, it may be safer and more manageable for a truss to be reversed so that the apex hangs down. On the other hand, when being offered up into its roof position, the truss should be upright and the eaves' joints should be the main lifting points. Laying trusses on their sides and pushing/pulling/see-sawing them across walls and scaffolding, etc., may make manhandling them easier, but is a completely unacceptable practice.

Mechanical Handling

When mechanical means are used, the trusses should be lifted in banded sets and lowered onto suitable supports. The recommended lifting points are at the rafter (top chord) or ceiling tie (bottom chord) *node* points (where the joints occur). Lifting single trusses should be avoided, but if unavoidable, a suitable spreader bar should be used to offset the sling-forces.

13.23.5 Providing Profiles for Gable Ends

Figure 13.48: Gable-end walls are usually completed, or partially completed, before the trussed-rafter roof

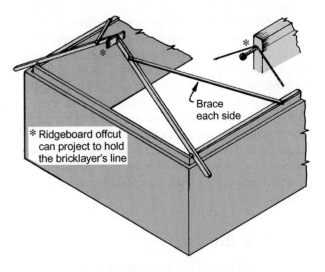

Figure 13.48 Profile erected for gable end

is erected. When this happens, a single trussed rafter frame or – if the trussed rafters are not yet on site – a pattern pair of common rafters, as illustrated, is fixed and braced up at each gable end to act as a profile for the bricklayer to use as a guide in shaping the top of the raking walls. If gable ladders (described later) are to project over the face of the wall, then the brickwork should be built up only to the approximate underside of the truss and completed after the ladders have been fixed in position.

13.24 ERECTION DETAILS AND SEQUENCE FOR GABLE ROOFS

13.24.1 Wall Plates and Restraint Straps

Wall plates for trussed rafters are jointed and bedded as already described for traditional pitched roofs. The next step is to mark the positions of the trusses at maximum 600 mm centres along each wall plate. This will indicate the clear areas for the positioning of the vertical restraint straps, which are now fixed.

13.24.2 Procedure

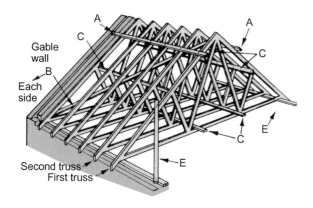

Figure 13.49 Erection of Fink trussed rafters

Figure 13.49: The erection procedure is open to a certain amount of variation, providing care is taken in handling and pre-positioning the trusses on the roof. Bearing that in mind, the following procedure is based on the notes given in the technical manual on trussed rafters, mentioned previously.

By using framing anchor truss clips (the recommended fixing, which does not rely on the high degree of skill required for successful *manual* skew-nailing) or by skew-nailing from each side of the truss with two 4.5 mm × 100 mm galvanized round wire nails

per fixing, the first truss is fixed to the wall plates, in a position approximately equal to the first pair of common rafters in a hipped end. This determines the apex for the diagonal braces, marked B. Stabilize and plumb the truss by fixing temporary raking braces E on each side, down to the wall plates.

Fix temporary battens A, on each side of the ridge and resting on the gable wall. Position the second truss and fix to the marked wall plate and to the temporary battens A, after measuring or gauging to the correct spacing. Proceed until the last truss is fixed near the gable wall. Now fix the diagonal braces B with two 3.35 mm × 65 mm galvanized round wire nails per fixing and then continue placing and fixing trusses in the opposite direction, braced back to the first established end. Finally, fix the braces marked C throughout the roof's length and fix horizontal restraint straps at maximum 2 m centres across the trusses onto the inner leaf of the gable walls, as illustrated previously in Figure 13.37(e).

13.25 HIPPED ROOFS UNDER 6 m SPAN

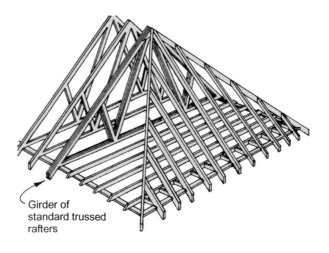

Figure 13.50 Hip-end construction for roof under 6.0 m span

Figure 13.50: As illustrated, the recommended hip-end for a roof of this relatively small span is of traditional construction. The main difference being that instead of the saddle board and hips being fixed to a ridge board and the first pair of common rafters, they are fixed onto a *girder truss* made up of manufactured truss rafters. This consists of two or three standard trussed rafters securely nailed together by the supplier (preferably), or fixed on site to a nailing pattern stipulated by the supplier.

13.25.1 Procedure

After the wall plates are jointed, bedded and set, the hipped ends are set out as already described for traditional roofing, the positions of the standard trusses marked and the vertical restraint straps positioned and fixed. The erection sequence starts with the fixing and temporary bracing of the girder truss, followed by the infill of the standard trusses and bracing. The hip ends are then constructed, using hip rafters and jack rafters of at least 25 mm deeper section than the truss rafters to allow for birdsmouthing to the wall plates. Ceiling joists – unlike those illustrated – may also run at right-angles to the multiple girder truss, supported on the girder truss by mini-hangers.

13.26 HIPPED ROOFS OVER 6 m SPAN

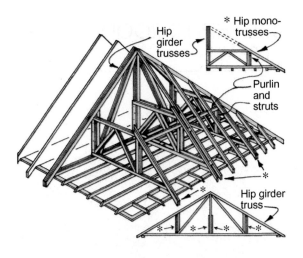

Figure 13.51 Hip-ended construction for roof over 6.0 m span

Figure 13.51: There are various alternative methods of forming hip ends in trussed rafter roofs. The example given here, therefore, is just one of the methods that deal with large spans. Assuming that the wall plates have been jointed and bedded and are now set, the erection procedure is as follows.

13.26.1 Procedure

Three special trusses, known as *hip girder trusses*, are – as mentioned for the previous hip end – securely nailed together by the supplier (preferably), or fixed on site to a nailing pattern stipulated by the supplier. The girder is fixed at the half-span (run) position and

infill ceiling joists are laid and fixed. In three positions, as illustrated, other special trusses rest across the ceiling joists and are fixed to them and to the vertical members of the girder truss. These secondary trusses are known as *hip mono trusses* and six are required at each hip end. Two in the centre are nailed together with rafter-thickness packings in between and fixed in position to straddle the central vertical girder member – and house a half-length *flying* crown rafter. The other hip mono trusses, nailed directly together in pairs, without packings, are fixed at the quarter-span position on each side of the girder truss and are splay-cut like jack rafters to fit the hip rafters. A short purlin and vertical struts, as illustrated, are fixed to the mono trusses. Hips and jack rafters, as before, must be at least 25 mm deeper to allow for purlin- and wall-plate birdsmouths to be cut.

13.27 ALTERNATIVE HIPPED ROOF UP TO 11 m SPAN

Figures 13.52(a)–(c): The most common construction for a hipped roof up to this span is referred to as a *standard centres hip system*. It is made up of a number

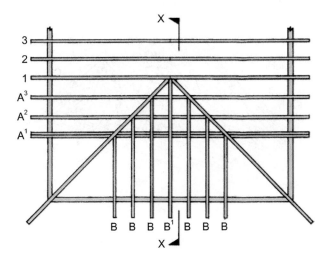

Figure 13.52 (a) Standard centres hip system

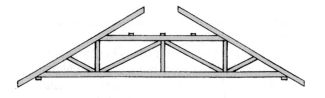

Figure 13.52 (b) Flat-top hip truss (in position A)

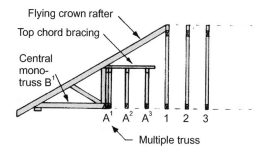

Figure 13.52 (c) Section X–X

of identical flat-top hip trusses A (Figure 13.52(b)) spaced at the same centres as the standard trusses, and a multiple girder truss of the same profile which supports a set of monopitch trusses B, set at right-angles to the girder. The corner areas of each hip are made up of site-cut jack rafters and infill ceiling joists attached to the hip girder. The flat tops (chords) of the hip trusses require lateral bracing back to the multiple hip girder.

13.27.1 Procedure

Assuming that the wall plates have been marked out for the hip end(s) and the truss positions, and that the vertical restraint straps have been fixed, a standard truss is first fixed at the half-span (run) position, labelled 1 in the illustration. The remaining standard trusses (2, 3 etc.) are then erected and braced. A ledger rail of 35 × 120 mm section – instead of a saddle board – is fixed at the apex of truss 1, at a height to suit the hip rafter's depth.

Next, the multiple girder truss A^1 is fixed, set in from the hip end by the span (length of bottom chord) of the mono truss B; then the two intermediate flat-top hip trusses, A^2 and A^3, are fixed and braced. Now a string line is set up, representing the *in situ* position of the hip rafters, and the flying rafters of the flat-top hip trusses are marked and cut back, allowing for the hip rafter's thickness. Next, the central mono truss B^1 is fixed, after its flying rafter has been trimmed and notched onto the ledger rail and its bottom end has been fitted into a prefixed truss shoe attached to the bottom chord of the girder. Now the hip rafters can be cut to length, birdsmouthed, notched and fixed in position. The remaining mono trusses B are then fitted and fixed into pre-fixed bottom-chord truss shoes, after trimming the flying rafter of each, ready for fixing to the hip rafter. Finally, the corner areas of each hip are completed with loose ceiling joists attached to the girder truss with mini hangers and loose infill jack rafters.

13.28 VALLEY JUNCTIONS

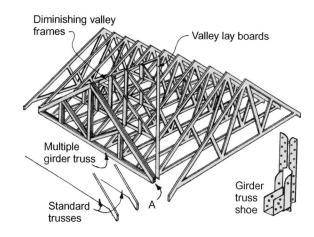

Figure 13.53 Valley junctions

Figure 13.53: Where a roof is so designed as to form the letter T in plan, a *valley set*, known as *diminishing valley frames* (as illustrated), can be fixed and braced directly onto the main trussed rafters in relation to lay boards or inset noggings. The noggings are required when the ends of the valley frames do not coincide with an underlying truss position. Lay boards or battens, running along the edges of the valley, are normally required by the roofer as a means of attaching the splay-cut ends of the tiling battens that are fixed at right-angles to the trusses.

13.28.1 Intersecting Girder Truss

At the intersecting point marked A, where the offshoot roof meets the trimmed eaves of the main roof – if no load-bearing wall or beam exists at this position – it will be necessary to have a multiple girder truss. This is to carry the ends of the trimmed standard trusses via hangers known as *girder truss shoes* (Figure 13.53). The girder is formed, either in the factory or on site, by nailing three intersection girder trusses together to a stipulated nailing pattern. Because of the heavy loads being carried, it may be necessary for the girder truss to have larger bearings in the form of concrete padstones.

13.29 GABLE LADDERS

Traditionally, when a verge projection was required on a gable end, this was achieved by letting the ridge board, purlins and wall plates project through the wall by the required amount – usually about 200 mm – to act as fixings for an outer pair of common rafters and the barge boards.

13.29.1 The Modern Method

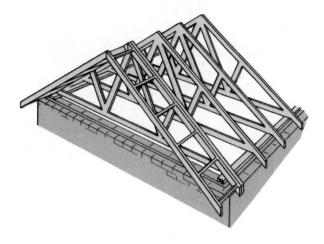

Figure 13.54 Gable ladders

Figure 13.54: As trussed-rafter roofs do not have purlins or ridge boards, the verge-projection is achieved by fixing framed-up assemblies, known as gable ladders, directly onto the first truss on the inside of the gable wall, as illustrated. The ladders, supplied by the truss manufacturer, are fixed on site by nailing at 400 mm centres and are subsequently lined with a soffit board on the underside and barge boards on the face side, after being built-in by the bricklayer. The usual practice is for the gable wall to be built-up to the approximate underside of the end truss, then completed after the ladders have been fixed.

13.30 ROOF-TRAP HATCH

When trusses are spaced at 600 mm centres, it should be possible to simply fix trimmer noggings between the bottom chords of the trusses to form the required roof-trap hatch. In other cases, when the trusses are closer together or a bigger hatch is required, it will be necessary – but not desirable – to cut the bottom chord of one of the trusses, as illustrated.

13.30.1 The Procedure

Figure 13.55: Typical details are shown for forming a trap hatch in the central bay of the trussed rafters. First, 35 mm thick framing timbers A are fixed to the inside faces of the ceiling joists that will be acting as

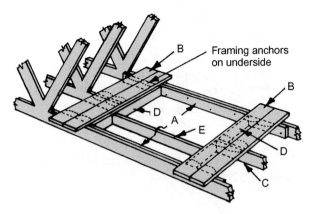

Figure 13.55 Forming the roof-trap

trimming joists. Next, two 150 × 38 mm boards B, spanning three trusses, must be fixed in position on each side of the proposed trap hatch before the central ceiling-joist chord is cut. The boards are fixed on the underside to the side of the ceiling-joist ties with bracket-type framing anchors and 31 mm × 9 gauge square-twisted sherardized nails, as per the manufacturer's instructions. The central chord C is then cut, and the opening trimmed with two trimmers D and, if required, an infill joist E.

13.31 CHIMNEY TRIMMING

Figure 13.56: Where the width of a chimney is greater than the normal spacing between trussed rafters, the trusses may have a greater spacing between them in the area of the chimney stack, providing the increased spacing is not more than twice the normal truss spacing.

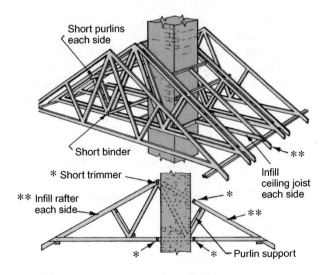

Figure 13.56 Chimney trimming

As illustrated, the non-truss open areas remaining on each side of the chimney stack are filled in with loose infill rafters, ceiling joists, trimmers, binders and short purlins strutted against the webs of the nearest standard trusses on each side.

As per the Building Regulations, the timbers should be at least 40 mm clear of the chimney stack. The infill rafters, which are nailed to the side of the infill joists and to the wall plate, should be at least 25 mm deeper than the trussed rafters to allow for a birdsmouth to be formed at the wall plate.

13.32 WATER-TANK SUPPORTS

Figure 13.57: For domestic storage of 230 or 300 litre capacity, the load is usually spread over three or four trusses. The details illustrated here are those recommended for a 300 litre tank within a Fink or W truss roof of up to 12 m span.

13.32.1 Spreader Beams

Two spreader beams of 47×72 mm section extend in length over the bottom chords of four trusses, as illustrated, up against the web on each side (close to the node points), sitting vertically on edge, not flat. The permanent longitudinal bracing, which is normally in this position, is offset in this area and fixed at the sides, up against the spreader beams.

13.32.2 Cross-bearers

Two cross-bearers of 35×145 mm section are now skew-nailed to the spreader beams, positioned at one-sixth the distance of the beam's bearing-length from each end – which is midway between the spacings of the first and second truss and the third and fourth truss in contact with the spreader beam.

13.32.3 Tank Bearers and Base Board

Next, two tank bearers of 47×72 mm section are skew-nailed across the first bearers, relative to the tank width, and a WBP plywood base board – not chipboard – is fixed to these. Like the spreader beams, the cross-bearers and tank bearers must sit vertically on edge, not flat.

13.32.4 Alternative Tank Support

If more headroom is needed above the tank platform, an alternative tank-bearer frame, the same or similar to that illustrated, may be used. This is made up of joist hangers and/or truss shoes, so the deeper cross-bearers, being parallel and between the trusses, can be dropped lower if required, on pre-positioned joist hangers. To allow for long-term deflection, there should be at least 25 mm between cross-bearers and the ceiling and the same between tank bearers and the ceiling ties of the trusses.

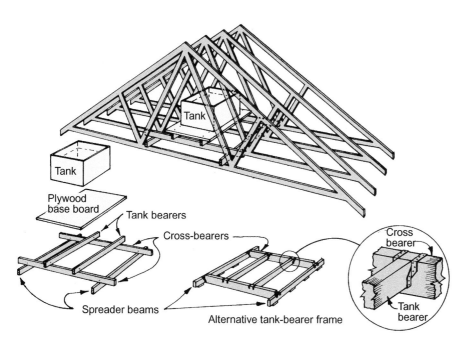

Figure 13.57 Water-tank supports

13.33 WORK AT HEIGHT REGULATIONS 2005 (AMENDED APRIL 2007)

These regulations from HSE (Health and Safety Executive) are only briefly outlined here, but are essential reading in their entirety for anyone working (or responsible for others working) at heights. In the construction industry, this covers any operation – not just roofing – where there is a risk of falling, whether it be from a low height or from a great height. The majority of fatal and serious injuries in industry are from falls – and over 53% of the serious injuries are from low heights.

The onus of responsibility is on the employer and/ or his representative (*the duty holder*), to carry out risk assessments and issue safety-method statements to all concerned with the work at height. By issuing method-statements, a share of the responsibility is passed on to employees (or sub-contractors), who must be responsible for their actions. The main key to ensuring safe workfing is identified as evaluating risks in a *hierarchy of risk assessment*. The order of the hierarchy is:

- AVOID working at height, if possible. For example, consideration might be given to prefabricating a truss rafter roof – or a section of it – at ground level on site and lifting and placing it onto the structure by crane. If, after assessment, work at height cannot be avoided, then:

- PREVENT a possible fall by erecting adequate scaffolding and work platforms with guard rails and toe boards, etc., to provide edge protection. Bear in mind that when erecting a roof, a scaffold around the outside does not prevent falls into the open well of the building. Alternatively, on small, isolated jobs, consider using a MEWP (mobile elevated working platform) – or a tower scaffold, instead of working from a ladder. Ladders might be the easy way, but they are not usually the safest way to do a job. If none of these are practicable, then:

- MITIGATE the effects of a fall by minimizing the fall-distance and thereby the consequences by providing safety nets just below the work area, or air bags on the floor level below. The safety nets must be installed by competent riggers.

In certain situations, additional measures may be necessary, such as providing fall-arrest equipment in the form of a safety harness – which must be securely attached to a strong anchorage point. To minimize trauma of suspension in a fall situation, a rescue plan should be part of the safety-method statement.

Although there is mention of a 2 metre rule applying before guard rails are legally required on a scaffold, risk assessment of falls is still required below this height and sensible, pragmatic precautions are expected to be taken. Mention is made of a *knee-high kick stool*, such as a traditional hop-up, not having to conform to the regulations, providing it is stable and in a good condition.

14

Erecting Timber-stud and Metal-stud Partitions

14.1 INTRODUCTION

Partitions are secondary walls used to divide the internal areas of buildings. Although often built of aerated building blocks, other materials, including timber and metal, are frequently used – especially above ground-floor level, on suspended timber floors, where block partitions would add too much weight unless supported from below by a beam or a wall.

14.2 TRADITIONAL BRACED PARTITION

Figure 14.1: This is only shown for reference and comparison with the modern stud partition illustrated in Figure 14.3 and the trussed partition seen in Figure 14.2. The partition was made up of 100 × 75 mm head, sill(s), door studs and braces – and 100 × 50 mm intermediate studs and noggings. The diagonal braces, which were bridle jointed to the door studs and sill, were included partly to give the partition greater rigidity against sideways movement; and partly to carry some of the weight from the centre of the partition down to the sill-plate ends, which were housed into the walls.

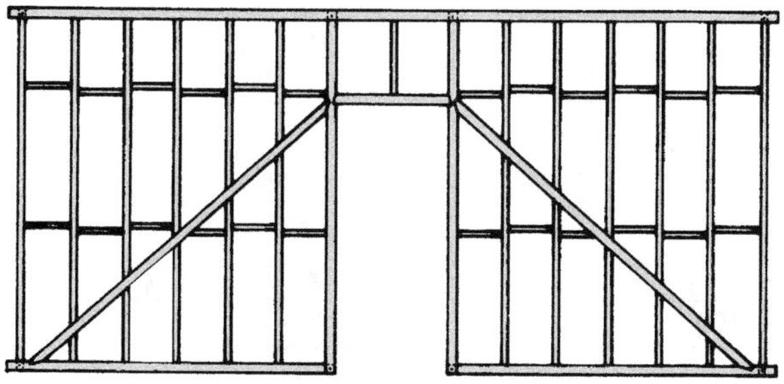

14.2.1 Jointing Arrangement

The main frame of this type of partition was through-morticed, tenoned and pinned (wooden pins or nails); the intermediate uprights (studs) were stub-tenoned to the head and sill; the door head was splay-housed and stub-tenoned to the door studs; the door studs were dovetailed and pinned to the sill; and the staggered noggings were butt-jointed.

14.3 TRADITIONAL TRUSSED PARTITION

Figure 14.2: As with the above braced partition, the trussed partition, with the advent of modern materials and methods, has been obsolete for many years. It was used for carrying its weight and the weight of the floor above. This should be taken into account before commencing any drastic alterations or removal on conversion works – as these partitions can still be found in old or grade I or II-listed buildings.

Note that timber partitions are now referred to as *stud* partitions, or *studding* (derived from old English *studu*, meaning post).

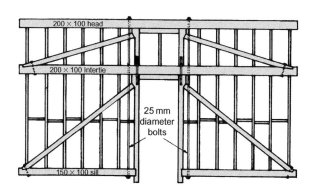

Figure 14.2 Traditional trussed partition (100 × 100 mm door posts, braces and straining heads; 100 × 50 mm studs and noggings)

Figure 14.1 Traditional braced partition

14.4 MODERN TIMBER-STUD PARTITIONS

Figure 14.3: Although these partitions can be made in a joinery shop and reassembled on site, with certain tolerances made for the practicalities involved, the common practice nowadays is to cut and build them up on site (*in situ*), piecemeal fashion. The detailed sequence of doing this is given below.

14.4.1 Sill, Sole or Floor Plate

Figure 14.3: The sill, labelled A on the illustration is cut to length and, if straight, can be used for setting out the floor position. Alternatively, its position can be snapped on the floor with a chalk line related to tape-rule measurements. The position of any door opening must be deducted from the sill plate setting-out. This deduction is an accumulation of

> door width (say 762 mm)
> thickness of door linings (say 28 mm × 2)
> a fitting tolerance (say 6 mm)
> door studs (50 mm × 2)

to give a total of 924 mm.

The plate is fixed to joists with 100 mm round-head wire nails, or to floor boards or panels with 75 mm nails or screws, or to concrete or screeded floors with nylon-sleeved Frame-fix or Hammer-fix screws, or cartridge-fired masonry nails or bolts, at approximately 900 mm centres.

14.4.2 Head Plate

Figure 14.3: The position for the head plate B can be fixed by plumbing up to the ceiling from the side of the sill at each end, either with a pre-cut wall stud and spirit level, a straightedge and spirit level, or a plumb bob and line, then by snapping a chalk line across the ceiling. The head plate is then cut to length, set out with intermediate stud-positions and propped up with two or three temporary uprights – purposely oversize in length – as illustrated in Figure 14.4. The plate is fixed to the ceiling joists with 100 mm round-head wire nails.

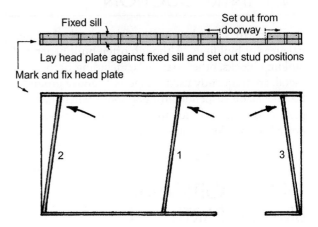

Figure 14.4 Marking and fixing the head plate

14.4.3 Wall Studs

Figure 14.3: The wall studs C should be marked to length either by a pinch-stick method (Figure 14.5(a))

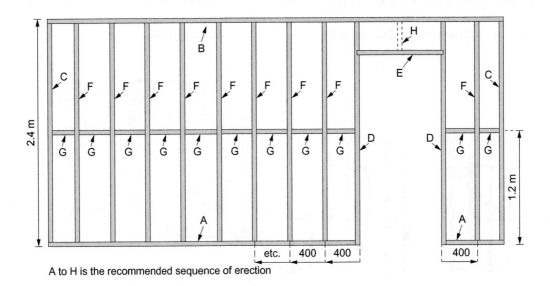

A to H is the recommended sequence of erection

Figure 14.3 Modern stud partition

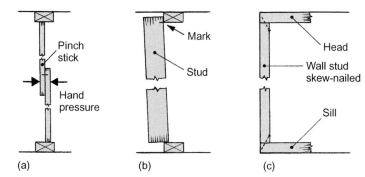

(a)   (b)   (c)

Figure 14.5 **(a)** Pinch stick method; **(b)** oversize stud method; **(c)** skew nail to plates

or by offering up a slightly oversize stud, resting on the floor plate and marked at the top, as indicated in Figure 14.5(b). Add 1–2 mm for a tight fit, then cut to length and fix to the block-wall with 100 mm Frame-fix or Hammer-fix screws. There should be at least three wall fixings and the ends of the stud should be skew-nailed to the plates, as shown in Figure 14.5(c).

14.4.4 Door Studs

Figure 14.3: The door studs D are now marked, as above, from floor surface to the underside of the head, tightening-allowance added, then cut, carefully plumbed and fixed. The base of each stud is fixed with two 100 mm round-head wire nails, driven slightly dovetail-fashion into the end of the floor plate and one central 75 mm nail skew-nailed into the floor. At the top, each stud is skew-nailed with three 75 mm wire nails, using the skew-nail technique indicated in Figure 14.6 (or, nowadays, may be skew-nailed with a Framing Nailer gun).

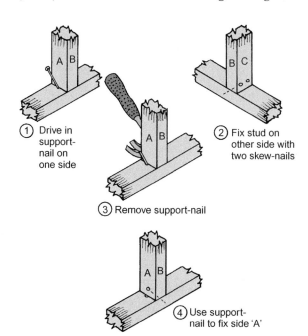

① Drive in support-nail on one side

② Fix stud on other side with two skew-nails

③ Remove support-nail

④ Use support-nail to fix side 'A'

Figure 14.6 Skew-nail technique

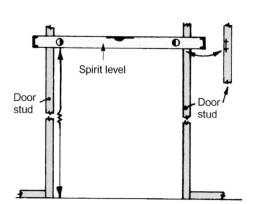

Figure 14.7 Mark head-housing

14.4.5 Door Head

Figure 14.3: Next, the position of the door head E is marked on the door studs, as illustrated in Figure 14.7. The measurement required for this is an accumulation of the door height (say 1.981 m) plus tolerance for floor covering (if carpet, say 15 mm) plus head-lining thickness (say 28 mm), giving a total of 2.024 m. This can be marked each side, but is best marked on one side of the door opening, then transferred across with a spirit level. If the head is to be butt-jointed and nailed (which is the most common trade-practice, see Figure 14.9(d)), cut to length to equal the width between studs at floor-plate level, or 924 mm (example measurement) as worked out in Section 14.4.1, and fix through the door studs with two 100 mm round-head wire nails each side. Alternatively, if to be housed (see Figure 14.9(b)) add 24 mm to the length, mark and cut the 12 mm deep door-stud housings each side *in situ*, working from a step ladder, slide the head into the housings and fix through the door studs with two 75 mm round-head wire nails each side.

14.4.6 Intermediate Studs

Figure 14.3: Studs F are next in sequence, cut to length for a tight fit as described for wall studs in Section 14.4.3. If the plates are not already set out for the intermediate stud positions, as suggested earlier, then these positions should now be marked out to suit the plasterboard sizes. The boards used are usually 2.4 m × 1.2 m and either 9.5 mm or 12.5 mm thick. These sizes dictate the spacing of the studs at either 400 mm centres (6 × 400 = 2.4 m) or 600 mm centres (4 × 600 = 2.4 m). The 9.5 mm thickness should only be used on studding spaced at 400 mm centres. The studs are nailed into position with three 75 mm wire nails to each abutment, using the skew-nail technique shown in Figure 14.6.

14.4.7 Noggings

Figure 14.3: These final insertions are short struts, G, that stiffen up the whole partition. If they are centred at 1.2 m from the floor, as shown (by measuring up at each end and snapping a chalk line), the joint of the plasterboard will be reinforced against the noggings. The noggings are cut in snugly and fixed by skew-nailing or end-nailing as indicated in Figure 14.8(a). To lessen the risk of bulging the door studs, the fixing of the noggings should be started from the extreme ends, working towards the door opening and being extra careful with the final nogging insertions. Note that intermediate stud H (above the door head) is optional on partitions of this limited height.

14.4.8 Studding Sizes

The timber used for studding can be of 100 × 50 mm sawn (unplaned) softwood, or ex. 100 × 50 mm prepared (planed to about 95 × 45 mm finish) softwood. For economy, 75 × 50 mm sawn, or ex. 75 × 50 mm prepared is sometimes used. Sawn timber is more traditional, but prepared timber is also used a lot nowadays to lessen the irregularities transferred to the surface material. To this end, *regularized* timber, machined-planed to a reduced, more constant sectional size, with slightly-rounded edges (arrises), is available and used for partitions nowadays.

14.4.9 Alternative Nogging Arrangements

Figure 14.8: Ideally, successive rows of noggings at 600 mm centres vertically (between the 1.2 m spacings shown in Figure 14.3) should be used to give greater rigidity and support to the plasterboard, although one row is normally sufficient. Points for and against the three alternative nogging arrangements are given below.

14.4.10 Straight Noggings

Figure 14.8(a): These can be positioned to reinforce horizontal plasterboard joints, but are not the easiest to fix. Various alternative methods of fixing are indicated by dotted lines denoting the nailing technique.

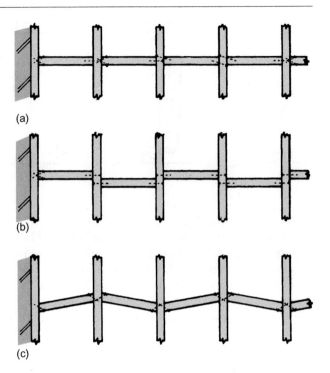

(a)

(b)

(c)

Figure 14.8 Alternative methods of nogging: **(a)** straight noggings; **(b)** staggered noggings; **(c)** herringbone noggings

The technique for skew-nailing (Figure 14.6) can be used here, with the support nail positioned under the nogging. Another technique, using a temporary strut, is shown in Figure 14.19.

14.4.11 Staggered Noggings

Figure 14.8(b): These cannot effectively reinforce the plasterboard joints, but as indicated, are easier to fix by end-nailing. Of course, if the plasterboard was placed on end, with vertical joints being reinforced on the intermediate studs, there would be nothing against this method.

14.4.12 Herringbone Noggings

Figure 14.8(c): These are positioned at an angle of about 10°, are easy to fix and achieve a tight fit, even with inaccurate cutting. If correctly positioned, they give about 50 per cent reinforcement to horizontal plasterboard joints. On the minus side, this method has a tendency to bulge the door studs.

14.5 DOOR-STUD AND DOOR-HEAD JOINTS

Figure 14.9: The strength of these joints is important, as any weakness, especially resulting in an upwards

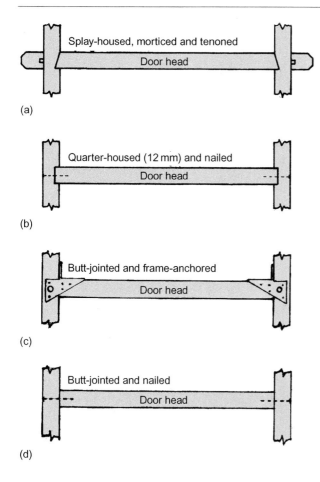

(a)

(b)

(c)

(d)

Figure 14.9 Door-stud and head joints: (a) too elaborate; (b) good compromise; (c) modern method (d) common method

movement of the door-head stud, can affect the door-lining head. Problems like this usually occur for two reasons: if a tolerance gap exists between the door-head stud and the lining-head; and if the door-head/door-stud joint is not strong enough to resist hammer-blows when the door-lining legs are being fixed at the top, or – more likely – when the door stop is being fixed to the underside of the door-lining head. Evidence of movement will appear as unsightly gaps to the corner housing-joints of the door lining. The risk of this happening can be avoided by using the joints shown in Figures 14.9(b) or (c) and by inserting packing or wedges in the gap between door-head and lining-head, directly above each door-lining leg (see also Chapter 6 which covers the fixing of door frames, linings and Doorsets).

14.5.1 Splay-housed, Morticed, Tenoned and Draw-bore Wedged

Figure 14.9(a): This traditional door-stud/head joint, although ideal for the job and strongly resistant to

displacement from timber which might twist and from lateral or vertical hammer-blows, is too elaborate and time-consuming nowadays.

14.5.2 Quarter-housed and Nailed

Figure 14.9(b): This is one of the recommended methods and is a good compromise between the other two extremes. The door-stud quarter housings (i.e. a quarter of the thickness) restrict twisting and upwards movement of the door-head and if the partition is well strutted with noggings, lateral movement should not be a problem.

14.5.3 Butt-jointed and Frame-anchored

Figure 14.9(c): This modern method is also recommended. It has all the virtues of the quarter-housed joint in restricting movement and is less time-consuming. The butt-jointed head can be nailed through the door-studs initially, then the framing anchors fixed at each end, or the framing anchors can be fixed in position on the head, the head located and fixed into the door-studs via the remaining framing-anchor connections. The anchors are recommended to be fixed with 3 mm diameter × 30 mm-long sherardized clout nails.

14.5.4 Butt-jointed and Nailed

Figure 14.9(d): This method is the most common trade-practice for attaching door-head to door-studs but, for reasons already stated, it is not the best. In the past, it was the method used on cheap work, which – it seems – has now become the norm.

14.6 STUD JOINTS TO SILL OR HEAD PLATE

Figure 14.10: There are four methods of jointing vertical, intermediate studs to the head plate and sill (floor) plate, as follows.

14.6.1 Stub-tenoned

Figure 14.10(a): The short tenons are morticed to half depth into the head and sill. This method involves too much hand work on site and is best suited to preformed partitions being made in the joinery shop, where machinery is available. Such partitions would be sent to the site in pieces, designed with length and

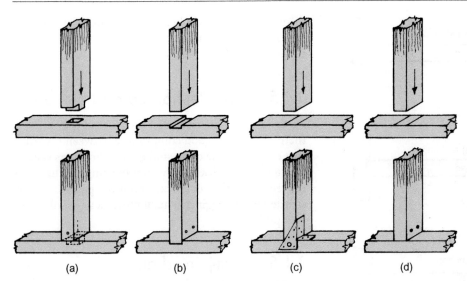

Figure 14.10 Stud joints to sill or head plate

(a)   (b)   (c)   (d)

height tolerances, ready for assembly and erection. This is useful in occupied premises, such as offices and shops, where the site work would be reduced.

14.6.2 Housed or Trenched

Figure 14.10(b): The housings are cut into a quarter-depth of the plate thickness and the studs are skew-nailed at each joint with two 75 mm round-head wire nails. This method can be easily handled on site; however, although housings are ideal for easier nailing, straightening and retaining any twisted studs, the method is rarely used nowadays because of the added time element.

14.6.3 Butt-jointed and Frame-anchored

Figure 14.10(c): This modern method, already referred to in Section 14.5.3 for door-head studs, has most of the benefits afforded by housing joints – with the exception of straightening out a stud already in a state of twist. The studs need to be fitted tightly and two framing anchors per joint – one on each opposite face-edge – are nailed into position with 3 mm × 30 mm sherardized clout nails.

14.6.4 Butt-jointed and Skew-nailed

Figure 14.10(d): Again, this is the most common trade-practice, not because it is the best, but because it is the quickest method of jointing. Originally used only on cheap work, this method is now widely used. The stud should be a tight fit, otherwise the relative strength of this joint is very much impaired. Three 75 mm wire nails should be used, two in one side, one in the other (as shown in Figure 14.6). This technique requires one support-nail to be partly driven into the

plate, at about 45°, on one line of the stud-position, so that when the stud is against it, the nail-head protrudes to the side. The stud is then fixed against the support-nail with two fixings, the support-nail removed with the claw hammer and used as the central fixing on the other side.

14.7 DOOR-STUD AND SILL-PLATE JOINTS

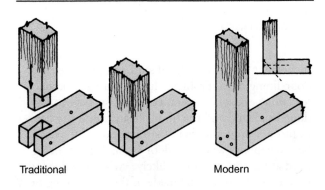

Traditional   Modern

Figure 14.11 Door-stud to sill-plate joints

Figure 14.11: Traditionally, these joints were dove-tailed and pinned (dowelled or nailed), as shown, to retain the base of the door-stud effectively. Studs of 100 × 63 or 75 mm thickness were used.

The present-day method uses door studs reduced to 50 mm thickness, resting on the floor and butted and nailed against the sill. As described previously, two 100 mm wire nails are driven slightly dovetail-fashion into the end of the plate and one centrally placed 75 mm wire nail is skew-nailed into the floor, as illustrated.

14.8 CORNER AND DOORWAY JUNCTIONS

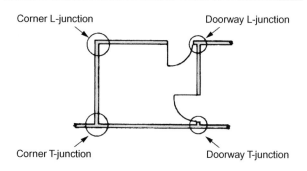

Figure 14.12 Corner and doorway junctions

Figure 14.12: This small-scale plan view of a room might be an uncommon layout for partitioning, but serves to illustrate the four junctions requiring different treatment. The best treatment each could receive would involve the use of three vertical studs, but other treatments shown here are sometimes used.

14.8.1 Corner L-junction

Figures 14.13(a)-(c): The plan view (a) shows the first choice of construction using three full-length corner studs to provide a fixing-surface on each side of the internal angle. As seen in the isometric view (b), short offcut blocks should be fixed in the gap existing

between two of the studs, to add rigidity and give continuity to the rows of noggings. For possible economy of timber and to achieve a similar result, a practical method of using *vertical* noggings is shown at (c). This could save a full-length vertical stud, as usually the noggings required can be cut from offcuts and waste material.

Alternative Methods

Figures 14.13(d) and (e): The first alternative (d) is reasonable, but would involve interrupting the partitioning operation to allow for plasterboard to be fixed to at least one side of the first partition, before the second partition could be built. The second alternative shown at (e) allows the whole partition to be built by providing a fixing-surface to each side of the internal angle, in the form of 50 × 50 mm vertical noggings fixed between the plates and the normal horizontal noggings. However, unless these vertical noggings are housed-in, there is a risk of them being displaced when plasterboard fixings are driven in – although modernday screwing would work.

14.8.2 Doorway L-junction

Figure 14.14: Where a doorway meets a corner L-junction, the problem of providing fixing surfaces on the internal angle is the same as before, and a similar treatment (a), using vertical noggings between normal horizontal noggings, is used instead of a full-length stud. The other methods, (b) and (c), are similar alternatives to those shown in Figures 14.13(d) and (e) the same considerations and comments apply.

*Vertical noggings

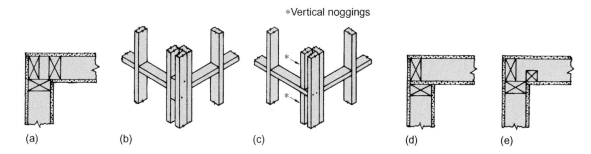

(a)   (b)   (c)   (d)   (e)

Figure 14.13 Corner L-junction

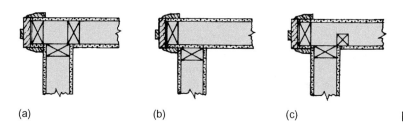

(a)   (b)   (c)   Figure 14.14 Doorway L-junction

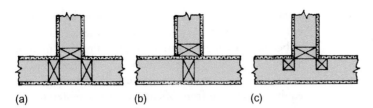

(a)　　　　　(b)　　　　　(c)

Figure 14.15 Corner T-junction

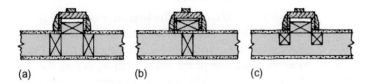

(a)　　　　　(b)　　　　　(c)

Figure 14.16 Doorway T-junction

Provision for Architrave

In good building practice, another consideration at this doorway junction, is that there should be provision for a full-width architrave on the side where the architrave is touching the adjacent wall. In practice, 3 or 4 mm less than the architrave width is provided on the inside angle to allow for eventual scribing of the architrave against an irregular plaster surface.

14.8.3 Corner T-junction

Figure 14.15: As illustrated in the plan-view details, this type of junction receives similar treatment in its three variations to those illustrated and described before.

14.8.4 Doorway T-junction

Figure 14.16: As seen in these illustrations, similar treatments apply in the same descending order. Another consideration in this situation is to pack out the lining (as shown), if necessary, to achieve full width architraves each side.

14.9 FLOOR AND CEILING JUNCTIONS

Figure 14.17(a): Although stud partitions do not normally present weight problems on suspended timber floors, certain points must be considered when a partition runs parallel to the joists and

- rests on the floor boards either in a different position to a joist below as at A, or
- rests in a position that coincides with a joist below, as at B (both situations inhibiting floor-board removal for rewiring, etc.);

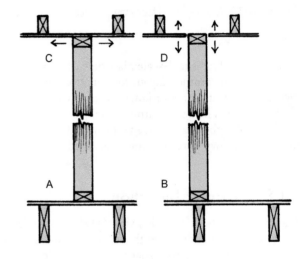

Figure 14.17 (a) Vertical sections

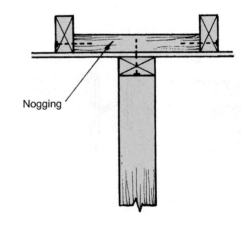

Nogging

Figure 14.17 (b) Head-fixings

- misses a joist required for head-plate fixings, C;
- creates a problem in board-fixings on each side of the head, if erected before the ceiling is boarded, as at D.

14.9.1 Creating Head-fixings

Figure 14.17(b): This shows a method of overcoming the lack of head-fixings by inserting 100 × 50 mm (or less) noggings between the ceiling joists at about 1 m centres. Where there is access above the ceiling joists, as in a loft, these noggings can be fixed before or after the ceiling is boarded; however, where there is no access above, as with floor joists that have been floored, then obviously the noggings must go in before the ceiling is boarded.

14.9.2 Double Ceiling-and-floor Joists

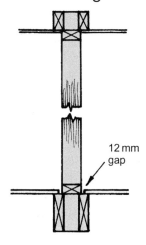

Figure 14.17 (c) Double ceiling-and-floor joists

Figure 14.17(c): This shows a way of overcoming all the previous issues, but uses more timber and requires extra work and careful setting-out at the joisting stage. Arrangements of double ceiling-and-floor joists, with support-blocks between, are set up. The blocks should be inserted at a maximum of 1 m centres and fixed from each side with 100 mm round-head wire nails, staggered as in Figure 14.17(d).

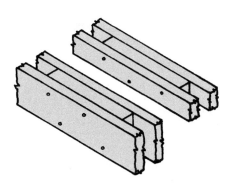

Figure 14.17 (d) Support-blocks

14.9.3 Creating a Beam Effect

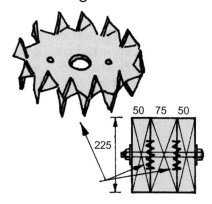

Figure 14.17 (e) Timber-connectors between joists

Figure 14.17(e): To achieve more of a beam effect with the double floor-joists – and so offset the disadvantage of a direct load – the support-blocks could be replaced by a continuous middle-joist, bolted into position with 12 mm diameter bolts, 50 mm round or square washers and 75 mm diameter toothed timber connectors at maximum 900 mm centres.

14.9.4 Partition across Joists

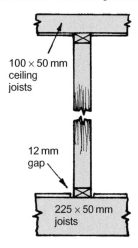

Figure 14.17 (f) Partition at right-angles to joists

Figure 14.17(f): This illustrates a situation that presents no problems, when the partition runs at right-angles to the floor and ceiling joists. The sill or floor plate is best fixed on the joists, to allow for expansion and contraction of the floor membrane – hence the 12 mm gap each side – but can be fixed on the flooring. Likewise, the head plate can either be fixed directly to the joists or to the boarded ceiling.

14.9.5 Fixing of Boards for Skimming

Figure 14.18(a): Plasterboards come in a variety of sizes, perhaps the most popular of these being 2.4 m × 1.2 m of 9.5 mm or 12.5 mm thickness. The illustration shows boards of 9.5 mm thickness, laid on edge, fixed to stud-spacings of 400 mm centres, in an arrangement suitable for skimming with finishing plaster, after reinforcing the joints with bandage or hessian skrim. Boards are fixed with 30 mm galvanized clout nails, at approximately 150 mm centres, or with BP (black phosphate) bugle-headed screws.

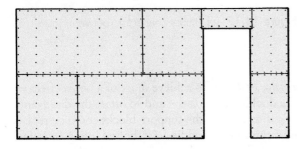

Figure 14.18 (a) Fixing boards for skimming

14.9.6 Fixing of Boards for a Dry-lining Finish

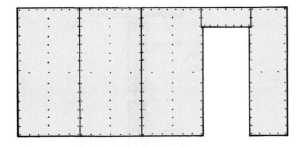

Figure 14.18 (b) Fixing of boards for a dry-lining finish

Figure 14.18(b): The illustration shows TE boards (tapered-edge plasterboards) of 12.5 mm thickness, laid on their ends to eliminate horizontal joints and fixed to stud-spacings of 600 mm centres, using either 38 mm galvanized clout nails, at approximately 200 to 300 mm centres – or BP (black phosphate) bugle-headed countersunk screws of a similar length. These achieve countersunk holes for filling, and reduce the risk of surface damage from hammer-blows. Decorators usually attend to the nail or screw holes and also fill, tape and refill the vertical joints within the indentation of the tapered edges, to achieve a finish.

14.9.7 Spacing of Studs

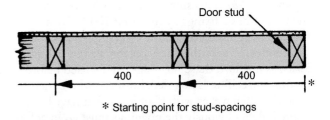

Figure 14.18 (c) Spacing of studs

Figure 14.18(c): As illustrated, it must be noted that the spacing of studs should start from the *edge* of the door-stud, to the *centre* of the intermediate studs, to achieve correct centres and full coverage of the door-stud edge by the board material.

14.9.8 Fixing Noggings

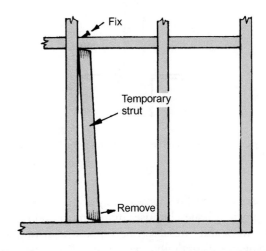

Figure 14.19 Alternative method of supporting noggings during fixing

Figure 14.19: As shown here, a temporary strut of 100 × 50 mm section, with bevelled ends to facilitate easy and quick removal, can be positioned to support a nogging during fixing with skew-nails, then lightly tapped at its base to remove it for the next fixing.

14.10 ERECTING METAL-STUD PARTITIONS

14.10.1 Introduction

As first mentioned in this chapter, metal-stud partitions, clad with 12.5 mm thick sheets of plasterboard, seem to be quite widely used nowadays for non-load-bearing partition-walls in dwellings and other categories of buildings. Their attraction seems to be that the relatively cheap metal components are light in weight, more stable than timber studs (which are susceptible to twisting and warping, etc.), easy to cut and fix – and are usually quicker and therefore cheaper to erect than timber stud partitions. The metal studs and tracks are made from pre-galvanized, folded mild-steel of varying thicknesses ranging from 0.5 mm to 0.9 mm. The studs, of which there are two types, have preformed holes in their inner webs (or partly formed holes, ready to be knocked out, if required) to accommodate service cables, etc., that might need to run through the partition; if the cable-holes are far apart – as they seem to be in the webs of some manufacturers' studs, the partition erector should try to keep them approximately in line, if possible – and, if used, the holes should be sleeved with rubber grommets.

The metal components for these partitions – because of their sectional shapes being likened to capital letters – are referred to as *U-tracks* for the *head-* and *floor-plates*, *H-* (or *I*) *studs* for the more common *intermediate studs*, and *C-studs*. The latter are used as *wall studs* and *door studs* – and as *end studs* when partitions are designed to change direction – thereby creating right-angled external corners.

14.10.2 Sectional Sizes and Lengths

Figures 14.20(a)(b)(c)(d): The sectional size of the metal studs and tracks that determines the stud-wall's thickness (without plasterboard cladding) seems to vary slightly between different manufacturers, but by only a few millimetres. My research showed five different wall-thickness sizes from one manufacturer, ranging from 50 mm to 146 mm, with sizes of 60 mm, 70 mm and 92 mm in between, to suit various heights and types of required partitioning. However, 70 mm and 92 mm seem to be the most likely sizes for the storey heights in domestic dwellings – and these sizes are available in lengths of 2.4 m, 2.7 m, 3.0 m, 3.6 m, 4.2 m and 5.0 m.

Special Tools Required

Apart from a pair of *leather work-gloves* (conforming to EN 388 Mechanical Hazards' Standards), as an essential requirement for handling and working with the dangerous cut-and-burred edges of these metal components, you will also require a large pair of *Straight-cut aviation metal-snips* (commonly referred to as *tin snips*) to cut the lengths of track and studs to size on site – and a *crimping tool*, used to speedily fix together two overlapping metal components. Alternatively, these metal-on-metal joints can also be

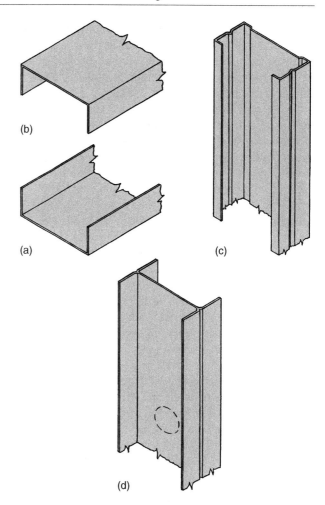

Figure 14.20 Isometric portions of: **(a)** metal U-track as a floor plate; **(b)** metal U-track as a head plate; **(c)** metal C-stud; and **(d)** metal H- (or I) stud

fixed with 13 mm wafer-head, self-tapping screws – as described further on. If you are the plasterboard-fixer, as well as the partition-erector, you might also need a cordless *Drywall Screwdriver*, with a screw-nose cone and a collated screw magazine for the bugle-headed screws.

14.10.3 Setting Out and Fixing

The setting-out procedure for constructing metal-stud partitions is basically the same as for timber-stud partitions. First, by careful measurement, its parallel position is pencilled on the floor at one end of its length and then the other. Next, a chalk line is pulled taut over these marks and snapped onto the floor. Alternatively, two overlapped lengths of metal U-track can be held (or clamped) together to act as a straight-edge – after sliding and extending them to relate to the pencil marks at each end – and intermittent or continuous pencil lines can be made on the floor. If there is to be a doorway in the partition, its position should now be marked on the floor, at right-angles to the partition-line. The opening should equal the door-width, plus twice the timber door-lining's thickness, plus a fitting tolerance of, say, 4 mm.

Next, the U-track floor plate is cut to length on each side of the marked doorway-opening. This is done by transferring the doorway's pencil mark up from the floor to the side of the U-track, squaring the mark across the underside, snipping the U-track sides down to the underside mark, then by bending the track-base downwards from the vertical cuts to allow the tin snips to complete the square-cut across the track. Each piece can now be screwed to the floor – and for boarded or chipboard floors, or similar decking material, the recommended fixings (to avoid the risk of piercing unseen, underfloor service pipes) are 13 mm wafer-head screws at 600 mm centres, but not more than 300 mm from the wall-ends.

When U-track sections are over 70 mm wide, the screw fixings should be staggered – and on each side of doorway openings (on all widths of track) more screws are required to offset the cantilevered weight and the repetitive opening-and-closing action of the door. So, at each doorway-end of the U-track floor plates, there should be four screw-fixings, two side-by-side (across the track) as close as possible to the door stud and two side-by-side at not more than 150 mm from the first pair of screws.

When fixing into sand-and-cement screeded and/or concrete floors, nylon-sleeved Frame-fix or Hammer-fix screws could be used – and (on this type of floor only) at least one manufacturer recommends that a strip of DPC (damp-proof course) material should be laid under the U-track floor plate. Although, I fail to see the need for this if it is known (or can be assumed) that the floor was constructed with a sandwiched DPM (damp-proof membrane) over its superficial surface.

If any floor upon which a metal-stud partition is to be built is found to be uneven and wavy, it is recommended that a timber floor-plate should be fixed below the metal U-track.

Figures 14.21(a)(b)(c)(d): To position the U-track head plate, the floor plate line can be transferred to the ceiling from marking plumb-lines on the walls at each end and snapping a chalk line onto the ceiling – or this line could be established with a simple laser level. Alternatively, one of the C-shaped wall-studs can be cut to size – after measuring its height against one of the abutting walls, from within the fixed U-track floor plate to the underside of the ceiling; and then by deducting, say, 4 or 5 mm for tolerance. Once cut to length, it should be tried into position at each end of the U-track and, if it fits, more C-studs can be

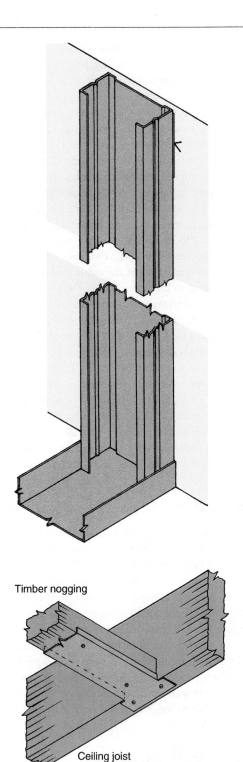

Timber nogging

Ceiling joist

Figure 14.21 **(a)** With the C-stud held in position, the wall is marked at the top on each side, after plumbing the stud laterally with a spirit level; **(b)** As illustrated, timber noggings, optionally supported by metal Fixing Channel with shallow, upturned edges (to give it strength), can be fixed between the ceiling joists. Note that the 0.7 mm thick galvanized steel has to be edge-snipped and flattened where it is to be fixed to the undersides of the timber joists, then fixed as indicated, with wafer-head screws.

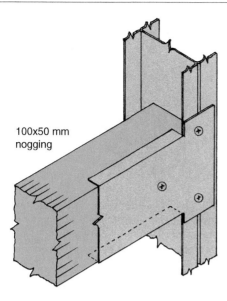

100x50 mm
nogging

Figure 14.21(c) Horizontal noggings, fitted between H(I)-studs, with optional metal Fixing Channel, as described at (b), fitted and fixed to the flanges of the studs with 13 mm wafer-head self-tapping screws, at not-more-than 600 mm centres.

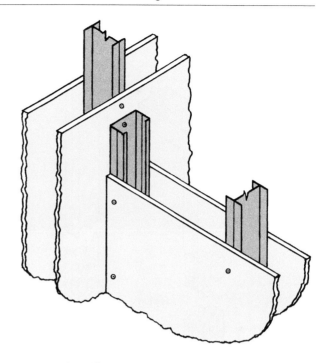

Figure 14.21(d) A pre-positioned H(I)-stud within a partition, ready to receive 25 mm screw-fixings at maximum 600 mm centres from the C-stud being fixed against the plasterboard at a tee-junction

marked and cut from it and set aside. Now fit one of the C-studs into the U-track channel at one end, up against the wall, as illustrated in Figure 14.21(a), and plumb it up with a spirit level before pencil-marking each side of its top position on the wall. Then repeat this procedure at the opposite end of the U-track floor plate. And between these short-length, double pencil-lines at the top of each opposite wall (depicting the position of the head plate), a pre-cut length of inverted U-track can now be propped (or held) up and fixed to the ceiling. The fixings can be with 38 mm screws into timber ceiling joists (if the joists are at right-angles to the partition), or, if the ceiling joists run parallel to the partition (and not directly above it), pre-fixed timber noggings will be required, as illustrated in Figure 14.21(b). And they can either be end-nailed to the joists (as in Figure 14.8(a)), or, as shown here, they can be screwed to pre-cut pieces of galvanized steel *Fixing Channel*, after the shallow, upturned edges have been snipped and flattened/hammered out, as indicated. Next, fit and fix each C-shaped wall-stud with screw fixings at approx. 600 mm centres, between the extreme screw fixings first-placed nearer the top and the bottom.

Note that if the C-shaped wall-studs are being fixed to the plasterboard surface of another metal-stud partition, similar horizontal timber noggings, or metal Fixing Channel, backed up with timber noggings, will be required, as illustrated in Figure 14.21(c), unless a pre-positioned H(I)-stud has been carefully positioned at this tee-junction, as illustrated in Figure 14.21(d).

14.10.4 Completing the Doorway Opening

With the U-track floor plates on each side of the opening, the U-track head plate and the C-studs against each wall now in position and fixed, the doorway opening can be completed. There are three ways of doing this, as detailed in Figures 14.22(a) to (d).

Figure 14.22(a): The first method is to use two pre-cut C-studs (already cut to the first wall-studs' length) as subframes for the timber door-lining. But they need an inner core, so full-length snug-fitting timber-inserts are slid into the confines of the C-shapes. Then these two infilled C-studs are placed into the downturned side-edges of the U-track head plate whilst being slid into the upturned side-edges of the U-track floor plate on each side of the doorway. The C-studs are plumbed with a spirit level and fixed at the base and the top, as illustrated, with 13 mm wafer-head screws on each side. Note that when the timber door-lining is eventually fitted and screwed into the opening, the screw-fixings must be long enough to achieve good fixings (after pre-drilling through the C-stud metal) into the two timber inserts.

Figure 14.22(b): The second method still uses two pre-cut C-studs, as before, placed in similar positions in the U-track head- and floor-plates, but, as illustrated, they are set in from the ends of the U-track

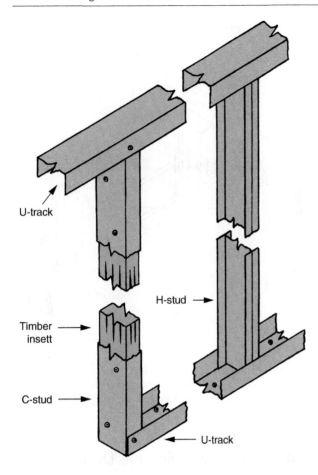

Figure 14.22(a) C-studs with snug-fitting timber inserts create subframe-fixings for the timber door-lining

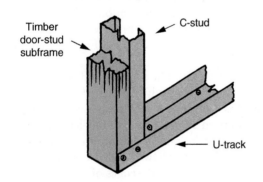

Figure 14.22(b) C-studs set back in the U-track and fixed to timber studs that act as subframe-uprights for fixing the timber door-lining

floor plate to allow timber studs (equal to the width of the C-studs) to act as subframe uprights for the eventual fixing of the timber door-lining. The subframe uprights must be screwed to the C-studs, or vice versa.

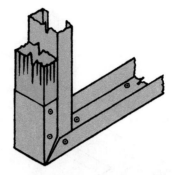

Figure 14.22(c) The U-track floor plates are mitre-snipped at their sides and turned up to be fixed to the timber-stud subframe, or to the alternative infilled C-stud. This arrangement strengthens the opening at floor-level

Figure 14.22(c): The third method also uses two pre-cut C-studs – and these either require snug-fitting timber inserts, as described at 14.22(a), or timber-stud facings, as described at 14.22(b). The only difference with this third variation, is to do with the U-track floor-plate's connection to the C-stud (or timber-stud) on each side of the door opening. Basically, the U-track is initially allowed to run over the doorway-opening floor-marks on each side by 150 to 200 mm, then the doorway-opening marks on each U-track are squared across the U-track's underside – and the sides of the U-track (in relation to the squared underside marks) are snipped at approx. 45° to allow the underside of the U-track to be bent up to form right-angled uprights. As illustrated, the short U-track uprights are fixed to the doorway C-studs or the doorway timber studs, depending on which method is used.

Figure 14.22(d): Whichever of the three methods is used, the next step to complete the door-opening is to pre-form a section of U-track to form a doorway-head. To do this, a section of U-track is cut 300 mm longer than the basic door-opening width already formed. When cut, lay it against the foot of the opening, so that 150 mm runs past on each side – and mark each U-track side with a pencil. Square these marks around the three sides of the U-track and snip them down on each side at each end. Bend the short *legs* down to form right-angles and position/slide the two-legged structure in between the door-studs to form the doorway head – at the correct door-plus-door-lining-head height. Check with a spirit level before fixing the side-legs on each side with 13 mm wafer-head screws. Note that, as illustrated, U-track door-heads usually require one short-length H-(I) stud to be in an unfixed position – ready to accommodate plasterboards overlapping the doorway opening.

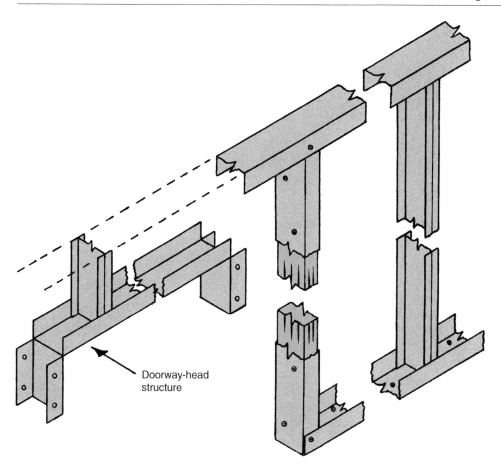

Doorway-head
structure

Figure 14.22(d) The
two-legged U-track
structure for forming the
doorway head

14.10.5 Positioning the H-(I) Studs and Fixing the Plasterboards

The intermediate H-studs are pivoted into the top and bottom U-tracks and positioned at 600 mm centres for 1200 mm wide plasterboards and 450 mm centres for 900 mm wide plasterboards. The studs upon which vertical jointing of two plasterboards occurs, are *not fixed* to the U-tracks – they are thus able to be adjusted sideways (up to the H-stud's indented centre line) to achieve a more precise central position at the plasterboard-fixing stage. But the studs in between jointed and/or end-fixed plasterboards are screwed at the top and bottom with 13 mm wafer-head screws driven through the sides of the U-tracks into the flanges of the H-studs, before the drylining stage begins.

Finally, plasterboards are usually end-cut between the floor and ceiling (after reducing the 'tight' measurement by about 15 mm) and fixed side-by-side to the H- and C-studs with 25 mm self-tapping, bugle-headed screws at approx. 300 mm centres. A joint gap of approx. 3 mm between boards is usually allowed – unless tapered-edge boards are used. When positioning the boards for fixing, they can be rested on small offcuts of plasterboard and/or levered up to the ceiling with purpose-made, double-ended timber *lifting-wedges* (also known as *rocker-wedges*), which are operated by foot.

15
Geometry for Arch Shapes

15.1 INTRODUCTION

Brick or stone arches over windows or doorways, in a variety of geometrical shapes, can only now be seen mainly on older-type buildings. Present-day design favours straight lines for various reasons, including visual simplicity, cost, and structural requirements in relation to new materials and design. Curved arches in domestic buildings have been replaced mostly by various types of lightweight, galvanized, pressed-steel lintels as shown in Figure 15.1(a).

However, arches should not be regarded as old-fashioned or obsolete and will still be required to match existing work on property maintenance, conversions and extensions. Also, some architects nowadays are using geometrical shapes in modern design.

Arches for internal beamed-openings in walls, finished in plaster, seem to be quite popular nowadays. But instead of these being formed traditionally with a structural brick-arch, the modern practice is to use a lightweight, galvanized steel *arch-former*, which is easily fitted and fixed at each end of the beam, ready for plastering.

However, whenever required, brick or stone arches are built on temporary wooden structures called centres, dealt with in the next chapter.

15.2 BASIC DEFINITIONS

Figure 15.1 (b):

- *Springing line*: a horizontal reference or datum line at the base of an arch (where the arch *springs* from).
- *Span*: the distance between the reveals (sides) of the opening.
- *Centre line*: a vertical setting-out line equal to half the span.
- *Rise*: a measurement on the centre line between springing and intrados.
- *Intrados* or *soffit*: the underside of the arch.
- *Extrados*: the topside of the arch.
- *Crown*: the highest point on the extrados.
- *Voussoirs* (pronounced *vooswars*): wedge-shaped units in the arch.

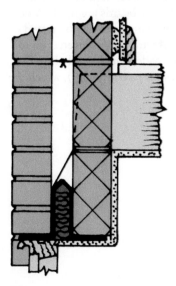 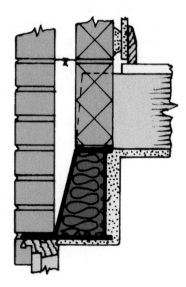

(a)

Figure 15.1 (a) Modern steel lintels

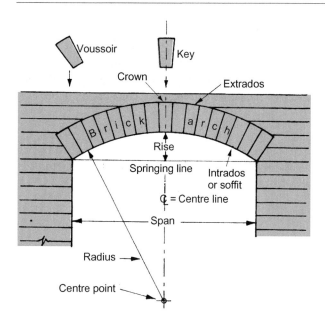

Figure 15.1 (b) Basic definitions

- *Key*: the central voussoir at the crown (i.e. the final insert that *locks* the arch structurally).
- *Centre*: the pivoting or compass point of the radius.
- *Radius*: the geometrical distance of the centre point from the concave of a segment or circle.

15.3 BASIC TECHNIQUES

Before proceeding, a few basic techniques in geometry must be understood.

15.3.1 Bisecting a Line

This means dividing a line, or distance between two points, equally into two parts by another line intersecting at right-angles. Figure 15.2(a) illustrates the

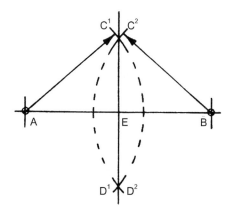

Figure 15.2 (a) Bisecting a line

method used. Line AB has been bisected. Using A as centre, set the compass to any distance greater than half AB. Strike arcs AC^1 and AD^1. Now using B as centre and the same compass setting, strike arcs BC^2 and BD^2. The arcs shown as broken lines are only used to clarify the method of bisection and need not normally be shown. Draw a line through the intersecting arcs C^1C^2 to D^1D^2. This will cut AB at E into two equal parts. Angles C^1EA, BEC^2, AED^1 and D^2EB will also be right-angles (90°).

15.3.2 Bisecting an Angle

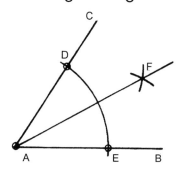

Figure 15.2 (b) Bisecting an angle

This means cutting or dividing the angle equally into two angles. Figure 15.2(b) shows angle CAB. With A as centre and any radius less than AC, strike arc DE. With D and E as centres and a radius greater than half DE, strike intersecting arcs at F. Join AF to divide the angle CAB into equal parts, CAF and FAB.

15.3.3 Semi-circular Arch

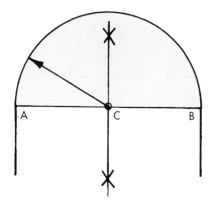

Figure 15.3 Semi-circular arch

Figure 15.3: Span AB is bisected to give C on the springing line. With C as centre, describe the semicircle from A to B.

15.3.4 Segmental Arch

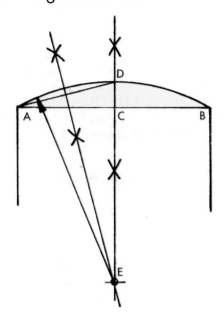

Figure 15.4 Segmental arch

Figure 15.4: Span AB is bisected to give C. The rise at D on the centre line can be at any distance from C, but less than half the span. Bisect the imaginary line AD to intersect with the centre line at E. With E as centre, describe the segment from A, through D to B.

15.3.5 Definition of Geometrical Shapes

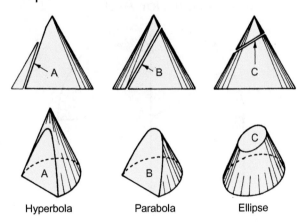

Figure 15.5 (a) Definition of geometrical shapes

Illustrated in Figure 15.5(a) is an explanation for some of the other arch shapes to follow.

Hyperbola

This is the name given to the curve A, produced when a cone is cut by a plane (a flat, imaginary sheet-surface) making a larger angle with the base than the side angle of the cone (e.g. for a 60° cone use a 70–90° cut).

Parabola

This is the name given to the curve B, produced when a cone is cut by a plane parallel to its side (e.g. for a 60° cone use a 60° cut).

Ellipse

This is the name given to the shape C, produced when a cone or cylinder is cut by a plane making a smaller angle with the base than the side angle of the cone (or a cylinder). The exception is that when the cutting plane is parallel to the base, true circles will be produced.

Axes of the Ellipse

ABCD = ellipse axes
ABC = semi-ellipse axes

AE or EB = semi-major axis
CE or ED = semi-minor axis

Figure 15.5 (b) Axes of the ellipse

Figure 15.5(b): An imaginary line through the base and top of a cone or cylinder, that cuts exactly through the centre, is known as an axis. The shape around the axis (centre) is equal in any direction, but when cut by an angled plane – to form an ellipse – the shape enlarges in one direction, according to the angle of cut. For reference, the long and the short lines that intersect through the centre, are called the major axis and the minor axis.

The axes on each side of the central intersection, by virtue of being halved, are called semi-major and

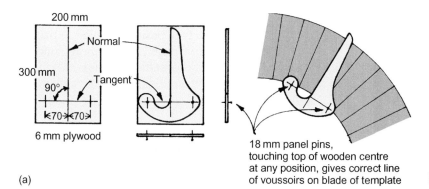

Figure 15.6 (a) Tangent-template

200 mm

←Normal

300 mm

90° Tangent

←70→←70→

6 mm plywood

(a)

18 mm panel pins,
touching top of wooden centre
at any position, gives correct line
of voussoirs on blade of template

semi-minor axes. The semi-elliptical arch is so called because only half of the ellipse is used.

15.4 TRUE SEMI-ELLIPTICAL ARCHES

True semi-elliptical shapes are not normally used for brick arches, as the methods of setting out do not give the bricklayer the necessary centre points as a reference to the radiating geometrical-normals of the voussoir joints. However, the problem could be solved by using a simple, purpose-made tangent-template, as shown in Figure 15.6(a). Therefore, the *true* semi-elliptical arch methods shown here might only serve to build a complete knowledge of the subject being covered, leading on to the methods favoured by bricklayers.

15.4.1 Intersecting-lines Method

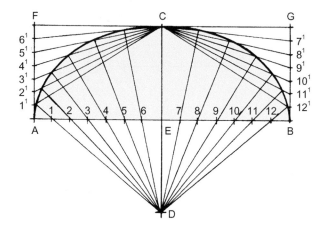

Figure 15.6 (b) Intersecting-lines method

Figure 15.6(b): Span AB, given as the major axis, is bisected at E to produce CD, a lesser amount than AB, given as the minor axis. Vertical lines from AB and

horizontal lines from C are drawn to form the rectangle AFGB. Lines AF, GB, AE and EB are divided by an equal, convenient number of parts. Radiating lines are drawn from C to $1^1, 2^1, 3^1$, and so on to 12^1; and from D, through divisions 1 to 12 on the major axis, to intersect with their corresponding radial. These are radials 1 to 1^1, 2 to 2^1, 3 to 3^1, and so on. The intersections plot the path of the semi-ellipse to be drawn freehand or by other means, such as with the aid of a rubber flexi-curve or with plastic French curves.

15.4.2 Intersecting-arcs Method

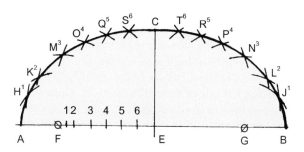

Figure 15.7 Intersecting-arcs method

Figure 15.7: Draw the major axis AB and the semi-minor axis CE as before. With compass set to AE or EB, and C as centre, strike arcs F and G on the major axis; these are known as the *focal points*. Mark a number of points anywhere on the semi-major axis between F and E; place the first point very close to F. Number these points 1,2,3, etc. Now with the compass set to A1, strike arcs H^1 from F, and J^1 from G. Reset compass to B1, strike arcs H^1 from G, and J^1 from F. Continue as follows:

compass A2, strike K^2 from F, L^2 from G
compass B2, strike K^2 from G, L^2 from F
compass A3, strike M^3 from F, N^3 from G
compass B3, strike M^3 from G, N^3 from F
compass A4, strike O^4 from F, P^4 from G

compass B4, strike O⁴ from G, P⁴ from F
compass A5, strike Q⁵ from F, R⁵ from G
compass B5, strike Q⁵ from G, R⁵ from F
compass A6, strike S⁶ from F, T⁶ from G
compass B6, strike S⁶ from G, T⁶ from F

These arcs plot the path of the semi-ellipse to be completed as before.

15.4.3 Concentric-circles Method

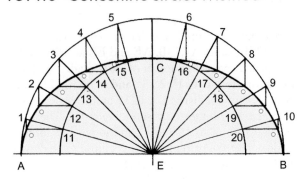

Figure 15.8 Concentric-circles method

Figure 15.8: Draw the major axis AB and the semi-minor axis CE as before. Strike semi-circles radius EA and EC. Draw any number of radiating lines from E to cut both semi-circles. For convenience, the angles of the radials used here are 15°, 30°, 45°, 60° and 75°, each side of the centre line CE. Draw vertical lines inwards from points 1, 2, 3, etc. on the outer semi-circle, and horizontal lines outwards from points 11, 12, 13, etc. on the inner semi-circle. These intersect at points °, which plot the path of the semi-ellipse to be completed as before.

15.4.4 Short-trammel Method

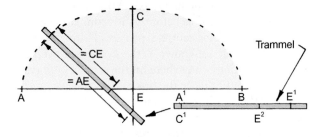

Figure 15.9 Short-trammel method

Figure 15.9: Draw the major and semi-minor axes as before. Select a thin lath or strip of hardboard, etc., as a trammel rod. Mark it as shown, with the semi-major

axis A¹E¹ and the semi-minor axis C¹E². Rotate the trammel in a variety of positions similar to that shown, ensuring that marks E¹ and E² always touch the two axes, and mark off sufficient points at A¹/C¹ to plot the path of the semi-ellipse to be completed as before.

15.4.5 Long-trammel Method

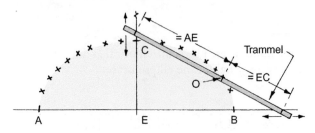

Figure 15.10 Long-trammel method

Figure 15.10: This is similar to the previous method, except that the semi-major and semi-minor axes form a continuous measurement on the trammel rod; the outer marks thereon move along the axes, while the inner mark, O, plots the path of the semi-ellipse. This method is better than the previous one when the difference in length between the two axes is only slight.

15.4.6 Pin-and-string Method

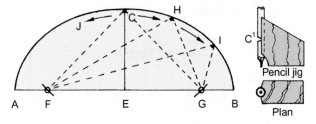

Figure 15.11 Pin-and-string method

Figure 15.11: This method uses focal points on the major axis. These are shown here as F and G, and either point equals AE or EB on the compass, struck from C to give F and G. This time, to describe the arch shape, drive nails into points F, C and G, pass a piece of string around the three nails and tie tightly. Make a pencil jig, if possible, and cut a notch in a pencil – as shown at C¹ – remove nail at C, replace with pencil and jig and rotate to left and right, as indicated at HIJ, to produce a true semi-ellipse.

15.5 APPROXIMATE SEMI-ELLIPTICAL ARCHES

These *approximate* semi-elliptical shapes, as previously mentioned, are preferred for brick or stone arches, as they eliminate the freehand flexi-curve, simplify the setting out and give the bricklayer definite centre points from which to strike lines for the radiating geometrical-normals of the voussoir joints.

15.5.1 Three-centred Method

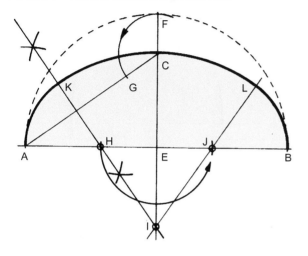

Figure 15.12 Three-centred method

Figure 15.12: Draw major axis (span) and semi-minor axis (rise) to the sizes required, as described before. Draw a diagonal line from A to C (the chosen or given rise). With centre E, describe semi-circle AB to give F. With centre C, strike an arc from F to give G. Bisect AG to give centres H and I. With centre E, transfer H to give J. Draw the line through IJ to give L. HIJ are the three centres. Draw segments AK from H, BL from J, and KL from I, to cut through the rise at C and complete the required shape.

15.5.2 Five-centred Method

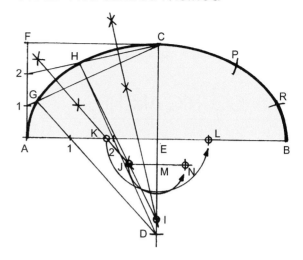

Figure 15.13 Five-centred method

Figure 15.13: Draw major and semi-minor axes as before. Draw lines AF and CF, equal to CE and AE, respectively. Divide AF by three, to give A12F. Draw radials C1 and C2. With centre E and radius EC, strike the arc at D on the centre line. Divide AE by three, to give A12E. Draw line D1 to strike C1 at G, and line D2 to strike C2 at H. Bisect HC and extend the bisecting line down to give I on the centre line. Draw a line from H to I. Now bisect GH and extend the bisecting line down to cut the springing line at K and line HI at J. IJK are the three centres to form half of the semi-ellipse. The other two centres are transferred as follows: with centre E, transfer K to give L on the springing line; draw a horizontal line from J to M and beyond; with centre M, strike an arc from J to give the centre N. To transfer the normals G and H, strike arc CP, equal to CH, and BR, equal to AG. To form the semi-ellipse, draw segments AG from centre K, GH from J, HCP from I, PR from N, and RB from L.

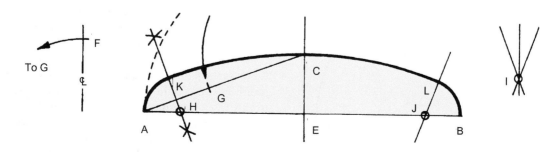

Figure 15.14 Depressed semi-elliptical arch

15.5.3 Depressed Semi-elliptical Arch

Figure 15.14: This arch uses a very small rise. The geometry is exactly the same as that used for the three-centred method explained in Figure 15.12.

15.6 GOTHIC ARCHES

15.6.1 Equilateral Gothic Arch

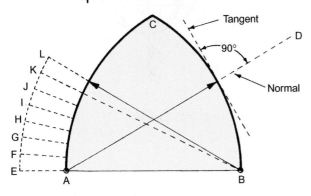

Figure 15.15 Equilateral Gothic arch

Figure 15.15: The radius of this arch, equal to the span, is struck from centres A and B to a point C. The line AD highlights a geometrical normal to the curve and a line at right-angles to this, as shown, is known as a tangent. Normals E, F, G, H, I, J, K, L, etc. are indicated by broken lines to form the voussoirs of the arch. Incidentally, points A, B and C of this arch, if joined by lines instead of curved arcs, form an equilateral triangle, where all three sides are equal in length and contain three angles of 60°.

15.6.2 Depressed Gothic Arch

Figure 15.16: This arch is sometimes referred to as an *obtuse* or *drop* Gothic arch. The centres for striking

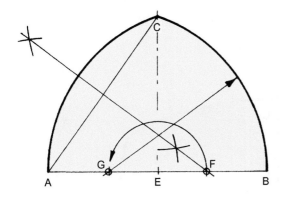

Figure 15.16 Depressed Gothic arch

this arch come within the span, on the springing line. Bisect the span AB to give the centre line through E. With compass less than AB, strike the rise at C from A. Alternatively, mark the chosen or given rise at C from E. Draw line AC and bisect to give centre F on the springing line. With centre E, transfer F to give centre G. Strike segments AC from F and BC from G.

15.6.3 Lancet Gothic Arch

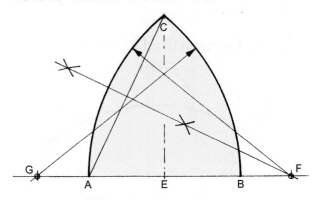

Figure 15.17 Lancet Gothic arch

Figure 15.17: The centres for this type of arch are outside the span, on an extended springing line. Bisect the span AB to give the centre line through E. With compass more than AB, strike the rise at C from A. Alternatively, mark the chosen or given rise at C from E. Draw line AC and bisect to give centre F on the extended springing line. With centre E, transfer F to give centre G. Strike segments AC from F and BC from G.

Note that line AC in the above is optional and need not actually be drawn.

15.7 TUDOR ARCHES

15.7.1 Tudor Arch – Variable Method

Figure 15.18: This method is best and can be used to meet a variety of given or chosen rises. The geometry is usually mastered when practised a few times.

Draw the span AB and bisect it to give an extended centre line through E and down. Mark the rise at C. Draw vertical line AF, equal to two-thirds rise (CE). Join F to C. At right-angles to FC, draw a line down from C. With compass equal to AF, and A as centre, transfer F to give G. With the same compass setting, mark H from C on line CI. Draw line from G to H and bisect; extend bisecting line down until it intersects with line CI to give centre I. Draw line from I,

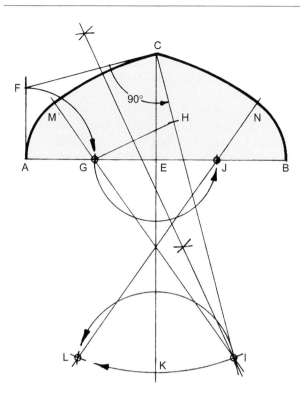

Figure 15.18 Tudor arch – variable method

15.7.2 Tudor Arch – Fixed Method

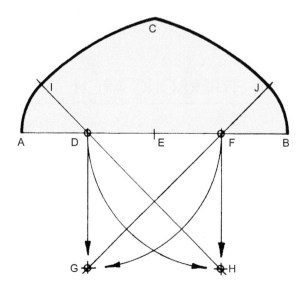

Figure 15.19 Tudor arch – fixed method

Figure 15.19: This method is simpler and can be used when the rise is not critical and the only information given or known is the span.

Draw span AB and divide by four to give DEF. Draw vertical lines down from D and F. With D as centre, transfer F to intersect the vertical line, giving G. With F as centre, transfer D to intersect the other vertical line, giving H. Draw diagonals from H and G, extending through D and F on the springing line. To complete, strike segments AI from D, IC from H, BJ from F and JC from G.

15.7.3 Depressed Tudor Arch

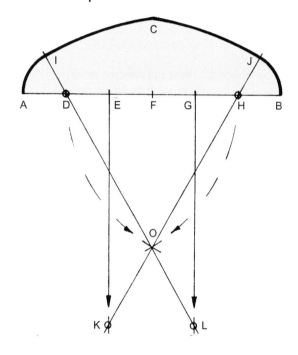

Figure 15.20 Depressed Tudor arch

Figure 15.20: Draw span AB and divide by six to give DEFGH. Draw vertical lines down from E and G. With centre D, transfer H down to O, and with centre H, transfer D down to O. Draw diagonal normals through DO and HO, extending down to intersect the vertical lines at K and L, and extending up past the springing line to I and J. To complete the arch, strike segments AI from D, IC from L, BJ from H and JC from K.

Note that division of the span can be varied to achieve a different visual effect, as can the angles of the normals at D and H, drawn here at 60°. For example, 75° would make the arch more depressed.

15.7.4 Straight-top Tudor Arch

Figure 15.21: Draw span AB and divide by 9. Mark one-ninth of span from A to give D, and one-ninth

extended through G on the springing line. With E as centre, transfer G to give J on the springing line. Again with E as centre, transfer I, through K, to strike arc at L. With K as centre, transfer I to give centre L. Draw line from L to extend through J on the springing line. To complete, strike segments AM from G, MC from I, CN from L and NB from J.

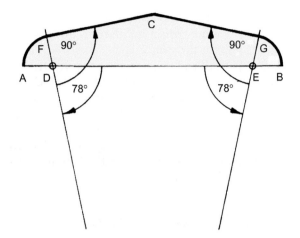

Figure 15.21 Straight-top Tudor arch

from B to give E. With a protractor, or a roofing-square containing a degree facility, set up diagonal normals passing through D and E at 78° to the horizontal. With centre D, strike arch curve AF, and with centre E, strike BG. From F and G, draw straight crown lines at 90° to the two normals, to intersect at key-position C.

Note that positions of the centres D and E can be varied to achieve a different visual effect, as can the angles of the normals at D and E, drawn here at 78°; however, the arch top or crown must always be tangential (at 90°) to the normals.

15.8 PARABOLIC ARCHES

15.8.1 Triangle Method

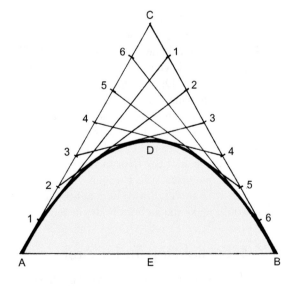

Figure 15.22 Parabolic arch – triangle method

Figure 15.22: Draw AB equal to the span, and vertical height EC equal to twice the rise. Join AC and CB to form a triangle. Divide each side of the triangle by a convenient number of equal divisions. This illustration uses seven divisions each side, numbered A, 1, 2, 3, 4, 5, 6, C, and C, 1, 2, 3, 4, 5, 6, B. Join 1 on AC to 1 on CB, 2 on AC to 2 on CB, and so on. These lines are tangential to the curve and give the outline shape of the parabola to be drawn freehand or by other means.

15.8.2 Intersecting-lines Method

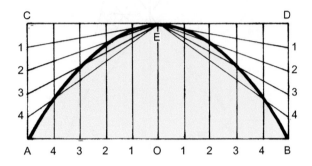

Figure 15.23 Parabolic arch – intersecting lines method

Figure 15.23: Draw a rectangle ABDC in which AB equals the span and AC equals the rise. Bisect AB to give vertical line EO. Divide OA into any number of equal parts, and OB, CA and DB, into the same number of equal parts as OA. Draw vertical lines up from horizontal divisions, and radiating lines from E to vertical divisions. The intersections thus formed, being base-vertical 1 (bv1) intersecting side-radial 1 (sr1), bv2 intersecting sr2, etc., on each side of the centre line, gives the outline shape of the parabola, to be completed as before.

15.9 HYPERBOLIC ARCH

15.9.1 Intersecting-lines Method

Figure 15.24: As before, draw a rectangle ABDC in which AB equals the span and AC equals the rise. Bisect AB to give vertical line EFO. Make apex E from F equal to half the rise (EF = FO/2). Divide OA, OB, CA and DB, as before and draw intersecting radials thus: base-radial 1 (br1) intersecting side-radial 1 (sr1), br2 intersecting sr2, etc., on each side of the centre line. The intersections give the outline shape of the hyperbola, to be completed as before.

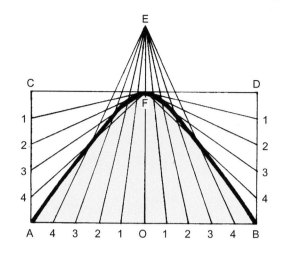

Figure 15.24 Hyperbolic arch

Note that, as with true semi-elliptical shapes (Figures 15.6 to 15.11), parabolic and hyperbolic arches, if constructed of brick or stone, requiring centres and normals for alignment of joints, present the same problems. To overcome this, the tangent-template (see Figure 15.6(a)) could be used.

16

Making and Fixing
Arch Centres

16.1 INTRODUCTION

The temporary wooden structures upon which brick arches are formed, are known as *centres*. They can be made in the joinery shop – taking advantage of a greater variety of machinery – or on site. The construction of the centre can be simple or complex, depending mainly on two factors: the span of the opening, and how many times the centre is to be used for other arch constructions.

16.1.1 Simple or Complex

For small spans up to about 1.2 m, the centre can be simple, of single-rib, twin-rib or four-rib construction. For spans exceeding 1.2 m, the centre becomes more complex, of multi-rib construction.

16.1.2 Practical, Working Compass

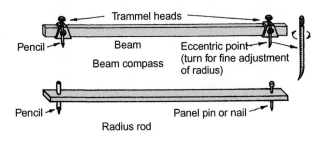

Figure 16.1 Radius rod and beam compass

Figure 16.1: A beam compass or a radius rod is required to set out the full-size shape of the centre, either directly onto the rib material (single and twin-rib) or onto a hardboard or similar setting-out board (four-rib and multi-rib centres), from which a template is made of the common rib shape. The beam compass consists of a pair of trammel heads and a length of timber, say of 38 × 19 mm section, known as a beam. To improvise,

a radius rod can be easily made, consisting of a timber lath with a panel pin or nail through one end, the other end drilled to hold a pencil firmly.

16.2 SOLID TURNING PIECE (SINGLE-RIB)

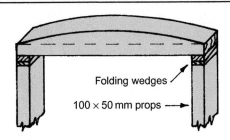

Figure 16.2 (a) Turning piece

Figure 16.2(a): Solid turning pieces are used for segmental arches with small rises up to about 75 mm. If the rise is too slight (say 10 mm rise, 900 mm span), then a beam compass or radius rod would not be practical for drawing the curve and a triangular trammel frame or trammel rod should be used. As illustrated in Figure 16.2(b), to make the trammel rod or frame, mark the required span AB and rise CD on the rib material, place a board against CB and mark and cut line A^1C^1. As shown separately, make a sawcut at C^1 to take a pencil. To mark semi-segment CA, position pencil at C^1 and push the trammel against the protruding nails at A and B, moving to the left. Reverse the trammel and move to the right to mark semi-segment CB.

16.2.1 Cutting the Segment

Figure 16.2(c): Ideally, the curve for the solid turning-piece is best cut with a narrow band-saw machine, or, if on site, with a jigsaw. Alternatively, as shown in

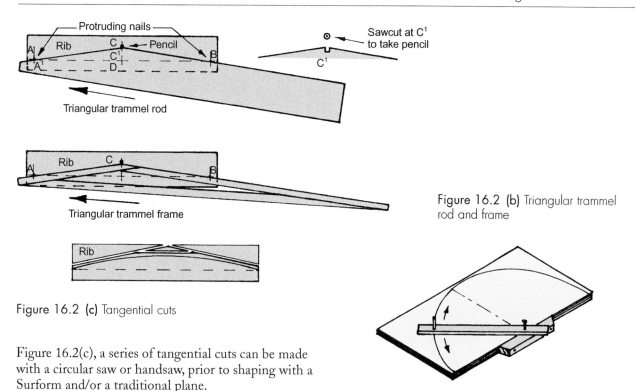

Figure 16.2 (b) Triangular trammel rod and frame

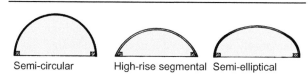

Figure 16.2 (c) Tangential cuts

Figure 16.2(c), a series of tangential cuts can be made with a circular saw or handsaw, prior to shaping with a Surform and/or a traditional plane.

16.3 SINGLE-RIB CENTRES

Figure 16.3 (a) Single-rib centres

Figure 16.3(a): These centres follow an unconventional method of construction, but are surprisingly strong and effective. Their strength is dependent upon the stress of compression achieved in bending the hardboard or plywood skin over the curved rib. For this reason, they are more suitable for semi-circular, high-rise segmental and semi-elliptical centres – in that order of diminishing suitability – and less suitable for Gothic, Tudor and low-rise segmental centres.

16.3.1 Marking and Cutting the Rib

Figure 16.3(b): The plywood rib, preferably of 18 mm thickness, is set out along the springing line to the span, minus the skin thickness each side, and marked with a radius rod (or beam compass), as shown. The shape is best cut – as before – by band saw, a corded or cordless jigsaw, or a corded or cordless reciprocating saw, finishing with a flat spokeshave, Surform and/or a smoothing plane.

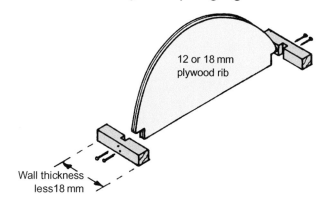

Figure 16.3 (b) Setting out

16.3.2 Adding the Springing Blocks

Figure 16.3 (c) Adding the springing blocks

Figures 16.3(c)–(e): Two 50 × 50 mm blocks, cut in length to the brick size of the arch, minus, say, 18 mm inset-tolerance, are half-housed centrally to fit housings cut into the springing points of the rib on each side and fixed with 63 mm oval, lost-head nails or countersunk screws (Figures 16.3(c) and (d)). The skin, of 3.5–4 mm hardboard or plywood, long enough to cover the curved shape of the rib and as wide as the block-length, must have the rib thickness pencil-gauged through the centre (Figure 16.3(e)).

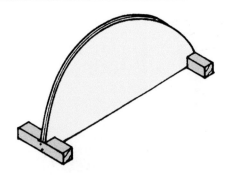

Figure 16.3 (d) Springing blocks in place

16.3.3 Fixing the Skin

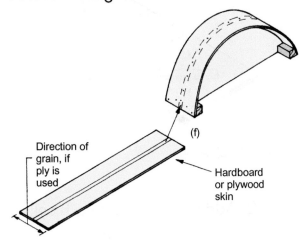

Direction of
grain, if
ply is
used

Hardboard
or plywood
skin

(f)

Figure 16.3 (e) Skin; (f) Skin in place

Figure 16.3(f): When finally fixing the skin to the springing blocks and the rib, using 25 mm panel pins or 32 mm small-headed clout nails, it is important to ensure that the skin is taut and is centred on the rib, following the gauge lines, as indicated. Note that the length of skin can be calculated, or measured directly from the rib shape by encircling the finished curve with a tape rule.

16.4 TWIN-RIB CENTRES

Figure 16.4(a): These centres are superior to the single-rib type and are suitable for all arch shapes. The ribs may consist of 12–18 mm plywood – or other sheet

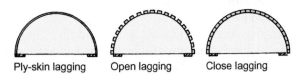

Ply-skin lagging   Open lagging   Close lagging

Figure 16.4 (a) Twin-rib centres

material like chipboard or Sterling board, if the arch is not being reused too many times. Bearers, of 75 × 25 mm or 100 × 25 mm section, are fixed to the underside of the centre, at each end of the springing line. These act as spacers between the two ribs and as bearers supporting the centre on the props.

16.4.1 Bracing-spacers

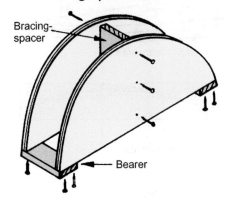

Bracing-
spacer

Bearer

Figure 16.4 (b) Bracing-spacer board and bearers

Figure 16.4(b): A bracing-spacer board, equal to the inside width of the centre, as indicated, can be inserted to keep the ribs initially parallel and square. The centre is then covered with a hardboard or plywood skin or, alternatively, with strips of timber known as *laggings*, illustrated in Figures 16.4(c) and 16.4(d). These can be placed close together, referred to as *close lagging*, or be spaced apart and referred to as *open lagging*, as indicated in Figures 16.4(d) and (e). Close lagging should be used

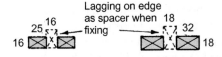

Lagging on edge
as spacer when   18
25 16   32
       fixing
16   18

Figure 16.4 (c) Lagging sizes for small spans

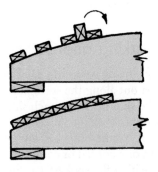

Figure 16.4 (d) Open and close lagging

Four-rib Centres   231

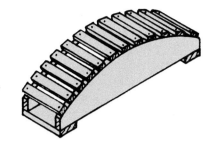

Figure 16.4 (e) Centre with open lagging

for gauged arches with tapered bricks (voussoirs), and open lagging for common arches with ordinary bricks used as parallel voussoirs.

16.4.2 Making the Centre

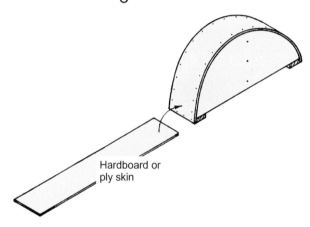

Figure 16.4 (f) Applying the skin

Figure 16.4(f): To make the centre, set out and form one rib as previously described. Use the first rib as a template to mark out and form the second rib. Cut the two 75×25 mm bearers in length to the brick size of the arch, minus 18 mm inset-tolerance, and fix to the underside of the ribs at each end with minimum 50 mm round-head or lost-head wire nails. Cut and fix the bracing-spacer (Figure 16.4(b)). Cut the hardboard or plywood skin to width (equal to length of bearers) and fix to the ribs, as indicated in Figure 16.4(f), or cut the laggings and fix to the ribs with 38 or 50 mm lost-head (or brad-head) oval nails.

16.4.3 Lagging Jig

Figure 16.4(g): This shows a plan view of a simple jig for cutting the lengths of lagging quickly. Two 50×25 mm battens are fixed to a bench, stool top or scaffold board, to form the jig. As indicated, a nail at one end positions the lagging strip without trapping

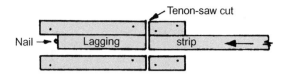

Figure 16.4 (g) Lagging jig

sawdust being pushed along from the right; the saw at the other end cuts the required length repeatedly after feeding in the strip towards the nail.

16.5 FOUR-RIB CENTRES

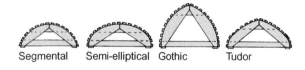

Figure 16.5 (a) Four-rib centres

Figure 16.5(a): These centres follow traditional methods of construction and are suitable for all arch shapes except low rise, which require only one or two ribs for small spans, as shown previously in Figures 16.2(a) and 16.4(e). Because of the time involved in making four-rib centres, they are less favoured than twin-rib centres using sheet material. As seen in elevation and section A–A in Figure 16.5(b), four ribs are required, two tie beams, two collars, two bearers, two optional struts, an optional brace and laggings.

16.5.1 Construction Details

Figures 16.5(b) and (c): Each pair of ribs is connected to a top collar and a bottom tie beam by clench-nails.

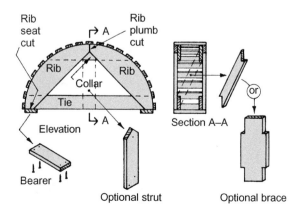

Figure 16.5 (b) Semi-circular centre

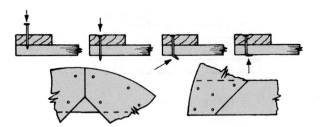

Figure 16.5 (c) Staggered clench-nailing

As indicated in Figure 16.5(c), when driven in, the nails purposely protrude by at least 6 mm and are then bent sideways and clenched over to secure the joint. For extra strength, optional struts can be fixed from the birdsmouth at the apex of the ribs, against the collar, to a central position on the face of the lower tie. The two bearers are fixed to the ends of the lower ties to line up the rib structures and act as bearers, supporting the centre when resting on the props and wedges.

16.5.2 Optional Skins

Hardboard or plywood skins can be used to cover the centre, in single or double thicknesses – or traditional laggings can be used, efficiently cut to length as previously described and fixed to the curved ribs, as before with 38 or 50 mm lost-head, or brad-head, oval nails.

Note that, if considered necessary, an optional brace, indicated in section A–A of Figure 16.5(b), can be used to achieve squareness in width and greater rigidity between the two rib structures.

16.5.3 Setting Out

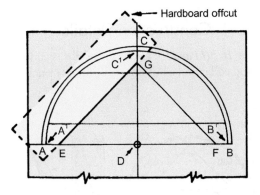

Figure 16.5 (d) The setting-out board

Figure 16.5(d): To produce the rib shape for this type of arch and also to ensure that the finished centre is true to shape, a setting-out board is required. The figure indicates the setting out for a semi-circular centre. The first line to be drawn is line AB on the

springing line, representing the span minus 2–3 mm (this reduction is a practical tolerance to allow for brick reveals being slightly out of square or irregular). Next, bisect line AB to produce the centre line CD. Then set the beam compass or radius rod to AD and describe the semi-circle ACB. Reset the compass or radius rod to AD minus the lagging thickness and describe the second semi-circle, producing A^1, C^1 and B^1, representing the rib curve. Now measure down at least 60 mm from C^1 to G, as the depth of the rib's plumb cut and measure inwards at least 60 mm from A^1 to E, and B^1 to F, being the length of the rib's seat cuts. Draw tangential lines EG and GF, representing the underside of the ribs. Draw two lines parallel to the springing, representing the top collar and bottom tie beam; these timbers should be about 100–150 mm wide × 25 mm thickness.

16.5.4 Marking and Cutting the Ribs

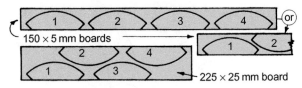

Figure 16.5 (e) Method of marking out the ribs

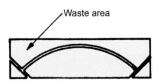

Figure 16.5 (f) Producing a hardboard template

Figures 16.5(e) and (f): The rib segments can be cut from a 150 × 25 mm or 225 × 25 mm board, as illustrated. A hardboard template, shown in Figure 16.5(f), facilitates the marking out of the ribs onto the board economically. To make the template, lay a piece of hardboard on the setting-out board, as shown by the dotted outline in Figure 16.5(d), strike the compass or rod from centre D to describe the quadrant A^1C^1, mark the plumb cut CG and the seat cut AE, then remove and cut to shape.

16.5.5 Marking the Collar and Tie

Figure 16.5(g): The collar and tie are laid in position on the setting-out board and marked with the compass or radius rod setting of A^1D, struck from centre D^1, squared up from D. As shown, a small temporary wooden block can be fixed to the tie, to retain the point of the radius rod or compass.

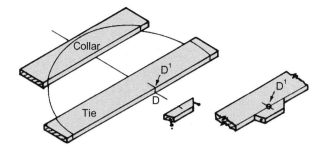

Figure 16.5 (g) Marking out the collar and tie

16.6 MULTI-RIB CENTRES

These centres also follow traditional methods of construction, simply because their multi-rib design allows a variety of configurations suitable for very wide arch-spans. They are mainly used for high-rise segmental, semi-circular and semi-elliptical arches. According to the span, they may have to be set out directly on the floor, or onto two or more setting-out boards placed side by side. The principles of setting out and producing rib templates are similar to those described for four-rib centres.

16.6.1 Semi-hexagonal Configuration

Figure 16.6(a): An elevation and sectional view of a semi-circular centre suitable for spans up to about 4 m are illustrated. The joints of each rib radiate from the centre point, comprising six sector-shapes with angles of 30° (6 × 30° = 180°). Given joint-lengths of about 60 mm, it follows that the underside of the ribs is tangential to the curve and forms two semi-hexagonal shapes. This is seen more clearly in Figure 16.6(b). For extra strength, struts should be added as shown.

16.6.2 Semi-octagonal Configuration

Figure 16.7(a): An elevation and sectional view of a semi-elliptical centre are shown (drawn by a three-centre method), suitable – as before – for spans up to about 4 m. The joint-lines radiate from the centre point, comprising eight sector shapes with angles of 22.5° (8 × 22.5° = 180°). (Alternatively, the joints and struts can radiate from the three centres of the semi-ellipse.) Measuring equal amounts of about 60 mm along each joint-line, to determine the line of the ribs, the underside of the ribs forms two distorted semi-octagonal shapes; true octagonal shapes occur with semi-circular centres, as seen more clearly in Figure 16.7(b).

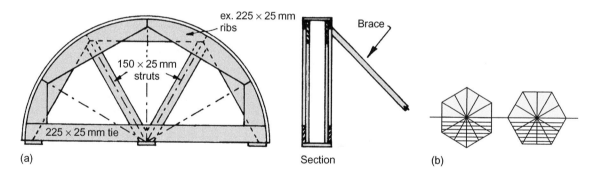

Figure 16.6 (a) Multi-rib semi-circular centre; (b) Semi-hexagons

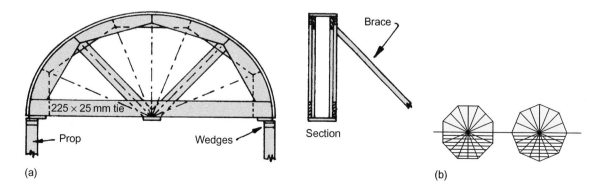

Figure 16.7 (a) Multi-rib semi-elliptical centre; (b) Semi-octagons

16.6.3 Props and Folding Wedges

Props are required to give support to the arch centre and its temporary load. The timber sizes and arrangements of these props will vary according to the size of the centre and arch to be supported. Folding wedges, under each end of the arch-centre and on top of the props, are required for three reasons. First, they can be adjusted to help set the centre's springing line and height to the correct level. Second, soon after the arch has been 'locked' by the insertion of the final key brick or voussoir – and without being left overnight – the wedges are very gently eased to drop the centre by about 3–4 mm, allowing the arch to settle without cracking. Third, when the centre is finally being removed, further easing and removal of the wedges will make the job easier and less hazardous. Of course, metal, telescopic props could be used instead of timber, but they cannot be stabilized easily with lateral strutting.

16.6.4 Prop Arrangements

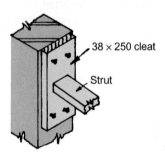

Figure 16.8 (a) Prop for small centres

Figure 16.8(a): An unconventional but effective method of temporary support for small centres is shown. Short-lengths of timber cleats are fixed on each reveal of the opening, nailed with heads protruding and strutted as shown. The strut is vital to this arrangement. Figure 16.8(b) shows a slightly modified traditional method, using 100×50 mm single props each side, with plywood gussets and 100×50 mm bearers – and another method is shown using double props each side. Both of these methods are for large spans and each should have some form of strutting.

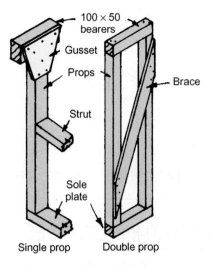

Figure 16.8 (b) Props for large centres

16.6.5 Wedge Shapes and Lagging Sizes

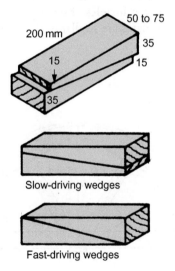

Figure 16.8 (c) Folding wedges

Figure 16.8(c): Folding wedges, as illustrated, should be 'slow-driving' (be cut with a shallow angle), as these are better for fine adjustments, non-slip bearing, and easing (slackening) prior to striking (removing) the centre. Lagging sizes for large centres are usually 25×38 mm or 25×50 mm.

17

Fixing Architraves, Skirting, Dado and Picture Rails

17.1 ARCHITRAVES

Figure 17.1(a): This is the name still used to describe the cover moulds of various designs and sizes that are fixed around the edges of door linings and door frames, etc., primarily to cover the unsatisfactory joint between the lining or frame within the opening and the plaster (or plasterboard) surface of the wall. Even though present-day architraves are usually plain in design, they add a visual finish to the opening. Architraves are referred to as being in sets. A set of architraves for a door opening consists of two uprights, called jambs (or legs) and a horizontal piece called a head. A door lining or frame normally requires two sets of architraves, one on each side of the wall.

17.1.1 Number of Sets Required

On a second-fixing operation, the architraves need to be fixed before the skirting, so that the skirting on each side of a doorway can be butted up against the

back of the architrave leg. To do the job efficiently, count the number of doorways in the dwelling and cut all of the sets required in one operation. For example, if there are five doorways, then ten sets of architraves are required, which is 10 heads and 20 legs. These are initially cut up squarely by chop saw, with an allowance in length for mitring. The length of the legs will be the door-opening height plus the width of the architrave-head plus about 30 mm allowance. The length of the heads will be the door-opening width plus the width of the architrave-leg each side plus about 40 mm allowance. Some of this allowance is required for the margins.

17.1.2 Margins

Figure 17.1(b): The architraves are always set back from the edge by a small amount, referred to as a margin. The margin is usually either 6 mm or 9 mm and whatever amount is decided upon or required, it must be consistently maintained around the opening. Inconsistent margins, often varying in size between the legs and the head and tapering from, say, 6 mm to 3 mm, are too often seen and reflect a very low standard of workmanship, spoiling the appearance of the finished job. The best way to establish a consistent margin, is to mark the required amount around the lining or frame's edge with a sharp pencil. It need not be a continuous line; broken lines of about

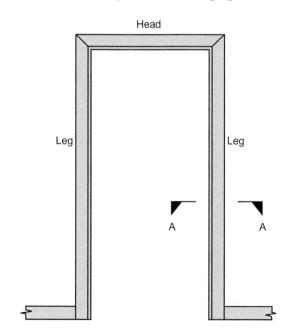

Figure 17.1 (a) Elevation of a set of architraves

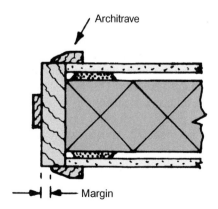

Figure 17.1 (b) Section A–A

50–150 mm length, marked at about 450 mm intervals will be enough to achieve a good margin, even if the architrave leg is 'sprung' out of shape and has to be pulled in or out before nailing.

17.1.3 Margin-template

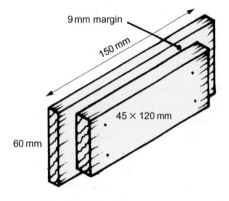

Figure 17.2 (a) A shouldered margin-template

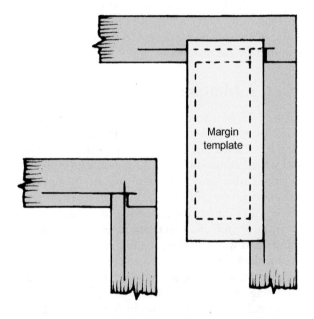

Figure 17.2 (b) Use of the margin-template

Figures 17.2(a) and (b): Measuring and marking these margin lines would be tedious and time-consuming, so they need to be gauged and this is best done with a *shouldered margin-template*. This can be made quite easily by nailing two pieces of timber together, as illustrated, with a 6 mm margin on one side and an alternative 9 mm margin on the other. The ends are also shouldered to allow the template to over-run when marking against each top corner. These over-running marks (Figure 17.2(b)) create an overlapping intersection for marking the leg and head mitres accurately. Holding the template with one hand, the pencil in the other, the template is held horizontally against the head and the corner and

marked, then slid to the mid-area and marked, then to the other corner and marked. On each side jamb, the template is held vertically against the head and marked (to create the intersection), then slid down to mark the mid-area, the bottom and intermediate positions between these three, making five marks on each side and three on the head. This method of marking should take no more than one minute per set of architraves.

17.1.4 Mitring and Fixing Technique

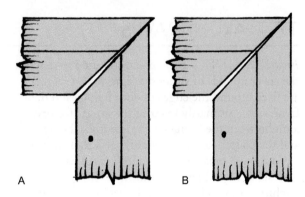

Figure 17.3 Ill-fitting mitres

Figure 17.3: Next, two sets of architraves can be placed near each doorway, ready for mitring and fixing. Starting with the first set, stand each leg in position (one at a time) and mark the inner point of the mitre from the margin-intersection point, onto the inner edge of each leg. These mitres are then cut on the waste side of the mark by chop saw and, without the need for marking, the left-hand mitre of the head is also cut. Next, the left-hand leg is fixed carefully to the margin marks, at top, bottom and centre positions, with three nails only, their heads left protruding. The left-hand mitred head is then tried in position. If the mitre is a good fit, then the head is held firmly while the right-hand mitre is marked at the intersection and then cut carefully. The right-hand leg is then fixed, again with three nails only, their heads left protruding. Next, the head is placed in position and the mitres are checked. If the appearance is as at A, then the head-mitres need easing with a plane, if as at B, then the leg-mitres need easing, which means releasing the provisionally nailed leg(s).

17.1.5 Final Fixing

If the mitres are a good fit, then the head can be fixed with three or four nails, 50–60 mm from the mitre-point at each end and one or two between. Also, the nailing of the legs can be completed with two more nails between the spacings of the nails already used, making seven nails in all on each leg, about 50–60 mm up the leg and the same distance from the mitre-point, then spaced approximately 300 mm apart. The nails used throughout are 38 mm oval nails, preferably the lost-head type, which are easier for punching in. The mitres are sometimes nailed through the top or side edge to achieve flushness on the face side and to close and hold the mitre. These mitre-fixings and the face-fixings nearest to the mitres, being six nails in all, are most likely to cause splitting and it is advisable to blunt their points to reduce this risk. This is done by holding the nail between forefinger and thumb, standing its head on a solid metal object or a concrete/brick surface and tapping the point with a hammer.

17.1.6 Mitre Saw or Mitre Box

Figure 17.4(a): Traditionally, mitres used in second-fixing operations were always cut in a purpose-made mitre box, but nowadays, as mentioned in the chapter on tools, mitre (chop) saws are quite commonly used. However, mitre boxes are still used and if one is to be made, it should be at least 600 mm long, have a solid timber base, as illustrated, of (ideally) 45 mm thickness and have plywood, MDF or Sterling board sides of 18 mm thickness. The width of the base and sides must be parallel and the base must be wide enough to take a piece of architrave lying flat with at least 12 mm tolerance. The sides must be wide enough to allow for attachment to the base with sufficient upstand to accommodate the skirting-height plus at least 10 mm tolerance. The sides are nailed to the base with 50 mm nails or screwed with 38 mm ($1\frac{1}{2}$ in) countersunk screws. The 45° mitres must be marked and cut carefully with a panel or other fine-toothed saw. Once the cut is about 10 mm deep, to avoid 'cutting blind', only let the saw run in the opposite cut while increasing the depth of the nearest cut, then turn the box around and repeat this operation as many times as necessary to control the plumbness of the cut.

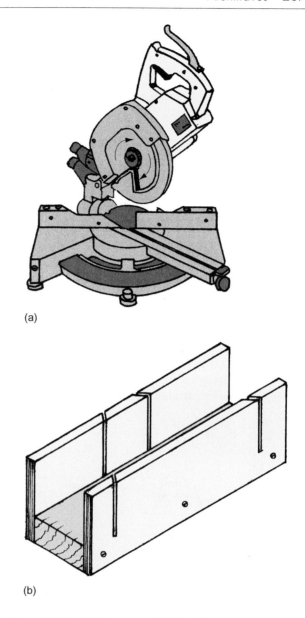

(a)

(b)

Figure 17.4 (a) Mitre saw; (b) Mitre box

17.1.7 Mitre Block

Figure 17.4(c): This is shown here for comparison with the mitre box and although architraves can be cut on it, the mitre block is more suitable for mitring smaller sections like glazing beads and quadrant moulding, cut with a tenon saw. When making one of these, more precise mitring will be achieved if, as illustrated, the solid top block is kept wide enough to give better control of the back-saw within the extended mitre cut. A piece of ex. 75 × 50 mm timber, about 450 mm long can be used, lying flat on a 150 mm wide base of 18 mm ply, MDF or Sterling board. The mitres are marked before screwing or nailing up through the base, so that the fixings can be placed strategically to avoid clashing with the 45° mitred saw cuts.

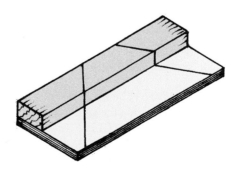

Figure 17.4 (c) Mitre block

17.1.8 Splicing

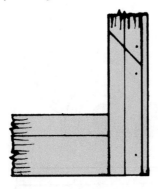

Figure 17.5 Leg-splicing

Figure 17.5: Ideally, each architrave member should be in one piece, but sometimes there is a need to use up the offcuts. The joining of two pieces is known as *splicing*, which is done at 45° across the face. Splicing should never be done on a head piece and only be done sparingly on the legs, as low as possible – well out of eye level – and the splice, as illustrated, should be cut to face downwards in the doorway. This tends to make it less obvious.

17.1.9 Scribing

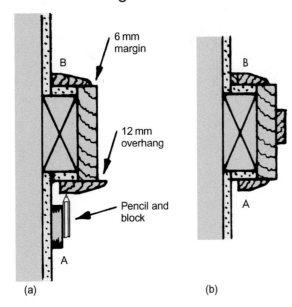

6 mm margin

B

12 mm overhang

Pencil and block

A

(a)   (b)

Figure 17.6 **(a)** Horizontal section showing scribing technique; **(b)** horizontal section showing A, gap and B, the infill

Figures 17.6(a) and (b): Where doorways are close to an adjacent wall, very often the architrave leg on that side touches the wall surface (B in Figure 17.6(a)) or, as illustrated at A in Figure 17.6(b), leaves an undesirable gap. In the first instance, the carpenter usually has to 'scribe' the architrave to the wall surface to achieve a good fit thereto and, at the same time, achieve the

required margin. The scribing technique for this is indicated at A, in Figure 17.6(a) and is described as follows:

1. With a minimum of temporary fixings, say two or three, fix the unmitred leg to the lining's edge and establish a constant overhang of any amount, but say 12 mm from the lining's face.
2. Add the margin required, say 6 mm, to the overhang-amount, making 6 + 12 = 18 mm. This is the amount to be scribed from the wall-surface. To mark the architrave-cut, most text books suggest that this can be done with a pair of dividers, which is theoretically possible, but not very precise in practice. A good practical way is to cut a small wooden block, equal in thickness to the scribe-amount (18 mm), but minus 3 mm for half the pencil's diameter, so that the pencil held on the 15 mm thick block measures 18 mm to the pencil point.
3. Holding the scribing block's edges between the second index finger and the thumb, place the forefinger on the projecting pencil and mark/scribe the architrave leg from top to bottom.
4. Release the leg from its fixings and very carefully rip down on the waste side of the line, achieving a slight undercut with the saw.
5. Try in position, then ease with a plane or Surform file, if necessary. Mark and cut the mitre.
6. Fix in position. In the case of any small, undesirable gaps between the leg and the wall's surface (as at A in Figure 17.6(b)), the standard of work would be improved if a small section of timber was glued to the back edge and the increased width of architrave was scribed as before, finishing up like B in Figure 17.6(b). Alternatively, if the wider skirting boards to be fixed next have an identical thickness and moulded edge, this can be used and scribed to the wall to avoid the time and labour of increasing the architrave width.

17.1.10 Double Architraves

Figures 17.7(a) and (b): In the case of two doorways close together, separated by a partition wall, the architraves often come very close to each other on the side of the double doorway and again, small, undesirable gaps can result, indicated in Figure 17.7(a). A small

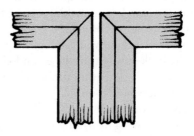

Figure 17.7 **(a)** Close doorways

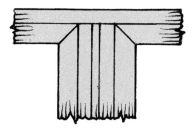

Figure 17.7 (b) Double architrave

section of timber can be glued in the gap, or as in Figure 17.7(b), a double architrave can be produced or built-up and glued together on site. This can be scribed or partly mitred into a double door-head placed across both doorways.

17.1.11 Storey-frame Architraves

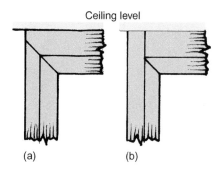

Ceiling level

(a)   (b)

Figure 17.8 (a) End-grain exposed; (b) end-grain concealed

Figure 17.8: Architrave legs that run up from floor to ceiling are required in the case of storey frames or linings. These are doorways that incorporate a glazed 'light' above the door transom. Very often, the architrave-head is required to be deeper to fit the ceiling without violating the head-margin. This means that the mitred legs will show unacceptable end-grain (Figure 17.8(a)), although the wider head can be partly mitred into the legs (Figure 17.8(b)).

17.1.12 Plinth Blocks

Figure 17.9(a): The need for these may only now be found on refurbishment works. Plinth blocks were used when deep, built-up skirtings, often involving one or two stepped face-boards, could not be mastered at the doorway by the relatively thinner architrave. Plinth blocks, therefore, are to accommodate the side abutment of any skirting member which would otherwise be unsightly in sticking out past the face of the architrave leg. The skirting was housed about 6 mm into the plinth-block side to offset any shrinkage across the block, and the base of the architrave leg

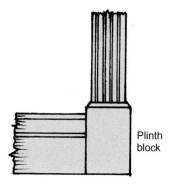

Plinth block

Figure 17.9 (a) Use of plinth blocks

was half-lap jointed and screwed into the back of the block, to make the block an integral part of the leg.

17.1.13 Cornice Blocks

Figure 17.9 (b) Use of cornice blocks

Figure 17.9(b): As with plinth blocks, cornices are not found in modern buildings. They were at one time used as an alternative to the mitred corners of a doorway, especially on higher class work where hardwood linings, architraves and skirting, etc. were being used. The cornice block, usually slightly thicker than the architrave, was often carved or routered out in a decorative way. The leg and head architraves were simply butted up squarely to it. The design of architraves used on jobs involving cornices were often, as illustrated in Figure 17.10, either fluted, or reeded.

17.1.14 Architrave Shapes and Sizes

Figure 17.10: Architrave sizes vary, but those quite commonly used are ex. 50 × 19 mm, ex. 63 × 19 mm and ex. 75 × 19 mm. On large jobs, architrave is usually ordered by *the metre run*, so it will arrive in random lengths, hence the occasional need for splicing. On small jobs, it may be obtained in specified lengths with head-lengths added together. A variety of modern and traditional shapes still in use are illustrated. Unshaped architraves, i.e. similar to the *single round edge* illustrated here, but minus the round edge, seem to be quite popular nowadays.

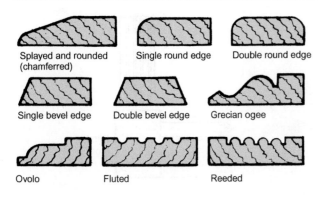

Figure 17.10 Architrave designs

17.2 SKIRTING

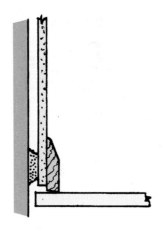

Figure 17.11 Skirting

Figure 17.11: Skirting is a protective board fixed at the base of plastered or plasterboarded (dry-lined) walls, which also covers the transition joint between the wall surface and the floor. The depth of the skirting boards, for visual balance, is usually at least 25 mm more than the width of the architraves used; and the shaped or moulded face-edge usually matches or is compatible with the architrave.

17.2.1 Scribing to the Floor

If the skirting boards are reasonably straight-edged in length and the floors are not uneven, it is theoretically possible to fix the skirting without having to fit it to the floor. Otherwise, it should be scribed. This is done by positioning a length of skirting, then laying a pencil on the floor touching its face and running a line along it. Next, the bottom edge is under-shot (meaning more planed off the back edge than the front) to the line with a smoothing plane or Surform, used diagonally across the edge, rather than along it, while the skirting is laying face-up on a saw stool, etc., and the

planing action is aimed at the floor. This technique allows the plane to remove any irregularities of the floor-line more easily.

17.2.2 External and Internal Angles

Figure 17.12: External angles are mitred, either with the mitre saw mentioned previously, or with the purpose-made mitre box, but internal angles should always be scribed. In effect, this means that the moulded profile of the skirting is cut from the end of one piece, A, to fit the moulded face of another piece, B, already fixed squarely into the internal corner of the adjacent wall.

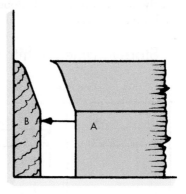

Figure 17.12 Internal scribe

17.2.3 Scribing Internal Angles

Figure 17.13: The professional technique for scribing an internal angle in a 90° corner is first to cut the end of the moulded board to be scribed with a 45° mitre cut A. This actually produces a sawn outline or profile of the skirting's moulded shape, regardless of whether it is a simple, i.e. modern or complex traditional design. Next, the scribe is produced by cutting in the waste area of the mitred profile with a coping saw, keeping very close to the outline with a slightly under-cutting angle, as indicated in B.

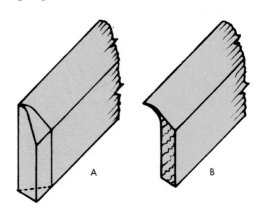

Figure 17.13 Scribing technique

17.2.4 Sequence of Fixing

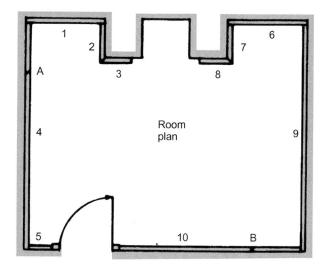

Figure 17.14 Sequence of fixing

Figure 17.14: Even though well-fitted skirting scribes may have been achieved, scribes should always face away from the doorway, so that any slight gaps are not looked directly into. This will always be possible, providing a certain sequence of fixing is followed, from 1 to 10 as illustrated in Figure 17.14. In the illustration, it can be seen that skirting-pieces 1 and 6 are square-ended at each end and pieces 4, 5, 9 and 10 have only one scribed end – as do pieces 2 and 7, which are also mitred. The message here is that you should always avoid having two scribed ends on one piece of skirting board – because it is much more time-consuming in demanding a higher degree of precision and skill for it to be successful. Avoiding double-ended scribes, therefore, is another good reason for adopting a sequence of fixing.

17.2.5 Splicing

As with architraves, lengthening-joints (splicing) should be minimal or non-existent. However, if they are unavoidable – perhaps because of the long length of a particular wall, or because there is an excessive amount of offcuts to use up – the splice, which is made with 45° cuts across the top edges, should be treated like the corner-scribes and always face away from the direct approach viewed from a doorway, as indicated in Figure 17.14 at A and B.

17.2.6 Mechanical Fixings

Each piece of skirting is fixed as it is fitted, except when a mitred corner is involved, when it is wise to fit the two mitred pieces before fixing one of them.

This way, any adjustments to either mitre can be made more easily. Mechanical fixings, such as nails or screws, should be made at approximately 600–800 mm centres. Nail fixings must be punched in and screws should be countersunk or – especially in the case of hardwood skirtings – counterbored and pelleted. Skirting fitted to timber stud partitions can be fixed with 63 mm lost-head type oval nails – and 63 or 75 mm cut clasp nails are still good fixings for skirting being fixed to walls built with receptive aerated building blocks. Walls built with dense concrete blocks or brickwork can be drilled for screw fixings. This is made easier nowadays by drilling directly through the *in situ* skirting with a masonry drill to take Fischer-type nylon sleeved Frame-fix screws.

17.2.7 Adhesive Fixings

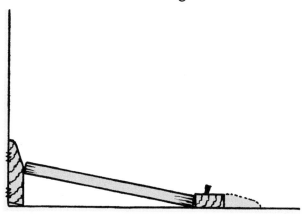

Figure 17.15 Temporary strutting

Figure 17.15: Present-day practices include fixing skirtings with so-called *gap-filling panel adhesives* such as Evo-Stik's Gripfill or Laybond's Connect. Gripfill is a solvent-borne, filled rubber resin and Connect is a solvent-free gap-filling adhesive specially formulated to provide good initial 'grab' or suction. Although either of these two adhesives may be used, the Laybond Connect is recommended when bonding to vertical surfaces. These adhesives are in 350 ml cartridge tubes which fit into silicone/mastic guns. Working speedily, one or two 6 mm diameter continuous beads should be applied to the back of the skirting board, 25 mm in from the edges, then it should be quickly placed and pressed (slid slightly, if possible) into position. A few temporary nails or pins may be needed if the skirting appears to come away from the wall in places. These should be left in overnight. Alternatively, as illustrated, small struts (maybe offcuts of skirting) can be wedged against the skirting, taking a foothold from short battens or offcuts fixed temporarily to the floor. If the floor has been overlaid with hardwood or plastic laminate flooring, temporary fixings could not be made and struts might need to be taken across to an opposite wall.

17.2.8 Skirting Shapes and Sizes

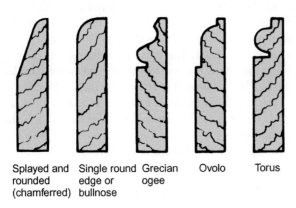

Splayed and rounded (chamferred) | Single round edge or bullnose | Grecian ogee | Ovolo | Torus

Figure 17.16 *Skirting designs*

Figure 17.16: Like architraves, skirting sizes vary, but those most commonly used are ex. 75 × 19 mm, ex. 100 × 19 mm and ex. 150 × 19 mm. If required, the thickness of these skirting boards can be increased to ex. 25 mm, but this would mean increasing the architrave-thickness. Again, like architraves, it is usually ordered *by the metre run*, so it will arrive in random lengths. A variety of modern and traditional shapes still in use are illustrated. Also of course, on certain jobs where simplicity is sought and moulded shapes are being avoided, plain square-edged boards may be used.

17.2.9 Dual Pattern Skirting

Figure 17.17(a)(b)(c): In recent years, dual pattern skirting of various depths has gained in popularity. The main advantage is that it offers a choice of moulded design on its alternate faces. Another advantage, not always realized, is that whatever edge touches the floor,

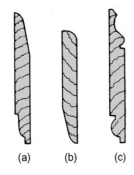

(a)   (b)   (c)

Figure 17.17 (a) ex. 25 × 175 mm chamferred/ovolo skirting; **(b)** ex. 25 × 150 mm bullnose/chamferred skirting; **(c)** ex. 25 × 175 mm Grecian ogee/Torus skirting

it has the effect of being 'undershot', by being relieved in thickness by the unseen moulding on its back face. This reduced edge-thickness makes it much easier to scribe – if necessary – to an uneven floor. However, there is also a downside: the lower face of the skirting is less solid (more so with the Grecian ogee, less so with the bullnose) and tends to cup slightly. This is sometimes quite noticeable on mitred external angles, which cannot be overcome by pinning, because of the absence of lower back-face material.

Finally, MDF is also used for moulded skirting nowadays. It has a good, smooth finish, is mitred and scribed satisfactorily, responds well to adhesive and screw fixings, but does not receive nails very well and offers more resistance to being scribed to uneven floors – requiring more effort and a sharp plane.

17.3 DADO RAILS AND PICTURE RAILS

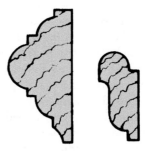

Figure 17.18 (a) Dado rail; **(b)** picture rail

Figures 17.18 (a) and (b): In recent years there has been a degree of revival of these traditional items, especially on home-improvement schemes. Basically, the same rules and techniques regarding mitring, scribing, splicing, fixing, etc., apply to dado and picture rails as already covered regarding the fitting and fixing of skirting. Dado rails of usually ex. 75 × 38 mm section, of whatever moulded design, are fitted and fixed around the walls of a room at about 900 mm up from the floor. Whatever height is used, the dado rails should be kept parallel to the floor, regardless of exact levels. Picture rails of about ex. 50 × 25 mm section, of whatever moulded design, but with the essential grooved or rebated top edge (as in Figure 17.18(b)) to hold the picture hooks, are fitted and fixed around the walls of a room, usually at the level of the top of the architrave-head of the room's doorway. Whatever height is used, the picture rails should be kept parallel to the ceiling, again regardless of levels.

18

Fitting and Hanging Doors

18.1 INTRODUCTION

Door-hanging is a vital part of second-fixing carpentry and requires a good standard of workmanship and speed. One to one-and-a-half hours is the established hanging-time for a lightweight internal door and two-and-a-quarter hours for a heavy external type or fire-resisting door. When hanging a door on a solid, rebated frame, this time includes fitting the door properly into the rebates – and on a door lining, it includes adjusting and fixing the planted door stops after the door has been hung. When more than one door is to be hung in the same locality, time will be saved by treating the fitting of locks and door furniture as a separate, secondary operation.

18.2 FITTING PROCEDURE

18.2.1 Removal of Horns

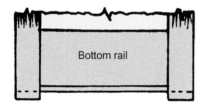

Figure 18.1 Increasing door-height

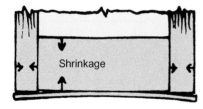

Figure 18.2 Bad practice

Figures 18.1 and 18.2: Traditionally, all doors (including flush doors) arrived on site with horns left on to give protection to the corners of the stiles. This also made it possible to increase the height of a door, if required, by gluing an additional piece of timber to the bottom rail,

between the horns (Figure 18.1). Without the horns, the additional timber would have to cross the opposing grain of the stiles, which is bad trade practice, not allowing for natural shrinkage (Figure 18.2).

18.2.2 Checking the Opening Size

Nowadays, with a few exceptions, horns are usually non-existent or very minimal in size. Although openings with non-standard height may only be met occasionally – usually on older-type property with odd-sized doors – it will be sensible on certain jobs other than new works, not to cut the bottom horns (if they exist) until the height of the door opening is checked. This can be done with a pinch rod or tape rule, although, if no major discrepancy is suspected, it will be quicker to hold the door against the opening and mark the top and side edges onto the frame, thereby gaining visual evidence of the amount to be planed off.

18.2.3 Checking the Hanging Side

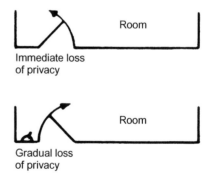

Figure 18.3 Hanging side

Figure 18.3: On which side the door hangs and whether it should open in or out, must be known. This information can be found on the *plan* views of the contract drawings, although sometimes a member of the site management team takes the responsibility of checking this out and marking the hinged side of the lining or frame with the letter H, at about one metre from the floor. Another guide, if needed, is that the lock edge of the door should be on the side nearest to the light switch at the side of the opening. Of course, on new contracts, this assumes that the electrician has put the switch in the correct position. On small contracts, where the builder or carpenter himself might decide on which side to hang the door, it may help to know that doors should give maximum privacy to a room while being opened (Figure 18.3). Another point to bear in mind on this subject, is that normally, external doors should open inwards, to avoid damage if caught by the wind in a partly open position.

18.2.4 Tools Required

The following tools are required to cover a variety of door-hanging jobs: pencil, tape rule, combination square, panel-type saw, size $5\frac{1}{2}$ jack plane and/or a $4\frac{1}{2}$ smoothing plane, optional power or cordless planer, one or two marking gauges, 32 mm bevel-edged chisel, claw hammer, optional mallet (for chisels with boxwood handles), large-sized birdcage awl with square tapered point, cordless screwdriver, portable router and hinge jig, bullnose rebate plane, or a shoulder plane with a removable front section (for easing rebates on solid timber frames), nail punch (for planted door stops), and a sharp marking knife or craft knife (for scoring across the plies or veneers of flush doors, if they are to be reduced in height).

18.2.5 Equipment Required

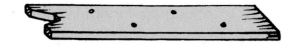

Figure 18.4 Extension stool-top

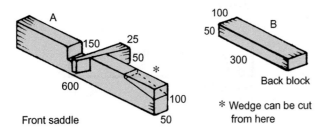

Figure 18.5 Saddle and block

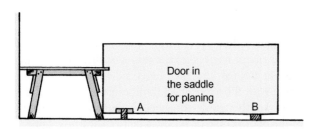

Figure 18.6 Door held firmly against saw stool

Figures 18.4–18.6: The equipment required for door-hanging can be as simple as one saw stool and an ex. 100 × 25 mm board of about 1 m length with a V cut in one end. The board, shown in Figure 18.4, can be screwed onto the top of the stool as and when required and helps to hold the door on its edge during the planing or *shooting-in* operation indicated in Figure 18.6. Additionally, a device known as a *saddle and block*, detailed in Figure 18.5, and easily made from two pieces of 100 × 50 mm timber, provides a very simple but effective way of holding the door firmly while shooting-in or cutting out the housings for the hinges, as indicated in Figure 18.6. A second saw stool is occasionally required to act as a trestle should it be necessary to lay the door across both stools to cut the horns or the bottom of the door.

18.2.6 Skilful Planing Requirement

When shooting-in the door, it is good practice to remove an equal amount from each side. This is done by judgement, rather than measurement. If the amount to be removed is in excess of a few millimetres, a power or cordless planer, if available, would save a lot of effort. However, to eliminate the unsightly rotary cutter marks that can be left on the door-edges, especially if the planer is pushed along too speedily, the edges should be finished off with a jack or smoothing plane. Although, unskilful hand planing – by not putting the correct pressure and momentum to the plane, or by not lifting off correctly (by raising the heel) – produces ridges and chatter marks which can be as unsightly as the pronounced rotary-cutter marks.

18.2.7 Closing Edge

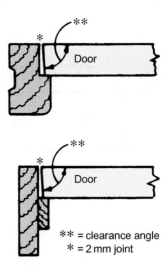

** = clearance angle
* = 2 mm joint

Figure 18.7 (a) Horizontal sections showing closing edge of door

Figure 18.7(a): The closing or lock edge of a door requires a slight angle of about 87–88° to clear the frame or lining's edge effectively. This should be achieved while shooting-in and not added afterwards. Again, this is done by judgement, rather than measurement. Although, until experience is gained, a sliding bevel could be set up and used for testing the edge while planing. The clearance angle in relation to the jamb's edge is indicated in Figure 18.7(a).

18.2.8 Clearance Joints

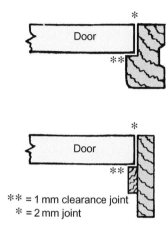

Figure 18.7 **(b)** Horizontal sections showing hanging edge of door

Figure 18.7(b): A consistent and unwavering 2–3 mm joint (gap) should be achieved around the door and frame or lining (side-edges and top) and the joint at the bottom should be a minimum 3 mm and a maximum 6 mm, unless extra allowance is required for floor covering, such as carpets. A two-pence coin is sometimes used as a feeler gauge for testing the top and side joints.

18.2.9 Arrises

After shooting-in the door and *before* screwing on the butt hinges, it is important to plane off the 'arrises' (sharp corner edges) from the top, bottom and side edges on both sides of the door. The appearance of this should be like a miniature chamfer, measuring no more than 1 to 1.5 mm across the 45° chamfer. Usually, one or two strokes with a smoothing plane will accomplish this. Sharp arrises on any joinery item can cut flesh and the paint or varnish film.

18.2.10 Planted Door Stops and Sunken Rebates

Figure 18.7(b): The door stop or the sunken-rebate shoulder must be 1 mm clear of the door on the

hanging side, to avoid a fault known as *binding*, which happens if these edges touch. The size of planted door stops varies in section between ex. 50 × 12 mm and ex. 38 × 12 mm, or ex. 50 × 16 mm and ex. 38 × 16 mm and requires fixing every 225 mm approximately, with 38 mm lost-head oval nails, staggered and punched under the surface by at least 1 mm. Solid frames with sunken rebates can present more problems than linings with planted stops, because the door must fit well into the rebates and if it – or the frame – is twisted, this will involve easing the shoulders of the rebates with a shoulder or bullnose plane.

18.2.11 Various Points to Note

1. When hanging hardwood doors, especially hardwood *flush* doors, or any door which may be easily surface-damaged, the door should be protected from being scratched or bruised by covering any door-bearing stools with dust sheets or soft fibreboard. For the same reasons, the door-side of the housing in the saddle block and the door-side of the wedge should be covered with masking tape.
2. When removing horns, or cross-cutting plywood or veneer-faced doors to reduce the height, *spelching out* of the fibres can be eliminated if the amount to be cut off is heavily scored with a sharp marking knife or chisel on both sides, then, after being cut very close to the line, is finished off by planing down to the knifed edge.
3. If more than 6 mm has to be removed from the bottom of a door, it is advisable to rip this off by saw. Any lesser amount should be planed, but first consider removing the cross-grain of the stiles, if practicable, with a fine saw, to make easier work of the planing, as indicated in Figure 18.8.

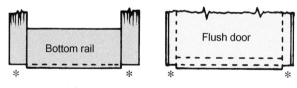

* = End-grain removed prior to planing

Figure 18.8 Reducing door height

18.2.12 Doors on Rising Butts

Figures 18.9 and 18.10: Doors being hung on rising butt hinges require the inner top edge of the door to be shot off on the splay. The amount of splay to be planed off is shown in Figure 18.9. Rising butt hinges are either left-handed or right-handed and a satisfactory way to identify which hand is needed is to name

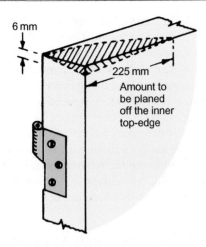

Figure 18.9 Rising butts

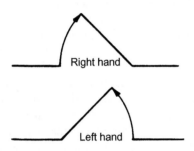

Figure 18.10 Handed hinges

the hand facing the door as it opens away from you, as shown in Figure 18.10.

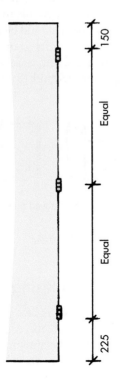

Figure 18.11 Hinge positions

18.2.13 Hinge Positions

Figure 18.11: There are no hard-and-fast rules about hinge positions, but they should always be less at the top than the bottom and must be clear of the end grain of any mortices (or their wedges) by at least 12 mm. Positions of 150 mm down to the top of the butt and 225 mm up to the bottom of the butt are the usual settings – but this may be governed by other doors already hung in the vicinity by others who used settings of 175 mm down and 250 mm up; in this case, their settings are followed. If three butts ($1\frac{1}{2}$ pairs) are specified or decided upon for a heavy door, the third butt is/was centred between top and bottom butt-positions, as in Figure 18.11, but nowadays the top of the mid-area butt is often specified to be about 150 mm below the bottom edge of the top butt.

18.2.14 Hinges and Screws

Butt hinges range in type and in size from 25 mm to 150 mm (75 mm and 100 mm being most commonly used on doors), and come in various kinds of metal. Screws should always match and may be recessed for Supadriv/Pozidriv, Phillips, etc., screwdrivers. Mild or bright steel butts must be painted in with the door, but brass or other non-ferrous metal butts, used on exterior or hardwood doors, should not be painted. The screw's head diameter must suit the countersinking in the butts and be either flush or slightly below the metal surface. On lightweight doors, 25 mm (1 in) screws are often used. On heavy doors, screws of 32 to 38 mm are used. The diameter of a screw's head is approximately twice the diameter of its shank, the latter of which is given nowadays on the manufacturer's screw-box label.

18.2.15 Knuckles In or Out

Figure 18.12: This illustrates different settings for butt hinges in relation to the knuckle part of the hinge, as follows.

(a) The butts are housed with a full knuckle projection. This is now standard practice to save time.
(b) The butts are housed-in to lessen the knuckle projection for aesthetic reasons, but, to enable the door to open beyond 90°, the centre of the knuckle must not be further in than the surface of the door. Although chiselled chamfers were used to accommodate the recessed knuckle, excessive time was added to the door-hanging operation and, therefore, this practice has not survived.
(c) In this example, only the knuckle on the frame side is partly housed-in to carry the door further

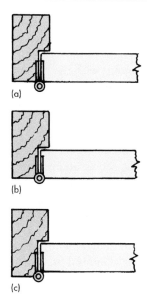

Figure 18.12 Knuckle projections

into the shoulder of the jamb, when a rebate is deeper than the door thickness (which is met occasionally). If ignored, an excessive clearance-joint between the shoulder and the door would look unsightly.

18.2.16 Marking the Hinge Positions

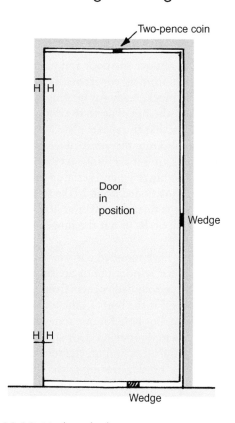

Figure 18.13 Marking the hinge positions

Figure 18.13: When the door has been shot-in and the clearance-joints have been achieved, set up the door in the opening as follows: insert an off-centre wedge, chisel or bolster under the door and lightly tighten up to a two-pence coin placed in the top, as illustrated, and then insert a small wedge at mid-height in the joint of the closing stile (the doubled-joint obtained this way can be tested, if necessary, with two two-pence coins held together and tried in several places (or slid along) above and below the wedge). Mark the hinge positions now, squarely across the frame or lining and the door. Alternatively, the butts could be already housed-in and screwed to the door before it is positioned in the opening and the unfolded leaves of the butts could then be marked onto the frame or lining.

18.2.17 Marking the Housings

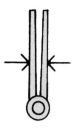

Figure 18.14 (a) Determining the housing-depth

Figure 18.14(a): The door is now removed from the opening and set up in the saddle and block, ready to be hinged. Place an open butt on the edge of the door, with the protruding knuckle inverted to rest against the door-edge, locate carefully to the correct side of the face mark and scribe around the leaf with a sharp pencil. Repeat this on each of the four butt-positions. Next, adjust the leaves of the butt until parallel (as illustrated), measure the overall thickness, deduct 2 mm for the joint, then divide by two to give the depth of housing required and set this on the marking gauge. Mark the butt-positions with this on the face of the door and the edge of the frame or lining. Then reset the gauge to the leaf-width of the butt and mark this on the edge of the door and the face of the frame or lining.

18.2.18 Cutting the Housings

Figure 18.14(b): Now cut the housings, using the chisel-chopping method, prior to chisel-paring. When leaning the chisel (at about 45–50°) for the cross-grain chopping action, the chisel's bevelled-edge should be on the side of the acute angle formed between the chisel and the timber. Note that if the housings are cut too deeply, the hinge-side joint will be lost and binding may occur between the door-edge and the face of

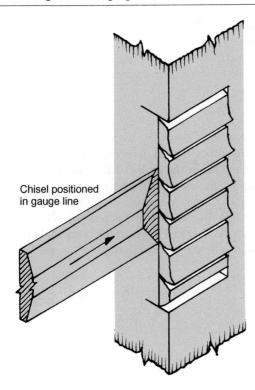

Figure 18.14 (b) Cutting the housings

the frame or lining. Also, if the wrong gauge screws are used and the countersunk heads protrude, binding may occur between the opposite screw heads.

18.3 HANGING PROCEDURE

Speed and skill in door-hanging is gained from the experience of repetition and it will assist if you adopt the set hanging procedure described below. (Note that the references are to a door-lining, but the procedure would also suit a doorframe.)

1. Check the opening size and hanging edge. Mark H on the lining face, mark T (for 'top') on the face of door (this is to avoid mistakenly changing the face-side during the shooting-in procedure) and remove horns, if necessary.
2. Offer the door up in position, judge the initial amount to remove from the edges and shoot-in.
3. Offer up again – if it now fits into the opening, concentrate on marking and planing the left-hand stile only, until a good fit is achieved against the lining-leg.
4. While keeping the door pressed or wedged against the left, lever or wedge up the door and check the fit against the underside of the lining-head. Remove the door and plane the top, if necessary, until another good fit is achieved.

5. Now concentrate on judging, marking and shooting-in the right-hand stile only, until (with the door pressed hard to the left) you are able to test for a double-joint with the aid of two two-pence coins held together.
6. If satisfied, re-establish the side and bottom wedges (Figure 18.13), after inserting a two-pence coin in the top joint. Now mark the hinge positions on the door *and* the lining-edge.
7. Dismantle the set-up and remove the arrises from all door edges *before* fixing the butts to the door.
8. Mark and cut out the housings for the butts, try the butts in lining-housings and make a pilot hole for one screw in each butt.
9. Screw butts to door, hang door and try closing and check the all-round fit.
10. Remove and adjust, if necessary, or finish screwing to the lining.
11. Fix top door stop (1 mm clear on hinge-side only), then closing-edge door stop, then hanging-edge door stop. Finally, punch in all nails.

18.3.1 Use of Portable Router and Hinge Jig

Nowadays, the door-hanging procedure is often speeded up by using a router and a hinge jig, both described and illustrated in sections 2.18 and 2.19 in Chapter 2. To take these into account here, the hanging procedure listed above would be changed as follows, after procedure number 5 had been completed:

6. If satisfied, remove the door and plane off the sharp arrises from all edges. Set the door up in the saddle and block, ready for hinging.
7. Now attach the hinge jig to the door lining, making sure that the swivel plate is at the top to establish the 3 mm top-joint allowance. Set up the router and rout out the hinge recesses.
8. Remove the hinge jig and set it up on the edge of the pre-positioned door. Make sure that the swivel plate is turned at 90° and hooked onto the top of the door. Rout out the hinge recesses and remove the jig.
9. Some butts have rounded corners to fit the routed recesses, but if standard butts are being used, then chisel out the rounded corners from the four (or six) recesses. Fit the butts in the lining-recesses and make a pilot hole for one screw in each butt.
10. Screw butts to door, hang door and try closing and check the all-round fit. Remove and adjust, if necessary, or finish screwing to the lining.
11. Fix top door stop (1 mm clear on hinge-side only), then closing-edge door stop, then hanging-edge door stop. Finally, punch in all nails.

19

Fitting Locks, Latches and Door Furniture

19.1 LOCKS AND LATCHES

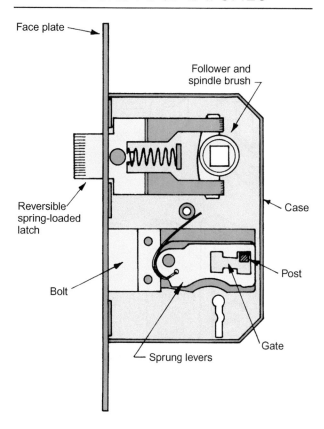

Figure 19.1 Mortice-lock mechanism

Figure 19.1: The types of lock used on dwelling houses are few and vary between locks, latches and combinations of these two. They may be mortice locks morticed into the door-edge, or various types of rim lock fixed on the inside face-edge of the door. The actual latch part of a lock is usually spring-loaded, quadrant shaped, or has a round-edged roller bolt, which holds the door closed (latches it) without locking it – unless it is a type of cylinder night latch, which requires a latch-key to open it on the cylinder-side of the door. The locking mechanism of a mortice lock is usually an oblong-shaped bolt which is shot in or out by inserting

and turning a key. The concealed part of the bolt, as illustrated, has a small metal post protruding from it, which must be moved through an open *gate* cut in the middle area of a specified number of sprung levers. This happens when the key lifts the levers, gains access to the edge of the bolt and moves it. The more levers a lock has, the greater the security. The quadrant-shaped latch is usually reversible to enable the hand of the lock to be changed from left to right, or vice versa.

19.2 MORTICE LOCKS

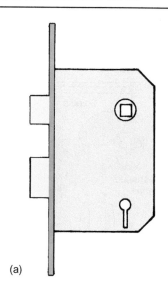

(a)

Figure 19.2 (a) Shallow mortice lock

Figure 19.2 (a) to (c): These locks vary in length (depth) between 64 mm and 150 mm. The deeper locks (Figure 19.2(b)) are of traditional size to receive door-knob furniture which consists of a spindle, two knobs (sometimes one of these is already attached to a metal spindle, the other being removable via a small grub screw), two rose plates, and two escutcheon plates (Figure 19.2(e)). The shallow-depth lock (Figure 19.2(a)) has its keyhole and spindle hole

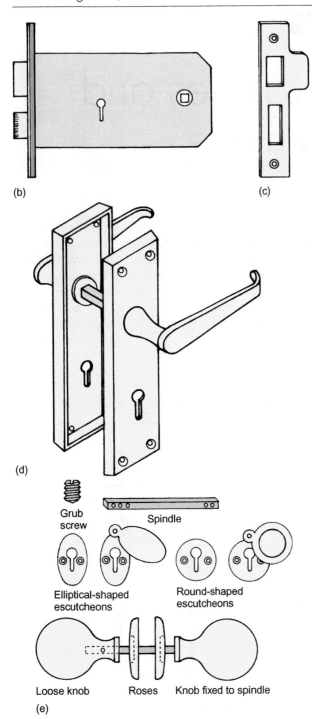

Figure 19.2 (b) Deep mortice lock; **(c)** striking plate (brass or chromed steel); **(d)** lock lever-furniture; **(e)** door knob furniture

vertically in line to receive a set of lock lever-furniture (Figure 19.2(d)). Both types of lock have a striking plate (Figure 19.2(c)) which is housed into the jamb to receive the projecting latch and the turned bolt. Shallow-depth locks should only be fitted with lever furniture, as door knobs come too close to the face of the jamb or lining on the door-stop side, resulting in

possible scraped knuckles. If door knobs are to be used, the lock should be at least 112 mm ($4\frac{1}{2}$ in) deep.

19.3 MORTICE LATCHES

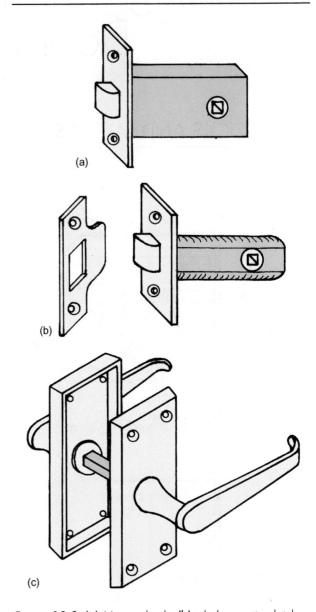

Figure 19.3 (a) Mortice latch; **(b)** tubular mortice latch; **(c)** latch lever-furniture

Figure 19.3: Mortice latches are used on internal doors not requiring to be locked. Two types are available: one, as in Figure 19.3(a), is oblong-shaped for morticing in to a 16 × 38 mm × 64 or 75 mm deep mortice hole; the other, as in Figure 19.3(b), is tubular-shaped for inserting into a drilled hole of 22 mm diameter × 64 or 75 mm depth. A mortice latch is always supplied with a striking plate and fixing screws, but requires separately a set of latch lever-furniture (without the keyhole), as shown in Figure 19.3(c).

19.4 MORTICE DEAD LOCKS

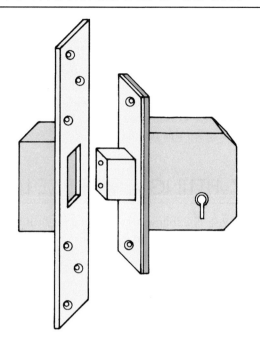

Figure 19.4 Mortice dead-lock and box-recessed striking plate

Figure 19.4: Mortice dead locks are for extra security and contain only a locking bolt – no latch. They are fitted to external doors in addition to a latch-type cylinder lock and are recommended by insurance companies to have five levers. The more expensive locks of this kind have a box-recessed striking plate or *keep* – and the brass bolt contains two hardened steel rollers to resist being cut with a hacksaw blade. Ironically, these locks can be more of a deterrent on the inside of a property than on the outside. The reason for this is that if burglars have gained access through a window – which is quite common – they like to leave by a door, which is less suspicious, easier and quicker than carrying stolen goods through a window. My preference with these locks is to make a keyhole only on the outside of the door, which is done to stop anyone deadlocking themselves in the property at night, in case there is a fire. The only door furniture required is supplied in the form of two escutcheons, one being a *drop escutcheon* with a pivoting cover plate that drops down to cover the keyhole on the inside of the door, as shown in Figure 19.2(e).

19.5 CYLINDER NIGHT LATCHES

Figure 19.5(a) and (b): This type of lock/latch is commonly used on front entrance doors and consists of the rim latch itself, as illustrated, with a turn-knob and a small sliding latch-button for holding the latch in an

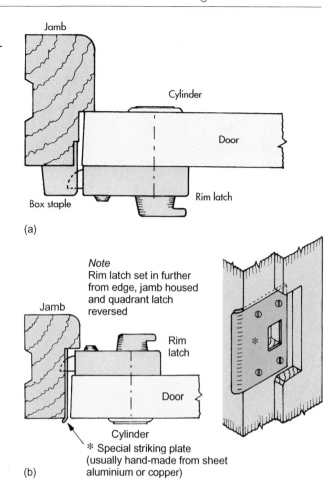

Figure 19.5 (a) Part horizontal section through normal inward-opening door; **(b)** uncommon outward-opening door

open or closed (locked) position, a cylinder with a bar that connects into the latch, a loose rose plate that provides a rim for the cylinder (or alternatively, a cylinder door pull), a back plate with connecting screws for securing the cylinder to the door and a box staple (also referred to as a *keep*) for receiving the striking quadrant-shaped latch. These locks are obtainable in standard sizes or narrow sizes. The latter is sometimes required on glazed entrance doors with narrow stiles. A standard cylinder night latch requires a 32 mm ($1\frac{1}{4}$ in) diameter hole to be drilled through the door at 61 mm in from the edge to the centre, to receive the cylinder. The narrow latch type requires the same size hole to be drilled, but at 40 mm in from the edge to the centre.

19.6 FITTING A LETTER PLATE

Figure 19.6: Measurements for the oblong aperture must be carefully taken from the letter plate (often referred to as a *letterbox*) and plotted on the outside *and*

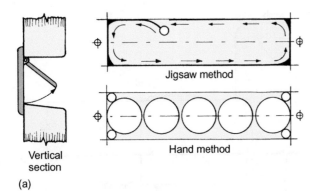

Vertical
section

(a)

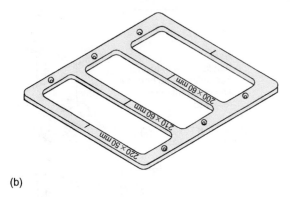

Jigsaw method

Hand method

the required area for the letter plate, a jig template such as the Trend model illustrated at Figure 19.6, is screwed in position. The router is set up and worked around the template to achieve a fast and precise finish. When complete, the template is unscrewed and the two screw holes become the centres for the small bolt holes to be drilled through the door to hold the letter plate. Certain makes and models of router might require a universal sub-base to accept the 30 mm guide bush needed for this template.

(b)

Figure 19.6 (a) Fitting a letter plate (two of three methods); **(b)** letter plate jig template/router method

the inside face of the door. This can best be done by marking a level centre line, as illustrated, on each face, to use as a datum line for the other measurements. This is also the line on which to mark the critical position of the 6 mm diameter holes to be drilled for the connecting bolts. If confident of the accuracy of your marking out, these holes are best drilled halfway through from each side. Slight misalignments midway can be overcome by the reaming effect gained by using a Sandvik combination auger bit.

The aperture can be cut out easily with a good jigsaw. Note that when reaching each corner, the saw is worked back and forth a few times to create space for the blade to turn through 90°. When cut, the hole is then cleaned-up with a wide bevelled-edge chisel and/or a Surform file.

Alternatively, a line of large-diameter holes can be drilled, with smaller holes drilled at each end to make it easier to cut the end grain with a sharp bevelled-edge chisel. By using this hand method, once the ends are chopped through from each side, the remaining timber above and below the large holes pares out quite easily with a wide bevelled-edge chisel. The exposed arrises of the aperture should be removed with a chisel, file and glasspaper.

Another method used nowadays is to cut out the aperture with a 12.7 mm ($\frac{1}{2}$″) collet plunge router. After pencilling vertical and horizontal centre lines in

19.7 FITTING A MORTICE LOCK

Figure 19.7: The following technique for fitting a standard mortice lock can – with the omission of the

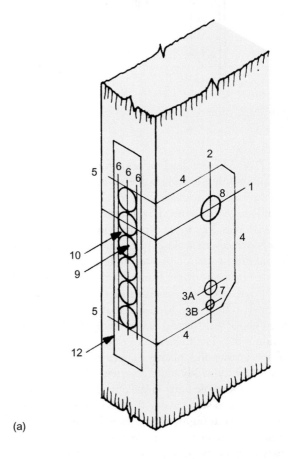

(a)

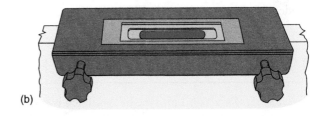

(b)

Figure 19.7 (a) Fitting a mortice lock – Hand method; **(b)** fitting a mortice lock – Router and jig method

spindle hole – be modified for fitting a mortice dead lock. The auger bits used can be power-driven or fitted into a ratchet brace. If power-driven, the drill must have a reversing facility and, ideally, a variable speed.

1. If no predetermined height exists, measure down half the door's height to mark the spindle level and square this around the door's edge to both sides.
2. Measure the lock from the outer edge of the face to the centre of the spindle hole and mark this on the door on each face with a pencil line gauged from the blade-end of the combination mitre square. Make a slight allowance for the undershot edge of the door.
3. Measure the lock vertically from the centre of the spindle hole to the top-centre and bottom-centre of the keyhole and mark cross-lines A and B on both faces of the door.
4. Hold the lock against the door, sight through the spindle hole to line it up with the spindle cross-mark, then mark lightly around the outer edge of the lock.
5. Square this outline across the edge of the door.
6. Set up a marking gauge and mark the centre of the door thickness, then reset the gauge to be 8 mm from the centre and mark a line on each side of the first gauge line.
7. Drill a 10 mm diameter hole at 3 A and a 6 mm diameter hole at 3B from each side of the door, to meet in the centre for greater alignment – and to avoid spelching out.
8. Drill a 16 mm diameter hole for the spindle, again from each side of the door.
9. As indicated, drill a series of 16 mm diameter holes close to each other for the lock-mortice. Use a depth gauge on the drill or bind masking tape around the auger bit to achieve a slightly oversize depth. Hold a flat rule or the blade of the combination square on the face of the door when drilling, to check on drill-alignment occasionally.
10. Clean out the mortice hole with, say, a 25 mm bevelled-edge chisel and a 16 mm firmer chisel. On mortices for deep locks, a so-called swan-neck chisel is sometimes necessary.
11. Complete the keyhole shape with a small chisel or a pad saw.
12. Try the lock in the mortice until, after easing, you achieve a slightly loose fit, then adjust it for central position edgewise and mark around the face plate with a sharp pencil. The edge-lines can be scored heavily with a marking gauge (or a *cutting gauge*) to reduce the risk of splitting the edges with a chisel, then, shallow chisel-chopping across the grain is carried out before hand-routering the face-plate depth, by judgement with the chisel held in a position similar to when it is being sharpened. Keep trying the lock until it

fits slightly under the surface. Then screw it into position.
13. Now close the door a few times until the latch marks the jamb's edge, or mark this position with a pencil, and square these marks across to relate to the striking plate's mortice hole. Position the striking plate onto the protruding latch and mark the plate's face where it protrudes past the face of the door. Now hold the striking plate against the marks on the jamb, with the edge-mark on the plate in line with the jamb's edge and mark around the plate with a sharp pencil.
14. Carefully chop out a shallow housing for the striking plate and try in position. Mark the outline for the latch and bolt mortice holes, remove the plate and drill/chop out for these. Fix the striking plate and try closing the door for easy latching and locking.

Another method used by some carpenters is to cut the mortice out with a heavy duty plunge router which, as before, has a 12.7 mm ($\frac{1}{2}''$) collet. A lock jig such as the Trend model illustrated at 19.7(b) is clamped to the door securely to guide the routing. Again, the router will require a 30 mm guide bush and a 12 mm Ø × 114 mm long-reach TCT cutter. This will cut up to a 70 mm deep mortice. Mortices for deeper locks will require finishing off with an auger bit and drill – or a twist bit and brace. The magnetized, interchangeable templates on this jig, allows both the mortice and the shallow face-plate recess to be cut.

19.8 FITTING DOOR FURNITURE

When fitting the door furniture, which is usually done after the doors have been painted or sealed, or – better still – has been done previously and then removed and replaced after painting or sealing, care must be taken with the vertical and lateral positioning and screwing of the furniture. This is because the 6.35 × 6.35 mm square-sectioned spindle sits quite loosely in the lock's spindle bush and if strained over to one extreme or the other by ill-positioned screws in the door furniture, binding can occur which may cause the latch and the lever handles or knobs to stick in the levered or turned position, without the springs being able to effect a self-return action. To avoid this, always feel for the correct position by gently moving the furniture from left to right and up and down – and by settling in the middle of these extremes. To help with this, on lock lever furniture, the keyholes can be sighted through for alignment. After fixing, always check lever handles or knobs for a smooth, easy movement and a self-returning action.

20

Fixing Pipe Casings and Framed Ducts

20.1 INTRODUCTION

Occasionally, according to the design of a property, there is a need to conceal vertical and/or horizontal pipes to improve the appearance of the room(s) that they pass through. These rooms are usually the bathroom or kitchen. The pipes may be 110 mm diameter soil pipes, 40 mm diameter waste pipes, or various small-diameter copper supply pipes. If the supply pipes are fitted with stopcocks, provision must be made for their access when the pipes are to be concealed. Basically, the two arrangements for concealing pipes are pipe casings and framed ducts. The studding for the latter can be in timber or, nowadays, in metal.

20.2 PIPE CASINGS

Figure 20.1(a)(b): Traditionally, solid timber of about 225 mm prepared width was used in the construction of pipe casings. The side casing had a beaded and rebated edge to receive a thinner timber casing. Vertical battens of about 50 × 25 mm section were fixed to the finished wall surfaces, carefully positioned laterally to ensure that the completed casing fitted squarely in the corner and was square in itself. Any under-achievement in this respect showed up badly at ceiling and floor levels. When working out the lateral position of the battens in relation to the width of each casing, allowances had to be made if it was decided that the casings would need to be scribed to the walls. Modern pipe casings are of similar construction, but the timber casings have been replaced by plywood or MDF of usually 12 mm thickness, with the rebated edge in the form of a planted, vertical corner batten, as at illustrated at (b).

20.3 FRAMED DUCTS

Figure 20.2(a)(b): Framed ducts/pipe casings take over from ordinary plywood or MDF casings when there are a greater number of pipes to conceal, or the concealment involves a more complex arrangement of vertical and/or horizontal pipework. The sawn

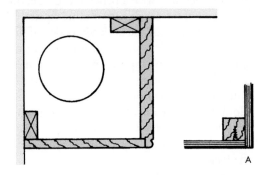

(a)   A

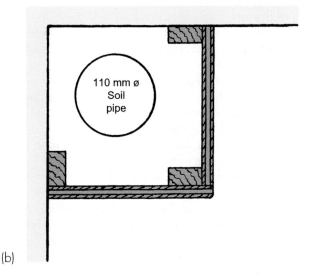

110 mm ø Soil pipe

(b)

Figure 20.1 **(a)** Horizontal section through traditional pipe casing; **(b)** Horizontal section through modern pipe casing

or prepared timber framing may be of 50 × 50 mm, 50 × 38 mm or 38 × 38 mm section. As illustrated, vertical or horizontal battens of 50 × 38 mm section are used to establish the ducting's position and anchorage to the walls or − in cases of horizontal ducting − to the wall and the floor or the wall and the ceiling. The other framing consists of a longitudinal corner-member and cross-noggings spaced at 600 mm centres. The framing is built up *in situ*, like stud partitioning. The butt-jointed noggings are skew-nailed to the longitudinal battens and nailed through the corner member at the other end. When

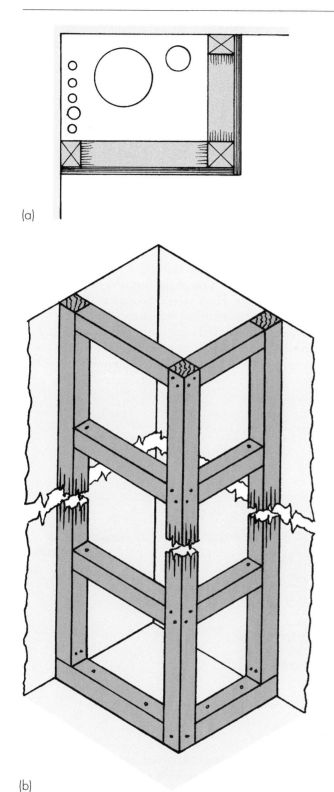

(a)

(b)

Figure 20.2 (a) Horizontal section through modern framed duct/pipe casing; **(b)** Isometric view of duct framing

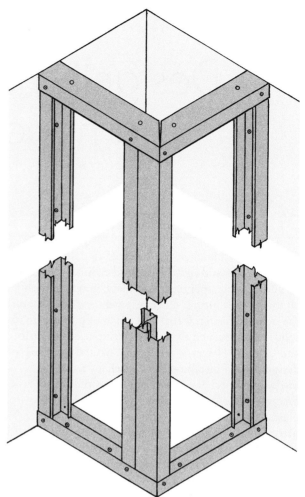

Figure 20.3 Framed duct formed with metal U-track and C-studs (see Chapter 14 for metal-studwork details)

the skeletal framework is completed, it is then covered with 6 mm or 12 mm plywood, MDF or plasterboard. If future access to a stopcock is required, the face panel(s) should be fitted with a small access panel.

21

Designing and Installing a Fitted Kitchen

21.1 INTRODUCTION

The first fitted kitchens are believed to date back to the late 1920s and mostly involved cabinet makers and second-fixing carpenters. Nowadays, however, with all the hi-tech fittings and equipment available – plus the manufacturing, fitting and fixing of purpose-made solid worktops such as granite, quartz/polymer resin, and Corian® ($\frac{1}{3}$ acrylic resin and $\frac{2}{3}$ natural mineral) – designing and installing fitted kitchens tends to be done by specialists. Nevertheless, if only for financial reasons, non cutting-edge kitchens using plastic-laminate and solid timber worktops still predominate – and a large number of kitchens are still installed by non-specialists. Anyone with good carpentry or DIY skills and a commonsense approach to design, is capable of successfully designing and installing a standard fitted kitchen.

21.2 ERGONOMIC DESIGN CONSIDERATIONS

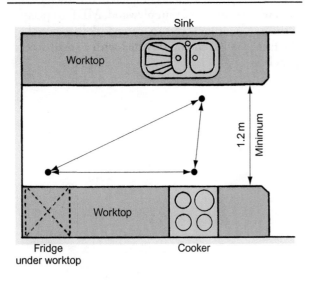

Figure 21.1 **(a)** Work Triangle and recommended minimum distance between worktops

Figure 21.1(a): In the early 1950s, researchers from Cornell University in New York, USA, conceived the idea of *the work triangle* as being the geometry determined by the positions of the sink, the refrigerator and the cooker. These were identified as the three main centres in a kitchen involving traipsing backwards and forwards during the cooking task. Study revealed that an ideal imaginary line joining the sink, fridge and cooker should measure no more than 20 ft (6.1 metres). It has also been established that the distance between opposing worktop-edges, as in a narrow, so-called galley kitchen, should be no less than 4 ft. (1.219 m).

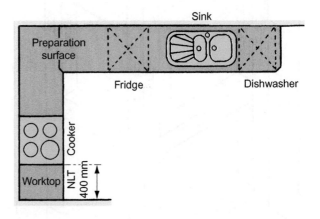

Figure 21.1 **(b)** Design considerations

Figure 21.1(b): Other design considerations that should be taken into account when juggling with the work triangle are:

- Have a good-sized food-preparation area between the sink and the cooker.
- Avoid putting the sink or the cooker in a corner. If unavoidable, keep them at least 400 mm from the return wall.
- Always have a worktop surface near, or over the fridge.

- Do not place the cooker too far from the sink.
- If intending to install an extractor hood that vents to the outside, try to keep the cooker or hob up against an exterior wall, or close to one to minimize ducting above the wall units.

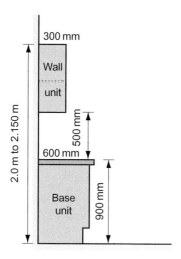

Figure 21.1 (c) Established ergonomic sizes

Figure 21.1(c): Although the ergonomic features of kitchen units are standardized to suit manufacturing purposes, there is room for some adjustment. As illustrated, standard base units, inclusive of worktop-thickness, should be about 900 mm high. Using details obtained from one manufacturer, their units can be lowered to 898 mm or raised to 923 mm to suit a person's height and working posture. Such adjustments are done by simply threading or unthreading the bottom section of the unit's adjustable legs. The plinth – that clips onto the front legs – is eventually reduced in width to suit the first example, but left in its full width to suit the second. Note that the latter does not allow for the plinth to be scribed to an uneven or un-level floor, as is usually necessary. These overall heights are arrived at as follows: 720 mm base unit + 38 mm worktop thickness + 140 mm minimum size of adjustable legs = 898 mm minimum; and 720 mm unit + 38 mm worktop + 165 mm maximum width of available plinth = 923 mm maximum.

However, such adjustments can create a problem if there is to be a free-standing cooker – instead of a hob – required to be flush to the worktop surface. Although modern cookers usually have screw-adjustable feet, their minimum and maximum adjustment would need to be checked out. Wall units should not be less than 460 mm, nor more than about 500 mm above the worktop. To put the cupboard shelves within easy

reach for most people, the ideal overall height from the floor is between 2 m and 2.150 m, as illustrated.

21.3 PLANNING THE LAYOUT

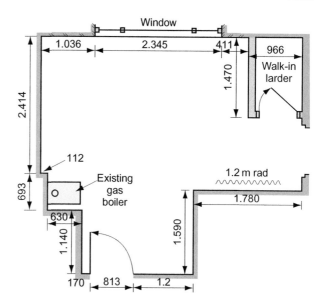

Figure 21.2 Rough-drawn measured survey of kitchen area

Figure 21.2: Although kitchen suppliers use CAD (computer-aided design) software nowadays to produce plans and three-dimensional views of a proposed kitchen, layouts can be produced from initial rough sketches and line drawings. Even if you have limited technical-drawing ability, simple scaled cutouts of the different sized units can be made and juggled with on a sheet of graph paper. If necessary, ideas for a kitchen design can be gained from magazines, kitchen showrooms or other people's homes. These ideas are borne in mind at the sketching and juggling stage. Sometimes all your ideas work out and an exciting design evolves – other times compromises have to be made. Apart from money, the size and shape of the room is the controlling factor in relation to the position of the door(s), window(s), radiator(s), etc. Therefore, the first stage is to make a rough freehand sketch and do a careful measured survey. The sketch shown above is reproduced from an actual kitchen refit completed in recent years. The original kitchen, apart from a sink and a walk-in larder, was practically non-existent.

Figure 21.3: Following on from the measured-survey sketch, ideas were juggled around and the final layout drawing evolved and is reproduced here. Note that this layout highlights practicalities such as the various small infill pieces required to achieve the U-shape arrangement of the base units and to infill

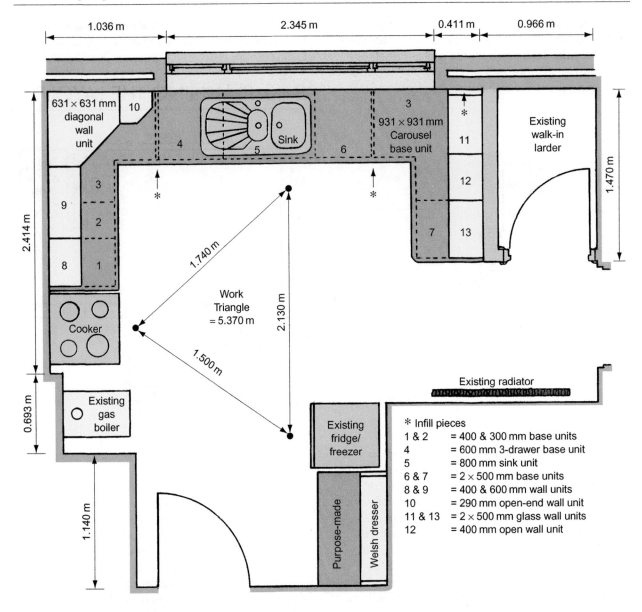

Figure 21.3 Final design layout of kitchen

the end of the number 11 wall unit. It was decided to keep the ventilated walk-in larder, which raised another practical issue. Wall units with glass doors and base units were to be fixed on the larder wall, which was of timber studwork and plasterboard. The base units would be secure enough fixed to this with modern cavity fixings, but to ensure the security of the wall units, a 100 mm strip of 12 mm plywood was fixed to the plasterboard surface, into the studs. This provided fixings for the metal mounting plates, as indicated in Figure 21.10. To accommodate the plywood strip, the back-edges of the wall units – that project beyond the groove for the back panel – had to be notched out and the suspension brackets had to be moved back into the units by 12 mm.

21.4 DISMANTLING THE OLD KITCHEN

If you are replacing an existing kitchen, then obviously the old one needs to be taken out. In doing this, use a sensible approach – not a sledge hammer! First, arrange for any integrated electrical wiring to be disconnected, the hot and cold water supply to be capped and any gas supply to be turned off and disconnected. After that, dismantle the fitted units generally in the reverse order of their assembly, i.e., cornices, pelmets, plinths, worktops, and so on. To speed things up, lever parts off with a crow bar instead of unscrewing everything tediously.

21.5 PRE-FITTING PREPARATION

If the composition of the walls is unknown to you, then check this out by tapping or piercing with a pointed awl, especially in a room not previously designated as a kitchen. Solid plastered walls are generally not a problem, dry-lined walls can be accommodated, but – as described in the paragraph under Figure 21.3 above – timber studwork and plasterboard may need to be overlaid with plywood strips, acting as fixing grounds. Other preparations are as follows:

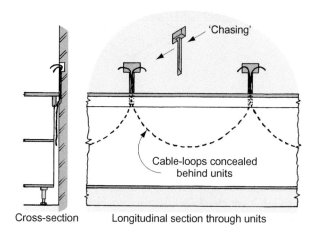

Figure 21.4 (a) Hit-and-miss 'chasing' above base units

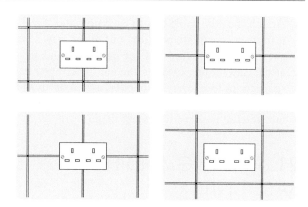

Figure 21.4 (b) Socket outlets related symmetrically to tiling

- *Figure 21.4(a)*: Now is the time to upgrade the electrics and install a new ring main (ring circuit), using 2.5 mm² two-core-and-earth cable to provide adequate double socket outlets above the worktops and in other places. These include electrical points for items such as a dishwasher, fridge and washing machine – and spurred, fused connection-units to an extractor fan, lighting in glass cabinets and lighting under the wall units, etc. If the cooker or hob is to be electric, then a separate circuit using 4 mm² two-core-and-earth cable must be run from a cooker control unit, back to a 30 amp fuse or a 32 amp MCB (miniature circuit breaker) in the consumer unit. Since changes in the electrical regulations came into force in January 2005, all of this work must now legally be done by a competent electrician – but 'chasing' (channelling) the walls to conceal the cables, can be tackled to lessen the specialist work (and, hopefully, the cost). As illustrated, this can be minimized by using a hit-and-miss technique for the outlets above the worktop, as the ring-main loops will be concealed within a void at the back of the cupboards.
- *Figure 21.4(b)*: If there is to be tiling on the back walls above the worktops (*splashbacks*) – which is quite common – the critical height and lateral

position of the socket outlets should be considered. The reasoning for this is partly to do with the ergonomics of handling the plugs – which is easier if the plugs are higher – but mostly to do with the symmetry of tile-joints related to socket-outlet positioning. Centre-line heights of outlets should be about 190 mm or more above the worktop. Being higher avoids clashes with sugar jars, etc.
- Drill through the outer wall and position the sink waste pipe, left poking through, ready for connection, and run the hot and cold water supplies, left poking up ready to enter the sink unit. These 15 mm Ø copper pipes should be capped off with so-called miniature valves. If a gas supply is required, this should also be installed at this stage.
- Any additional or new lighting required in the ceiling, should now be fitted.
- Metal mounting boxes should be screwed into each outlet-recess cut in the walls. Where tiles are to be fixed, the boxes can protrude by that thickness; (this should be taken into account from the outset, to reduce the cutting-depth into the wall). Oval-shaped plastic conduit should be fixed in the cable-chases, enabling easier future rewiring to be done. The conduit should run from a rubber grommet in the box and be side-fixed in the chases with small, galvanized clout nails. After the ring-main cabling has been looped through the plastic conduit, all the making good of plasterwork can be completed.
- The room should now be decorated, the ceilings finished, the walls and woodwork left ready for the final coat.
- Now clear the room and open the flat packs carefully, starting with the carcases. Line the floor with the cardboard to create a working area, examine each item for damage and start assembling the base units, minus the doors and drawers. (Manufacturers usually provide illustrated Assembly and Installation Guides for their range of units.)

Attach the adjustable legs to each unit and stand it in its approximate position, out of the way. Then assemble the remainder of the base units and set aside. Assembling carcases can continue if you have spare room for storage, otherwise assemble and fix in the order of base units, tower units, wall units, etc. You are now ready to start fixing.

21.6 FITTING AND FIXING BASE UNITS

Now to continue with the installation of the kitchen illustrated in Figure 21.3. In order to set out the position of the socket outlets required above the worktops in relation to the tiling joints, it was necessary in the early stages of preparation to mark the exact height of the worktop around the walls. In this case, the height was determined by the fact that the existing one-year-old electric cooker was to be reused. Its height of 910 mm was set on the wall as a datum measured up from its designated position, which was then carefully levelled and pencil-marked around the three walls as a continuous reference line.

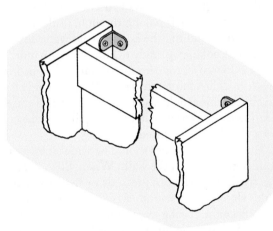

Figure 21.5 (a) Base-unit cheeks fixed to wall

Figure 21.5(a): With a few exceptions, base units are fixed to the wall via two small metal brackets supplied with the units. As illustrated, these are screwed close to the top of the side-cheeks on the inner back edges. If the wall surfaces are 'out of plumb' or not very straight in their length, the back edges of the units may need to be scribed to the irregular shape before the brackets are fixed. Sometimes, small pieces of thin packing strategically placed will suffice. Prior checking with a straightedge and spirit level will confirm whether this additional work has to be done.

Figure 21.5(b): As illustrated, the only other fixings required to hold the units are the two-part plastic

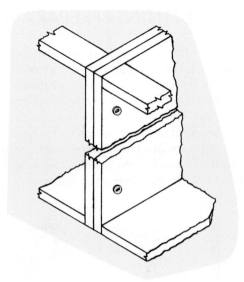

Figure 21.5 (b) Base-unit cheeks joined near front with two plastic connecting screws

connecting screws. These join the front edges of each unit's side-cheeks together. The male-and-female plastic screws require 8 mm Ø holes to be drilled through both cheeks whilst they are held firmly together with a pair of small G cramps. One fixing is positioned near the bottom, one near the top, clear of any hinge positions or drawer runners, etc. If necessary – according to what drill you are using – clamp a piece of flat waste material on the blind side of the cheeks and drill into it to avoid 'spelching' (breaking the edges around the exit holes).

Referencing the kitchen layout in Figure 21.3 again, having assembled the base units, screwed on the legs firmly and adjusted them to the approximate height, it was time to level up, adjust and scribe them to the wall. Starting with the left-hand carousel unit, the five wall-to-wall base units under the window were fitted (the fifth being another carousel unit in the right-hand corner against the larder wall). Only minimal scribing was required, mostly to do with the slightly bell-bottomed protrusion of the lower regions of traditional plaster-work, affecting the level of the units at right-angles to the wall. This was easily put right by planing about 4 mm tapered scribes off the lower back edges of the cheeks. Before repositioning each unit, the metal fixing brackets were screwed in place, as illustrated in Figure 21.5(a). With repeated reference to the spirit level, final adjustments were made to each unit's four adjustable legs. These adjustments also had to relate to the worktop reference line, which had to be 38 mm (worktop-thickness) above the top of the units.

Figure 21.5(c): The five base units now ready for fixing were composed of a 931 mm carousel unit each end, a 600 mm 3-drawer unit, a 800 mm sink unit and a 500 mm drawer-line unit, all adding up to 3.762 m

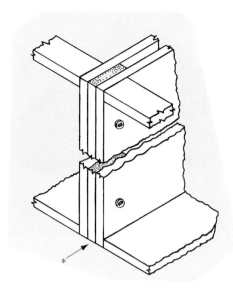

Figure 21.5 (c) * 15 mm laminate-edged infill strip between each carousel and adjacent unit

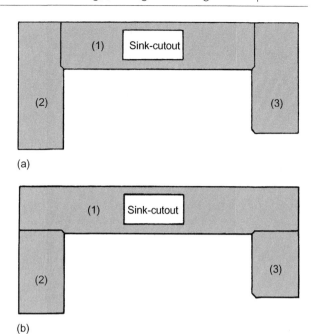

(a)

(b)

Figure 21.6 (a) Worktop (1) shown jointed at each end is impractical and should be avoided (b) More practical jointing arrangement

in relation to a room-width of 3.792 m. This meant that an infill of 30 mm was needed – which was known and accepted at the design stage. As illustrated above, this was made up of two 15 mm strips of melamine, edge-veneered to match the units. These were screwed to the face-side front of each carousel cheek and eventually connected to the adjacent units with the two-part plastic connecting screws supplied with the units. After marking the walls through the holes in the fixing brackets, the walls were drilled and plugged (see Chapter 3 on plugging, if necessary), all the two-part connecting screws were fitted and connected and the units were screwed back to the walls.

Using a similar procedure to that described above, the remaining base units on the side walls were fitted and fixed. This comprised a 300 mm drawer-line unit coming away from the left-hand carousel, finishing with a similar 400 mm unit up against the cooker. Only one 500 mm drawer-line unit was needed up against the right-hand carousel.

21.7 CUTTING, JOINTING AND FITTING WORKTOPS

Although guide-notes usually list the fitting of worktops after the wall units, I have always preferred to fit them before, without the restrictive, head-butting obstruction of the units above. However, once fitted they should be covered with the thick cardboard from the packaging before the work on the wall units is started.

Figure 21.6(a)(b): Laminate worktops are usually 3 m long, but some manufacturers can supply 4 m lengths as well. When fitting around two or more

corners in a kitchen, avoid having connecting joints parallel to each other at each end of one length of worktop, as in illustration (a) above. It is not ideal from a practical, fitting point of view – especially if a number of 'biscuits' are to be inserted in the joints. A better arrangement, shown in illustration (b) and used in the example kitchen, allows more control in fitting worktops (2) and (3) separately. Each of these arrangements requires 1 × 3 m and 1 × 4 m worktops. The 4 m length is cut in two for arrangement (a) and the 3 m length is cut in two for arrangement (b). Although small gaps between the worktop and the wall will be mastered by tiles or cladding, it may be necessary occasionally to scribe to the wall.

Figure 21.7(a): Because worktop (1) in Figure 21.6(b) was butted to the wall at each end, the square angle of the walls had to be checked and any major

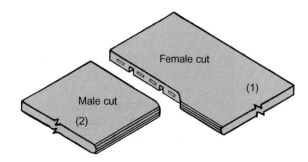

Figure 21.7 (a) Routed joint between worktops (1) and (2), showing biscuit and panel-bolt slots

deviation transferred to the worktop prior to cutting. Cutting can be done with a hardpoint handsaw or from the underside with a portable circular saw fitted with a 60 or 80 toothed TCT blade. Because circular-saw teeth revolve upwards towards the front of the saw's base, cutting from the underside minimizes damage to the laminate surface. After being fitted with a slight tolerance, worktop (1) was then removed and positioned for routing.

With the location bushes set up on a work-top-jig such as the Trend Combi 651, illustrated and described in Chapter 2, the jig was G-cramped into position and the first female joint on the left-hand side was cut. This had to be done in at least three cuts of increasing-depth to remove the bulk of the waste, before the finishing cut was made. Using a 1300 watt (minimum requirement) plunge router with a 12.7 mm ($\frac{1}{2}''$) collet and a 30 mm Ø guide bush, the first three cuts of varying-depth were run against the tolerance-side within the oversized guide slot of the jig. This left a small surplus edge on the joint-side of the jig which was routed off in one full-depth pass on the fourth finishing cut. After resetting the jig on the right-hand side of the worktop, the second female joint was cut in a similar way and the worktop was put back in position.

Worktop (2) was then sawn to length with an allowance of about 20 mm for finishing each end. The jig was then set up and a similar method of routing, as described above, was used to produce the first male end. Once routed, the worktop was laid in position and the quality of the joint carefully checked. If not a perfectly parallel fit, the jig would need adjusting before re-cutting the male end again. If the joint was satisfactory, the worktop could be marked for trimming the length to finish squarely against the cooker. The jig and router were used again to square this off. Finally, worktop (3) was cut to a provisional length and the whole jigging, routing and checking procedure repeated. This time, though – because exposed right-angled edges can be a dangerous height for children – the outer corner was removed by using the male-shaping side of the jig to produce the partly-rounded 45° shape.

Figure 21.7(b): Before removing the worktops to rout out the underside for the T-shaped recesses which house the panel bolt-connectors, and to cut the edge slots for the elliptical-shaped wooden biscuits, a pencil mark was run around the inner rails and cheeks of the sink unit to determine the position of the cut-out for the sink. Then the worktops were removed carefully – so as not to damage the routed edges – turned upside down, the jig set up again and used to form the bolt recesses. Three or four slots for No. 20 size biscuits were cut in each joint with a biscuit jointer – although, by using a biscuit-jointing cutter, this could have been done with the router.

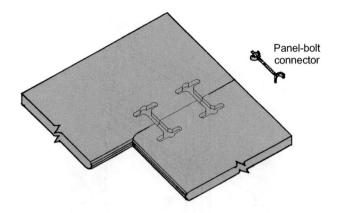

Figure 21.7 (b) 'T' or Bone-shaped recesses routed out to take panel-bolt connectors

21.7.1 Sink Cut-out

First, the pencilled outline of the inner base unit was squared up onto the laminate surface and the manufacturer's cardboard sink-template was related to it to mark the cut-out. Then the oblong shape was cut out with a jigsaw, after drilling a small entry hole. I learnt recently that an alternative trade-practice used nowadays for removing this oblong shape is to use a portable circular saw. Working from the underside, the cut-out is marked from the template or sink unit in relation to the marked outline of the unit. A portable circular saw (with a fine TCT blade) is set up with a limited blade-projection to slightly pierce through the laminate surface. The saw is carefully 'dropped on' to each of the four lines to cut the outline shape.

'Dropping on' is a technique for piercing the material in mid area. It can be dangerous and should only be practised by experienced woodworkers. If attempted, the back of the saw base should be pivoted against a temporary batten fixed to the material to prevent a dangerous, so-called *kickback*. It should be appreciated that by working a *circular* saw into each *vertical* corner, the cut-out will still be held by uncut corners below. This gives an advantage in stopping the cut-out from falling/breaking out prematurely and damaging the laminate edges. When the worktop is turned face up, a hardpoint handsaw is used in the corners to finish the job. Before the sink is finally fitted, the edges of the cut-out should be sealed with two or three coats of varnish – or at least receive a thorough spreading of silicone sealant.

21.7.2 Final Fixing

Understandably, plumbers usually like to fix the taps to the sink before it is fixed in position, but some plumbers also prefer to fix the sink as well as the taps before the worktop is placed and fixed in position. So, having previously agreed to the plumber's request, he was called in before the work proceeded. After the sink was installed, the long worktop was repositioned and fixed. This was done by screwing up through the plastic KD Fixit blocks previously fixed to the inner top surfaces of the base-unit cheeks – and through shank holes drilled through the manufacturer's partly drilled holes in the front rails.

Before worktops (2) and (3) could be fixed in the same way as worktop (1), they had to be joined together with the panel bolt-connectors. (Two per joint are usually supplied, but some fixers prefer three on 600 mm joints.) Before fixing these, though, the exposed extreme-ends of each worktop were veneered with laminate edge strip and carefully trimmed off with a Stanley (Craft) knife and a fine flat file.

With the male end of worktop (2) lined up to the female recess in worktop (1), but sitting back by about 150 mm, the three or four biscuits were tapped into the male-end slots. Working speedily now, with the panel-bolt connectors and a ratchet ring-spanner at the ready, a good-quality brand of clear silicone sealant was gunned along the routed female-recess and spread thinly with a spatula to within about 8 mm of the laminate surface. Following on quickly, the coloured jointing compound was then applied to the untreated 8 mm margin that had been left clear of the surface edges. (The coloured compound was used like this because it is not easy to spread and sets quickly; therefore the less applied, the quicker the assembly time.) Finally, the joint was brought closely together, the bolts inserted on the underside and quickly tightened with the ratchet spanner.

During this process, frequent checks were made regarding the flushness and closeness of the join (sometimes, a slightly fat or thin biscuit can upset the surface flushness. For this reason, some kitchen installers omit the biscuits completely – although the surfaces can be checked by putting the joint together dry, with the biscuits in place, before committing yourself to the final fit). With the aid of a wooden spatula, the squeezed-out compound was scraped off the worktop surface and cleaned up immediately with the cleaning solution provided – and any squeezed out droppings inside the unit were also cleaned up. This jointing procedure was repeated to join worktops (1) and (3) together before the final screw fixings were made on the underside.

21.8 FIXING THE WALL UNITS

Figure 21.8(a)(b): As there was no so-called tower units in the example kitchen used here, which would have been fixed next to determine the overall height,

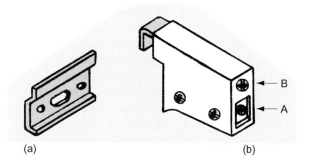

Figure 21.8 **(a)** Metal mounting-plate; **(b)** adjustable suspension-bracket

it was time to fix the wall units. These had already been assembled previously, minus the doors. After cutting and adapting the cardboard packaging to lay on the worktops for their protection, horizontal lines were pencilled around the wall to establish the top of the units. This equalled the units' height of 720 mm + 460 mm minimum height above the work-top. The line also acted as a datum for plotting the critical positions of the suspension-bracket mounting plates. These metal plates are fixed to the wall before the unit's integral suspension brackets are hooked upon them and adjusted for final position. By turning clockwise or anti-clockwise (with a non-powered screwdriver only) screw 'A', indicated above, is for vertical adjustment up or down and screw 'B' is for horizontal adjustment towards the wall or away from it.

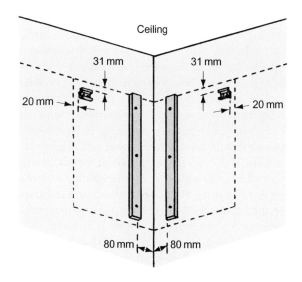

Figure 21.9 Positions for mounting plates and vertical battens (grounds) for corner wall-unit

Figure 21.9: Starting with the 631 × 631 mm diagonal corner unit, it was first held in position, up against the datum line, and each side cheek was marked on the wall, as illustrated. The two metal mounting plates were marked, plugged and screwed at 31 mm down from the top and 20 mm in from each side. The loose battens (supplied), acting as vertical grounds, were also fixed to the wall at 80 mm from the corner. The unit was then hung on the mounting plates via the integral suspension brackets and adjusted vertically to the datum line and a spirit level before being screwed back to hold its position. Finally, the unit was screwed to the grounds with the 8 × 16 mm flange-headed screws (4 each side) supplied.

The two wall units on the left of the corner unit, a 600 mm 2-door and a 400 mm single door, were fixed in a similar way. Each was fitted to a pair of metal mounting-plates positioned at 31 mm down from the datum and 20 mm in from each side as before – but instead of loose vertical rails, these units had a fitted horizontal rail, positioned low down behind the back panel. This needed to be drilled twice for final fixing to the wall with the $2\frac{1}{4}''$ (57 mm) × 8 CSK screws supplied.

The final unit on this side of the kitchen was the 290 mm open-end display unit on the right of the corner wall-unit. Because of its solid construction, mounting plates and suspension brackets were not needed this time. It was simply screwed to the wall at the top and bottom with two $2\frac{1}{4}''$ × 8 CSK screws supplied – and attached to the corner unit with four $1\frac{1}{4}''$ (32 mm) × 6 CSK screws supplied. These final four screws were concealed by being drilled and fixed from inside the cupboard unit, near the hinge positions. The exposed wall fixings were concealed with plastic KD cover caps.

The side-cheek attachment of the display unit near the hinge positions was also required on the other adjoining cheeks of the units. These fixings, though, were made with two-part plastic connecting screws – similar to the connections made on the base units, illustrated in Figure 21.5(b). Generally speaking, because of slight irregularities in the shape of most wall surfaces, it is better to make the cheek-connections – with the aid of a pair of G-cramps – before the units are screwed back to the wall through their lower, concealed rails.

Figure 21.10: On the other side of the kitchen now, against the larder wall, the three remaining wall units were fixed by using the same methods and techniques described above. However, to keep the symmetrical alignment of door-positions above in line with the door and drawer-positions below, a 31 mm infill was required against the left-hand wall, as indicated in the kitchen layout of Figure 21.3. This came about because the end of the 931 mm carousel base unit was 31 mm wider than the end of the 400 mm wall unit that had been joined to the 500 mm unit with glass-doors. Although infill pieces of limited size can be attached to the end of the unit, Figure 21.10 indicates an alternative method of fixing infill pieces to the wall.

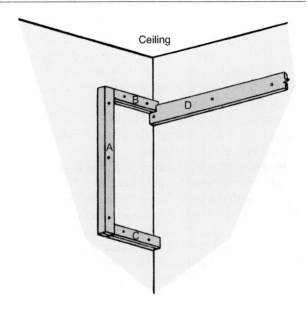

Figure 21.10 31 mm infill strip 'A' fixed to left-hand wall. 'B' and 'C' infill strips are optional. 'D' = 100 mm × 12 mm plywood strip fixed to stud partition to provide fixings for the metal mounting-plates

21.9 ADDING FINISHING ITEMS

Apart from finalizing the plumbing, electrics and decoration, the remaining finishing jobs in a logical sequence are: Fix manufactured backboard-splashbacks, or tiles, etc., around worktop-walls; Fix pelmets under wall-units (optional, but necessary if under-cupboard striplights are to be installed); Fix cornices above wall-units (also optional); Fix gallery rails to display units (if such units have been fitted); Scribe and reduce plinths to fit between underside of base units and the floor, minus an additional 2 mm to accommodate the plastic, plinth sealer-strip. Fit and fix plinths as per manufacturer's guide notes; If Cushion-close drawers have been chosen, fix the drawer-fronts and fit the drawers; If Standard drawer-box drawers are being used, assemble these and fit them; Fix the hinges to the doors and hang the doors; After all drawers and doors have been fitted and aligned carefully, and not before, fit the dummy drawer fronts where required, as per the guide notes; Finally, remove the backing adhesive from the small plastic buffers and fit them to all doors and drawers – and fit the following where required: hinge-hole cover caps, hinge-plate hole cover caps, wall-fixing screw cover caps, 8 mm-hole cover caps and cam cover caps.

22

Site Levelling and Setting Out

22.1 INTRODUCTION

Setting out the position of new buildings in relation to the required levels is a very responsible job and it must be borne in mind that mistakes can be costly. Apart from the upset to client and contractor, the local authorities can be unforgiving. Disputes over this can hold up a contract indefinitely and it is not unheard of for a partly-formed structure to be demolished and re-sited in the correct position. On large sites – especially those that involve a complexity of detail and the layout of roads *and* buildings – site engineers or surveyors are usually responsible for setting out and establishing the various levels. However, on small to medium-sized projects, such as the building of a few dwellings or a detached garage, the site levelling and setting out is usually done by the builder. This chapter covers the basic information required to set out the foundations, walls, drain-levels and site levels of a detached dwelling.

22.2 ESTABLISHING A DATUM LEVEL

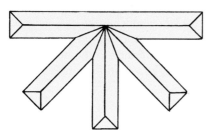

Figure 22.1 Ordnance Bench Mark (OBM)

Figure 22.1: It is essential in all site-levelling operations that the various levels required should have reference to a fixed datum. Wherever possible this should be the Ordnance Bench Mark (OBM), which is a chiselled-out arrow head with a horizontal recess above it, as illustrated above. OBMs are carved in stone blocks, usually found set in the walls of public buildings and churches. The inverted horizontal centre line over the arrow is a reference height set above the Ordnance datum by the Ordnance surveyors.

The original *Ordnance datum* was established as the *mean sea-level* at Newlyn in Cornwall. The value of OBM readings are recorded on Ordnance Survey maps related to block plans of built-up areas and can usually be viewed or obtained from a local authority's Building Control Department. Where there are no OBMs in the vicinity of a building site, an alternative, reliable datum – such as a stone step, plinth of a nearby building, or a manhole cover in the road – must be used.

22.2.1 Levelling Equipment Required – The Tripod Level

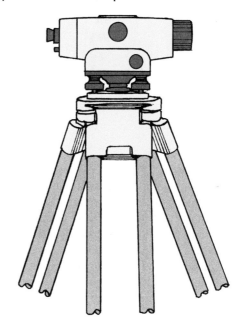

Figure 22.2 A typical automatic level

Figure 22.2: There are a variety of levelling instruments available and ranking high among these are the modified and improved versions of the popular traditional *dumpy level*. These instruments include the *quickset, precise,* and *automatic levels.* The automatic level represented above, operates by means of integral self-levelling glass prisms. Once the instrument has been attached to the tripod and levelled by means of a circular centre bubble, it is ready to be used. The principle upon which it works is that the observer's view of the horizontal hairline – that bisects the vertical hairline – across the lens establishes a theoretical horizontal plane, encompassing the whole site when the head of the instrument is pivoted through a 360° circle. This *theoretical plane* is referred to as the *height of collimation*, from which various levels can be measured and/or established. These levels are taken from a measuring rod known as a levelling staff.

22.2.2 The Levelling Staff

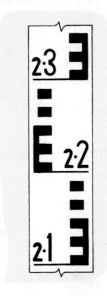

Figure 22.3 The 'E' staff

Figure 22.3: Metric staffs with five telescopic sections for easy carrying and use, are now available in aluminium or fibreglass. They can be extended to 5 m from a closed length of 1.265 m. The British Standard model indicated here is often referred to as the 'E' staff, for obvious reasons. Each graduated metre is marked with alternating black and red printing on a white background. Basically, the letters 'E' and the entire staff are divided into centimetres by virtue of each red or black square and each white space between the coloured squares equalling 10 mm. Thereby each 'E' has a vertical-value of 50 mm and the coloured squares and white spaces between each 'E' also equal 50 mm. A numbered value is printed at the bottom of each 'E', which is thereby expressing incremental readings of a decimetre (100 mm). Readings of less than 10 mm are estimated and can be quite accurate after a little practice. Clip-on staff-plumbing bubbles are available and advisable.

22.2.3 Terms Used in Levelling

Collimation Height: The height of a level's theoretical viewing plane above the original Ordnance datum established at Newlyn in Cornwall.

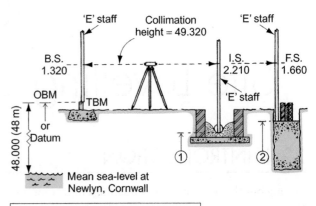

Figure 22.4 Graphic example of terms used in levelling

Backsight (B.S.): The first staff-reading taken from an instrument, usually on the datum.
Datum: A solid and reliable fixed point of initial reference, such as an OBM or a manhole-cover in the road.
Foresight (F.S.): The last staff-reading before an instrument is repositioned.
Intermediate sight (I.S.): All readings taken other than Backsight and Foresight.
Ordnance Bench Mark (OBM): As described above.
Temporary Bench Mark (TBM): This is a temporary datum transferred from the OBM and set up on site at a reduced level – or to the same level as any alternative datum used, such as a manhole-cover in the road – from which the various site levels are set up more conveniently. The datum peg, either a 20 mm Ø steel bar, or a 50 × 50 mm wooden stake, is usually set in the ground, encased in concrete. For protection, it should be strategically positioned to the front-side boundary of the site and have a confined fence-like guard built around it.
Reduced Level (R.L): Any calculated level position above the original Ordnance datum, as illustrated in Figure 22.4.

22.3 SETTING OUT THE SHAPE AND POSITION OF THE BUILDING

22.3.1 Setting-out Equipment Required – Setting-out Pegs

Figure 22.5(a): Setting-out pegs are usually of 600 mm length and made from 50 × 50 mm sawn softwood,

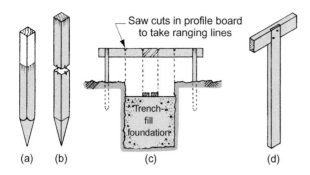

Figure 22.5 **(a)** Setting-out peg; **(b)** profile-stake; **(c)** profile board set up to give lateral position of foundation-trench and brickwork; **(d)** a boning rod

pointed at one end to enable them to be driven into the ground. The upper third should be painted with white emulsion for easy recognition. If preferred, pegs can be bought ready-made nowadays, with pyramidal or conical points.

22.3.2 Stakes

Figure 22.5(b): These are similar to setting-out pegs, but they are not painted and are about 900 mm long. They are driven into the ground in pairs and a profile board is nailed to their upper faces.

22.3.3 Profiles

Figure 22.5(c): These consist of ex. 150 or 100 mm × 25 mm boards nailed horizontally across a pair of stakes driven into the ground. (Corner profiles with three stakes were used originally for hand-digging, but unless using unreasonably long profile boards, come too near to the trenches and inhibit the mechanical digger.) There are two uses for profiles. 1) They can be set up – well clear of the intended trench lines – without any regard to being perfectly level and used for setting out the lateral position of the brick walls and the foundations. This involves marking the wall thickness and the foundation width on the board's face, as illustrated, and highlighting these marks with nails or saw cuts; Or, 2) if the tops of the profile boards (of which there may be at least eight) are set up level in themselves and in relation to each other, they can be used with a *boning rod* for sighting the levels of trench- and foundation-depths.

22.3.4 Boning Rods

Figure 22.5(d): As illustrated, these resemble a 'T' shape and are made to varying calculated lengths according to the reduced level required. The simple but effective technique requires one person to hold the boning rod

at any position along the trench, whilst another person sights across the two relevant profiles to see whether the top of the rod indicates that the trench levels or foundation pegs are too high, too low or correct. Traditionally, the rod was made from good quality prepared softwood of about ex. 75 × 25 mm section. The cross piece was usually jointed to the rod with a tee halving.

22.3.5 Surveying/Measuring Tapes

Fibreglass tape rules for setting out are now very popular, although traditional steel tapes with an improved finish are still available. The steel is either coated with white polyester and the printing protected with an additional top coat, or – for heavy-duty work – the steel is coated with yellow nylon and the printing protected with an additional coat. Tapes of 10, 20, 30 or 50 m are available, all with loop and claw ends and a fold-away rewinding handle. A sturdy, pocket tape-rule, such as the Stanley Powerlock® rule, 8 m long with a 25 mm wide blade, will also be required in the setting out.

22.3.6 Optical Squaring Instruments

Optical instruments such as digital theodolites, with an integral laser-plummet facility that emits a vertical beam from its tripod position, down onto the corner setting-out peg, can be used to establish 90° right angles required at foundation level – but they are extremely expensive for use on small to medium-sized sites. However, less expensive laser squares are available. Also, when the longest side of a right-angle in the setting out does not exceed about 20 m, the angle can be set out by using either a simple method of geometry or a method of calculation. Practical techniques for setting out right angles reliably by these methods are given in sections 22.3.9 and 22.3.10.

22.3.7 Ranging Lines

Ranging lines for initial setting out can be obtained in 50 m and 100 m reels of 2.5 mm Ø braided nylon, coloured orange or yellow for high visibility.

22.3.8 Establishing the Building Line

The first essential operation in setting out the position of the building is to establish *the building line* (sometimes referred to as *the frontage line*). As the term suggests, this line determines the outer wall-face of the building and is usually given on the site plan as a parallel measurement from the theoretical centre line of an existing or proposed road in front of the property.

Building-line positions are determined initially by the local authority, so it is imperative that they are

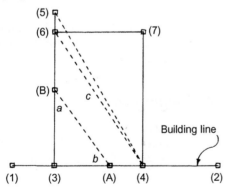

Figure 22.6 Building line related to existing properties and/or to the centre line of the road

adhered to. When a property is being built in between two existing dwellings, as illustrated in Figure 22.6, the building line can often be established by simply fixing a taut line to the wall-face on each side.

Alternatively, two setting-out pegs – numbered (1) and (2) in Figure 22.7 – are used to fix the position. They are driven into the ground and left protruding by about 100 mm on each side of the site, well clear of the flank-wall positions. Their centres are set to the given dimension of the building line from the centre of the road. After finally checking the embedded position of the pegs to the centre of the road again, 65 or 75 mm round-head wire nails are driven-in to correct any slight lateral deviation, with their heads left protruding by about 25 mm to hold a ranging line.

22.3.9 Setting Out from the Building Line

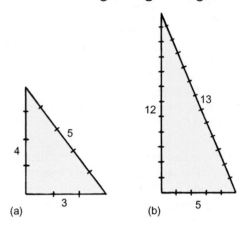

Figure 22.7 Two methods of establishing a right angle from the building line

Figure 22.7: After straining and tying a ranging line to the nail-heads of pegs (1) and (2), peg number (3) is measured in from the left-hand side boundary and driven-in centrally to the line to establish the corner (quoin) of the flank wall. The overall width of the building is then measured from this and peg number (4), illustrated, is driven-in. As before, wire nails are

driven into the pegs to determine exact positioning and are left protruding to hold a ranging line. The next step is to set out a right angle.

22.3.10 Forming a Right Angle

Figure 22.8 (a) $4^2 + 3^2 = 5^2$; (b) $12^2 + 5^2 = 13^2$

Figures 22.7 and 22.8(a)(b): The right angle to form the first flank wall can be made by using either a method of calculation, illustrated in Figure 22.8(c) and described below; a simple method of geometry such as the 3:4:5 method illustrated above at (a), or the equally simple 5:12:13 method illustrated at (b). These two ancient geometric formulas that preceded Pythagoras' theorem, enable right-angled triangles to be formed with integral sides, i.e, sides which can be measured in whole numbers or equal units. Providing the units remain equal and are set to the prescribed ratio, right angles of various sizes can be formed easily by changing the value of the unit.

When using either of these methods, choose a size of unit that will form as large a triangle as possible in relation to the setting out. For example, a 1.5 m unit to a 3:4:5 ratio = 3 × 1.5 = 4.5 m base line, 4 × 1.5 = 6 m side line and 5 × 1.5 = 7.5 m hypotenuse. This is indicated in Figure 22.7, where peg (B) is set up to a 3:4:5 triangle from pegs (3) and (A). Alternatively, the whole base line between pegs (3) and (4) in Figure 22.7 can be divided to provide equal units. The result of such a method is indicated between pegs (3), (4) and (5). The advantage with this is that the ranging line between pegs (3) and (5) is long enough to set up the rear of the building at peg (6). The setting up of peg (7), to complete the outline of the building, can either be set up in a similar way to peg (6), or its position can be determined by parallel measurements. Either way, once the four main pegs are established, diagonals (3) to (7) and (4) to (6) should be measured for equality to confirm that the setting out is truly square.

Method of Calculation

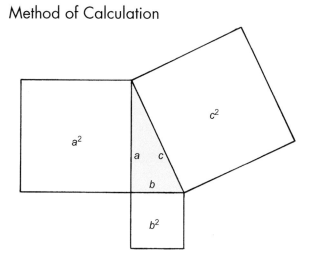

Figure 22.8 (c) Pythagoras' theorem: $a^2 + b^2 = c^2$

Figure 22.8(c): Finally, if preferred, the right angle can be formed by treating pegs (3), (4) and (6) as a triangle and using a method of calculation known as Pythagoras' theorem, where the square on the hypotenuse *c* of a right-angled triangle is equal to the sum of the squares on the other two sides *a* and *b*, i.e.

$$a^2 + b^2 = c^2$$

Thereby it is possible to find the length of the hypotenuse if the lengths of sides *a* and *b* are known. Once the sum of the *square* on the hypotenuse (represented by the superscript (^{2}) after the number, meaning that the number is to be multiplied by itself) has been worked out, the square root of that sum will give the length of the hypotenuse *c*, i.e.

$$\sqrt{c^2} = c$$

With reference to Figure 22.7 again, if the width of the building was 8.750 m between pegs (3) and (4) and the depth was, say 12.500 m between pegs (3) and (6), to find the length of the hypotenuse, *c* thus enabling peg (6) to be squared to peg (3), the following sum would be used:

$$c^2 = 8.750^2 + 12.500^2$$
$$(c^2 = 8.750 \times 8.750 + 12.500 \times 12.500)$$
$$c^2 = 76.562 + 156.25$$
$$c^2 = 232.812$$
$$c = \sqrt{232.812} \ (c = \text{the square root of } 232.812)$$
$$c = 15.258 \text{ m}$$

22.3.11 Squaring Technique

Figure 22.9(a): Whatever method of triangulation is used from the three variations given above, the precise pin-point positioning of peg (5), peg (6), or peg (B) is critical. It is usually done with two tape rules, each looped over the nails in the two relevant building-line pegs and strained over the site to meet at their intersecting apex point, as in Figure 22.9(a). Initially, this will be done to establish the position for the peg, then repeated to pin-point the nail position. Intermediate support, such as bricks-on-edge (if the pegs are about 100 mm above ground), should be used under the tape if there is too much sagging. If only one tape rule is available, a brick-on-edge can be set up at the apex point and adjusted for position after separate measurements are made to the base pegs and marked on the brick. The apex peg is then related to this mark and driven-in. To pin-point the apex with a single rule, bisecting arcs can be made on the peg,

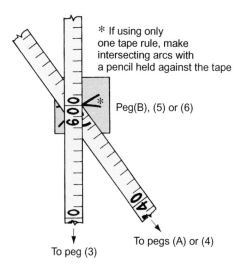

Figure 22.9 (a) Squaring technique with tape rules

as indicated in the Figure above. If ever a peg is found to be slightly out of position when pin-pointing, rather than move it, drive another peg in alongside.

Figure 22.9(b): When looping the end of a tape rule over the centralized nail in a peg, it should be realized that about three or four millimetres are gained or lost every time you read or set a measurement. This is

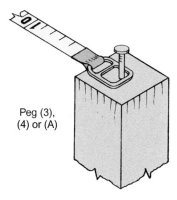

Figure 22.9 (b) Tape rule looped over RH wire nail

partly due to the inaccuracy of most end-loops and to the fact that half the nail's shank-diameter is over-sailing the centre point. With this in mind, the allowance that needs to be added to every *looped measurement*, or taken off of every *looped reading* can be worked out. 75 mm round-head wire nails have a 3.75 mm Ø, so nearly 2 mm is lost or gained here. Now check the tape with another rule and add any loop inaccuracy to this and you have your + or − allowance.

22.3.12 Positioning and Marking the Profile Boards

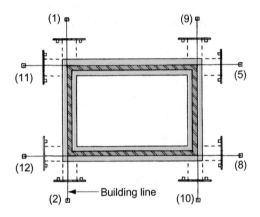

Figure 22.10 (a) Positioning of profiles, showing additional setting-out pegs (8) to (12)

Figure 22.10(a): As mentioned previously and illustrated above, the profiles for lateral positioning of the walls and foundation-trenches must be kept well back to allow access for the mechanical digger. The 900 mm stakes are driven-in to about half their length and the boards are fixed reasonably level with 50 or 65 mm round-head wire nails. The boards should be fixed on the side of the stakes that takes the strain when the ranging lines are stretched out.

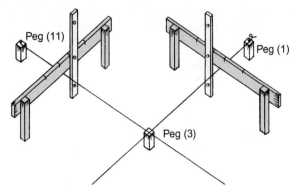

Figure 22.10 (b) Levelling up to profiles from setting-out lines

Figure 22.10(b): Bearing in mind the relative heights of the profile boards (about 450 mm) above the

setting-out pegs and lines (about 100 mm), as illustrated, it can be seen how the face-of-brickwork lines are transferred to the profiles. However, it is a delicate task and care should be taken in steadying a spirit level against the face of the board and adjusting it until a plumb position coincides with the ranging line below. Once marked with a sharp carpenter's pencil and squared onto the top edge, this acts as a datum from which the wall thickness and foundation width are marked. These are established by shallow saw cuts or by fixing wire nails into the top edge. Note that as indicated in Figure 22.10(a) and (b), additional setting-out pegs are required at points (8) to (12) to allow the ranging lines to be extended for marking the profile boards.

22.3.13 Sloping Sites

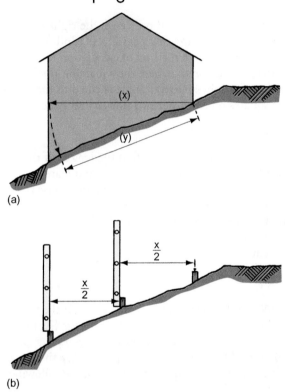

Figure 22.11 (a) Horizontal measurements on a sloping site; (b) stepped measurements via setting-out pegs and a spirit level

Figure 22.11(a)(b): Finally, sloping sites must be mentioned. Although these will not affect the triangulation methods of setting out a right angle, they can affect the horizontal measurements of walls. As illustrated, the given dimension of a wall (X) can be considerably reduced geometrically to produce (Y) if measured down a slope. To combat this, stepped measurements would be required at ground level, as indicated.

Appendix: Glossary of Terms

The terms and other technical names listed here for explanation, are relevant to those used in this book only – not to the industry as a whole. For continuity, some terms are explained in the appropriate chapter and may or may not be repeated here.

Aggregate: Stone, flint and finer particles used in concrete.

Apron lining: A horizontal board, covering the rough-sawn vertical face of a trimmer or trimming joist in a stairwell.

Architrave: A plain or fancy moulding, mitred and fixed around the face-edges of door openings, to add a visual finish and to cover the joint between plaster(board) and door frame or lining.

Arris, arrises: The sharp corner edges on timber or other material.

Ashlar: A square- or oblong-shaped hewn stone.

Ashlaring: Ashlars laid to a stretcher-bond pattern to form a stone wall; or the appearance of ashlars formed on the quoins of brick- or sand-and-cement-rendered walls.

Ashlering or ashlers: Vertical timber studs fixed in an attic room from floor to rafters, to partition off the lower, acute angle of the roof slope.

Balusters: Lathe-turned wooden posts, fixed between the handrail and string capping or handrail and landing nosing, as part of the balustrade of a staircase.

Baluster sticks: As above, but only square posts with no turning.

Balustrade: The barrier or guarding at the open side of a staircase or landing, comprising newel posts, handrail and balusters (spindles).

Bare-faced tenon: A tenon with only one shoulder.

Bearer: A timber batten, usually ex. 50 × 25 mm, that supports a shelf.

Bearing: The point of support for a beam, lintel or joist.

Bed, bedding or bed joint: A controlled thickness of mortar – usually 10 mm – beneath timber plates, bricks and blocks.

Birdsmouth: A vee-shaped notch in timber, that is thought to represent a bird's mouth in appearance.

Bits: Parallel-shank and square-taper shank tools that fit a power drill and/or a carpenter's ratchet brace for drilling and countersinking.

Block partitions: Partition walls built of aerated insulation blocks, usually measuring 440 mm long × 215 mm high × 100 mm thick.

Bow: A segmental-shaped warp in the length of a board, springing from the wide face of the material.

Boxwood: A yellow-coloured hardwood with close, dense grain – still used in the manufacture of four-fold rules and chisel handles, to a limited extent.

Brace: 1. A diagonal support. 2. A tool for holding and revolving a variety of drill bits.

Brad head: The head of a nail (oval brad) or awl (bradawl) whose shape is scolloped from the round or oval to a flat point.

Breather membrane: Water-resistant, breathable (vapour permeable) fabric material manufactured from high-density polyethylene. Available in 50 m rolls, 1.5 m and 1 m wide. Used in recent years for protecting roofs and timber-framed walls from external elements, whilst improving the energy efficiency and thermal targets in a building by making ventilation unnecessary in both cold and warm-decked constructions.

Bullnose step: A step at the bottom of a flight of steps, projecting past the newel post, with a quadrant-shaped (quarter of a circle) end.

Burr: A sharp metal edge in the form of a lip, projecting from the true arris of the metal.

Butt-joint: A square side-to-side, end-to-end or end-to-side abutment in timber, without any overlapping.

C-studs: Metal-stud partitioning, likened to capital letter C (because of its sectional shape) and used for wall-studs and/or door-studs.

Casement: Hinged or fixed sash windows in a casement frame.

Centimetre: One hundredth of a metre, i.e. ten millimetres (10 mm).

Chamfer: An equal bevel (45° × 45°) removed from the arrises of bearers or slatted shelves.

Chase, chased, chasing: Rough channels or grooves cut in walls or concrete floors to accommodate pipes, conduits or cables; or cut in the face of mortar beds to take the top-turned edge of apron flashings.

Chord: A reference to trussed-rafter rafters (top chord) and trussed-rafter ceiling-joist ties (bottom chord).

Chuck: The jaws of a drill or brace.

Cladding: The clothing of a structure in the form of a relatively thin outer skin, such as horizontal weather-boarding or tiles.

Cleats: Short boards or battens, usually fixed across the grain of other boards to give laminated support to the join.

Clench-nailed: Two pieces of cross-grained timber held together by nails with about 6–10 mm of projecting point bent over and flattened on the timber, in the direction of the grain for a visual finish, or across the grain for extra strength.

Coach bolt: This has a thread, nut and washer at one end and a dome-shaped head and partly square shank at the other. The square portion of shank is hammered into the round hole to stop the bolt turning while being tightened.

Common brickwork: Rough brickwork or brickwork built with cheaper non-face bricks, to be plastered or covered.

Concave: Shaped like the inside of a sphere.

Concentric: Sharing the same centre point.

Conduit: A metal pipe for housing electrical cables; although nowadays plastic and fibre tubes are mostly used.

Convex: Shaped like the outside of a sphere.

Corbel or corbelling: A structural projection from the face of a wall in the form of stone, concrete or stepped brickwork, to act as a bearing for wall plates and purlins, etc.; straight or hooked metal corbel plates, being the forerunner of modern joist hangers, were used at about 1 m centres, projecting from the face of a wall, to support suspended wall plates.

Course: One rise of bricks or blocks laid in a row.

Cramp: 1. Sash cramp – a tightening device for holding framed timber components together under pressure, usually while being glued. 2. Frame cramp or tie – a galvanized steel bracket holding-device, fixed to the sides of frames and bedded in the mortar joint.

Cross-halving: A half-lap joint between crossed timbers.

Cup or cupping: Concave or convex distortion across the face of a board, usually caused by the board's face being tangential to the growth rings.

Datum: A fixed and reliable reference point from which all levels or measurements are taken, to avoid cumulative errors.

Decimetre: One tenth of a metre, i.e. 100 millimetres (100 mm).

Dihedral angle: The angle produced between two surfaces, or geometric planes, at the point where they meet. For example, two vertical surfaces meeting at right-angles to each other, produce a dihedral angle of 90°, but incline the surfaces from their vertical state, to represent a hip or valley formation, and the dihedral angle thus produced is different, according to the degree of inclination.

Door joint: The necessary gap of 2–3 mm around the edges of a door for opening clearances.

Dot-and-dab: A plastering technique used for the dry-lining of solid walls, whereby the wall is trued up with small *dots* of fibreboard, followed by the plasterboard being adhered with *dabs* of plaster between the dots.

Dovetail key: The locking effect of a dovetail, or nails driven in to form a dovetail shape.

Dowel: A round wooden (usually hardwood) or metal pin.

Draw-bore pin: A front-tapered wooden dowel, driven into an offset hole drilled (separately) through a mortice and tenon joint, to pull up the shoulder-fit and permanently reinforce the joint.

Easing: 1. Removing shavings from an edge to achieve a better fit. 2. Concave and convex shaping of the top edges of stair strings (especially on the two right-angled wall strings containing tapered steps in a quarter-space turn).

Eaves: The lowest edge of a roof, which usually overhangs the structure from as little as the fascia-board thickness up to about 450 mm, where rainwater drainage is effected via a system of guttering and downpipes.

Eccentric point: The bent portion of a trammel-head pin, which causes the axis (centre) of the pin to move, when rotated, in an eccentric orbit, even though the pin-pointed position remains concentric. This allows fine adjustments to be made to the trammel distance without altering the trammel head itself.

Facework or face brickwork: Good quality face bricks, well-laid to give a finished appearance to the face of walls.

Fair-faced brickwork: Common brickwork, roughly pointed and bagged (rubbed) over with an old sack.

Fillet: A narrow strip of wood, rectangular or triangular in section, usually fixed between the angle of two surfaces, as a covermould.

Firring: Building up the edges of joists, with timber strips which may be parallel or wedge-shaped, to achieve a level, a sloping or a higher surface when boarded or covered in sheet material.

Flange: The bottom or top surface of a steel I beam or channel section.

Flashing: A lead or felt apron that covers various roof junctions.

Fletton: An extensively used common brick, named after a village near Peterborough and made from the clay of that neighbourhood.

Floating: see Rendering.

Flush: A flat surface, such as a flush door, or in the form of two or more components or pieces of timber being level with each other.

Gablet: The triangular end of a roof, known as a gablet when separated from the gable wall below, as in a gambrel roof.

Glue blocks: Short, triangular-shaped blocks, glued – and sometimes pinned – to the inside angles of steps in a wooden staircase and other joinery constructions.

Going: The horizontal distance, in the direction of flight, of one step or of all the steps (total going) of a staircase.

Grain: The cellular structure and arrangement of fibres, running lengthwise through the timber.

Green brickwork: Freshly laid or recently laid brickwork, not fully set.

Groove: A channel shape sunk into the face or edge of timber or other material.

Grounds: Sawn or planed (prepared) battens, which may be preservative-treated, used to create a true and receptive fixing surface.

Gullet: The lower area of the space between saw teeth.

Gusset plate: A triangular-shaped metal (or timber; usually plywood) joint connector.

H-studs: Metal sections used in partitioning as intermediate studs – and likened to capital letter H (or I), because of their sectional shape.

Half-brick-thick wall: A stretcher-bond wall, where bricks are laid end-to-end only, in one thickness of brick.

Hardcore: Broken brick and hard rubble used as a substrata for concrete oversites.

Hardwood: A commercial description for the timber used in industry, which has been converted from broad-leaved, usually deciduous trees, belonging to a botanical group known as angiosperms. Occasionally, the term hardwood is contradictory to the actual density and weight of a particular species. For example, balsa wood is a hardwood which is of a lighter weight and density than most softwoods.

Heel: The back, lower portion of a saw or plane.

Hone: Sharpen.

Housing: A trench or groove usually cut across the timber.

I-studs: Metal sections used in partitioning as intermediate studs – and likened to capital letter I (or H), because of their sectional shape.

Inner skin: The wall built on the dwelling-side of a cavity wall, usually constructed in blockwork.

Inner string or wall string: One of the two long, deep boards that house the steps at the side of a wooden staircase, being on the side against the wall – or a lesser-width board used in formwork (shuttering) to support the ends of the riser boards that abut the wall.

In situ concrete: Concrete units or structures cast in their actual and final location, controlled by _in situ_ formwork (timber or metal shuttering).

Jamb: The name given to the side of a door frame or window frame.

Joists: Structural timbers that make up the skeleton framework in timber floors, ceilings and flat roofs.

Kerf: The cut made by a saw during its progress across the material.

Kicker(s): Longitudinal support battens used at the base of shuttering panels.

Knots: Roots of a tree's branches, sliced through during timber conversion. Healthy-looking knots are known as _live_ knots, and those with a black ring around them are likely to fall out and are known as _dead_ knots.

Knotting: Shellac used for sealing knots (to stop them _bleeding_ or exuding resin) prior to priming. Shellac is derived from an incrustation formed by lac insects on the trees in India and nearby regions.

Lag or lagged: Wrapped or covered with insulation material.

Landing nosing: The narrow, projecting board, equal in thickness and shape to the front-edge of a tread board, that is fixed on all top edges of the landing stairwell. It is often rebated on the underside to meet a reduced-thickness of floor material.

Lignum vitae: Dark brown, black-streaked hardwood with extremely close grain. It is very hard and dense, about twice the weight of British elm, grown in the West Indies and tropical America.

Lintel: A concrete or metal beam over door or window openings.

Mirror plates: T-shaped, thin (brass or brass-coated, etc.) metal plates, with three fixing holes, traditionally used for fixing mirrors to walls.

Mitre: Usually a 45° bisection of a right-angled formation of timber (or other material) members – but the bisection of angles other than 90° is still referred to as mitring.

Muzzle velocity: The speed of a nail in the barrel of a cartridge tool.

Newel posts: Plain or lathe-turned posts in a staircase, usually morticed and jointed to the outer string and the handrail tenons and attached to the floor at the lower end and to the landing trimmer at the other. The newel posts assist in creating good anchorage of the staircase at both ends, as well as providing stability and strength to the remainder of the balustrade.

Noggings: Short timber struts, usually between studs or joists and rafters.

Normal: The geometrical reference to a line or plane at right-angles to another, especially in the case of a line radiating from the centre of a circle, in relation to a right-angled tangent on the outside.

Nosing: The projecting front-edge of a tread board past the face of the riser, reckoned to be not more than the tread's thickness.

Open-riser stairs: Stairs without riser boards.
Outer skin: The wall built on the external side of a cavity wall.
Outer string(s): One of the two long, deep boards that house the steps at the open side of a wooden staircase, away from the wall – or a wider board or panel used in shuttering to form the ends of fluid-concrete steps, that includes the depth of the waist-thickness at the open side of a concrete stair, away from the wall.
Oversite: An *in situ* concrete slab of 100 mm minimum thickness, laid over hardcore on the ground as part of the structure of this type of ground floor.

Paring: Chiselling – usually across the grain.
Pellets: Cork-shaped plugs for patching counterbored holes when involved in screwing and pelleting.
Perpends: Perpendicular cross-joints in brickwork or masonry.
Pilot hole: A small hole made with a twist drill or brad-awl to take the wormed thread of a screw.
Pin or pinned: Fixed with wooden-dowel pins, but more commonly the reference is to fixing with nails or panel pins.
Pinch rod: A gauge batten for checking internal distances for parallel.
Pitch: 1. The angle of inclination to the horizontal of a roof or staircase. 2. Repetitive, equal spacing of the tips or points of saw teeth or other equally spaced objects.
Pitch-board: A purpose-made triangular template for setting out stair strings.
Plant: Equipment.
Planted mould or stop: Separate mouldings or door stops, pinned or fixed by nails to the base material.
Plate: 1. A horizontal timber that holds the ends of vertical or inclined timbers in a state of alignment and framed spacing, as in the case of roof wall-plates and stud-partition sill plates and head plates. 2. Metal components such as striking plates for latches and letter plates.
Plumb or plumbing: Checking or setting up work in a true, vertical position.
Plumb cut: The vertical face of a cut angle.
Pocket screwing: Screws which are angled or skew-screwed into shallow niches and shank holes drilled at an angle through the (usually) hidden face of the timber being fixed – an example being the top riser and nosing piece of a stair.
Precast concrete: Concrete units cast in special mould boxes in a factory or on site, but not cast in their actual and final location.
Primed: Painted with the first coat of paint (priming) after being knotted.

Profile: 1. A horizontal board attached to stakes or pegs driven in the ground, across the line of an intended foundation strip – one at each end, set well clear of the digging area, has saw cuts or nails in the top edge of the board to mark the foundation and wall positions. When initially digging or building, ranging lines are set up across the boards to establish the required positions. In the case of mechanical digging, a thin line of sand is trickled vertically beneath the lines to mark the trench position, then the lines are removed and reinstated later when the building work is to be started. 2. Any object or structure acting as a template in guiding the shape of something being made or built.

Quadrant: 1. A right-angled sector shape, equalling a quarter of a circle. 2. A small, wooden bead of this shape.

Rebate: A return or inverted right-angle removed from the edge of a piece of wood or other material.
Rendering and/or floating coats: Successive coats of coarse plaster built up to an even and true surface for skimming.
Resin bonded: This is usually a reference to the cross-laminates of plywood having been bonded (glued) with synthetic resins. According to the type of resin used, the plywood may be referred to as moisture resistant (MR), boil resistant (BR), or – better still – weather and boil proof (WBP).
Retaining wall: A wall built to retain high-level ground on a split-level site.
Reveals: The narrow, return edges or sides of an opening in a wall.
Rise: The vertical distance of one step or of all the steps (total rise) of a staircase.
Riser: The vertical face or board of a step.
Runners: Sawn-timber beams, used in formwork.

Sarking: Roof boarding or sheeting material and/or roofing felt.
Scribe; scribing: Techniques used in joinery and second-fixing carpentry for marking and fitting mouldings against mouldings, or straight timbers against irregular shapes or surfaces.
Seat cut: The horizontal face of a cut angle.
Set, setting: 1. The alternate side-bending of the tips of saw teeth. 2. The chemical setting action that brings initial hardening of glue, concrete, mortar or plaster. 3. A coat of finishing plaster (see Skimming).
Shank: The stem or shaft of a tool or screw.
Shank hole: A small hole made with a twist drill to take the stem or shank of a screw.
Sheathing: Close-boarding or sheet material such as plywood or Sterling OSB, fixed to vertical framing (studs) as a strengthening-skin and a base for cladding with weather-boarding or tiles.

Sherardized: Ironmongery (hardware), such as nails and screws, coated with zinc dust in a heated, revolving drum and achieving a penetrated coating, claimed to be more durable than galvanizing.

Shuttering: Temporary structures formed on site to contain fluid concrete until set to the required shape; shuttering is also known as *formwork*.

Skew-nailing: Nailing at an angle of about 30–45° to the nailed surface, through the sides of the timber, instead of squarely through the edge or face.

Skimming or setting coat: The fine finishing plaster, traditionally applied to ceilings and walls in a 3–5 mm thickness and trowelled to a smooth finish.

Slatted shelves: Shelves formed with open-sided slats/battens (usually of ex 50 × 25 mm softwood) to allow air to pass through in an airing cupboard.

Soffit: The underside of a lintel, beam, ceiling, staircase or roof eaves' projection.

Softwood: A commercial description for the timber used in industry, which has been converted from needle-leaved, usually coniferous evergreen trees, belonging to a botanical group known as gymnosperms. Occasionally, the term softwood is contradictory to the actual density and weight of a particular species. For example, parana pine is a softwood that is quite heavy and dense, like most hardwoods.

Spall, spalling: A breaking or flaking away of the face material of concrete, brick or stone.

Span: 1. Clear span – the horizontal distance measured between the faces of two opposite supports. 2. Structural span – for design calculations, is measured between half the bearing-seating on one side to half the bearing on the other. 3. Roof span – measured in the direction of the ceiling joists, from the outer-edge of one wall plate to the outer-edge of the other.

Spindles: In carpentry and/or joinery terms, this is an alternative name for balusters and therefore refers to lathe-turned wooden posts, fixed between the handrail and string capping or handrail and landing nosing, as part of the balustrade of a staircase.

Spotting: Marking a trowel-line through a sliver of trowelled mortar when setting out walls and partitions on concrete foundations or oversites.

Spring or sprung: Warping that can occur in timber after conversion and seasoning, producing a *sprung, cambered* or segmental-shaped edge adjacent to the wide face of the material. Joists and rafters should be placed with the sprung edge uppermost.

Stretcher: 1. The temporary timber batten at the base of a door frame or lining that stretches the legs apart to the correct dimension until the fixing operation takes place. 2. The long face of a brick.

Strut: A timber prop, supporting a load vertically, horizontally or obliquely.

Stub tenon: A shortened tenon, usually morticed into its opposite member by only a half to two-thirds its potential size.

Stuck mould or rebate: Moulded shapes or rebates cut into the face of solid timber members.

Studs, studding, stud partitioning: Vertical timber posts.

Tamp, tamped, tamping: A term used in concreting, referring to the level surface being zig-zagged and tamped (compacted) with a levelling board. The tamping is effected by bumping the board up and down as it is moved across the surface.

Tang: A square-taper shape at the end of a round-shanked tool.

Tangent: This is a line that lays at right-angles to another line – known geometrically as a *normal* – that radiates from the centre of a circle.

Toe: The reference in carpentry to the front of a saw or plane.

Tonk Strips: Patented metal strips, with closely-spaced cutout slots that receive separate, metal lugs placed in them, to support adjustable shelves.

TRADA: Timber Research and Development Association.

Tread: The horizontal face or board of a step.

Twist: 1. Warping that can occur in timber after seasoning and conversion, producing distortion in length to a spiral-like propeller-shape. 2. Distortion in a framed-up unit caused by one or more of the members being twisted, or by ill-formed corner joints.

U-track: Metal-stud partitioning, likened to capital letter U (because of their sectional shape) and used for the head- and floor-plates.

Viability graphs: Diagrams in this book, devised to give a quick reference to regulation-compliance of stair-design in relation to the 2R+G rule.

Voussoir: A tapered brick in a gauged brick arch.

Waist-thickness: The critical measurement of concrete between a stair's soffit and the internal angle between the tread and the riser, measured at right-angles to the soffit.

Waling(s): Substantial horizontal timber(s) used to align/bind/support temporary formwork (shuttering).

Warp: Distortion of converted timber, caused by changing moisture content (see Bow, Spring, Twist, Cup and Wind).

Web: The connecting membrane between the flanges of a steel I beam or channel section.

Wind, winding: These terms are the equivalent of Twist and Twisting. The expression *in wind* means twisted and *out of wind* means not twisted.

Index

Abbreviations xiv, 6
Adhesive fixings
    flooring panels 91
    skirting 241–2
    stairs 118, 119, 120
Adjustable-strap cramps 108
Aggregate 276
Air bags 202
Alternating-tread stair 129, 135
American projection 3
Angle
    (corner) ties 155, 173
    fillets 181, 182
Angle-of-lean 50
Angular perspective 5, 6
Apron
    flashing 181, 188
    lining 112, 126, 271
Arch centres
    four-rib 231
    multi-rib 233
    single-rib 229
    twin-rib 230
Arches
    depressed
        Gothic 224
        semi-elliptical 224
        Tudor 225
    equilateral Gothic 224
    lancet, Gothic 224
    segmental 220
    semi-circular 219
    semi-elliptical 221–4
    straight-top Tudor 225–6
    Tudor 224, 225
Archformer 218
Arch geometry (methods)
    concentric circles 222
    five-centred 223
    fixed 225
    intersecting arcs 221
    intersecting lines 221
    long-trammel 222

pin-and-string 222
short-trammel 222
three-centred 223
variable 224, 225
Architrave 235–9, 271
    grounds 112
Arris 51, 271
Ashler studs 188, 190, 271
Auger bits 20
Axes of ellipse 220
Axonometric projections 5

Back gutters 194, 195
Backing
    bevel 160
    line 159
Backsight (BS) 266
Baffle boards (roof ventilation) 185
Balusters 114, 127–8, 271
Balustrade regulations 129, 130–137
Balustrades, fixing 125–8
Barge boards 158
Barrier, vapour 92–3, 181–7
Battened floor 92
Battens
    ground 73, 111–3, 122
    tile or slate 158
Beads
    clip-in 79
    shuffle 78
Beam
    spreader 201
    steel 86, 96
Beam compass (trammel) 228
Bearers
    cross 201
    stair 125
    tank 201
Bedding 177, 271
Bench mark 115
Bevelled housing 95
Bevel formulas (roofing) 168
Bevel, sliding 11

Binders 156
Binding (doors) 245
Birdcage awl 22, 23
Birdsmouth
    rafter 160, 271
    saw stool 48–51, 271
Biscuits, wooden 29, 262
Bisecting 219
Bitumen DPM 92
Blind tenon 95
Block plan 7
Blocks
    locating 78, 79
    setting 78, 79
Boarding, diagonal 184
Boards
    back-gutter 195
    barge 158
    fascia 157
    floor 90, 91
    lay 156
    ridge 153–6
    saddle 153–4
    soffit 157
    valley 157, 158
Boning rods 267
Brace
    diagonal 66, 196, 197
    ratchet 20
Bracing 195, 197
Bracing spacer 230
Bracket anchors 81, 82
Brackets, cradling 96, 157
Bradawls 23
Brads
    cut floor 34
    Maxi 31
Breather membrane 184, 186, 271
British Standards (BS)
    (743) 64
    (1192: Part 1: 1984) 1
        (Part 3: 1987) 1
    (1282: 1975) 92
    (4978) 87
    (5250) 185
    (5268: Part 2: 1991) 87
        (Part 5) 184
    (5395: Part 1: 1977) 131
        (Part 2: 1984) 136
Building Regulations, the
    Approved Document A 99
        (A1/2) 87
        (B: Fire safety) 132
        (F2) 182, 183
        (K, 2010) 129
        (L: Conservation of fuel and power) 88

(M: Access and facilities for disadvantaged
    people) 132
(Part C) 88, 93
(Part J) 90
Tables A1 and A2 (floor joists) 87, 95
A17 to A22 (roof joists) 180
(N: Glazing – Materials and Protection) 79
Bullnose step 114, 117, 119, 271
Butts (doors) 246

Cabinet projection 5
Cambered edges 100
Cap
    acorn 128
    ball 128
    mushroom 128
Capping
    landing 126
    string 126
Carpenter's pencil 8
Cartridge
    guns (tools) 31
    strengths 32
Cat walk 156
Caulking tool, wooden 78, 79
Cavalier projection 5
Cavity
    tray 181
    walls 86, 181
Ceiling joists 153
Celotex GA2000 187
Centre bob 10
Centres, arch 228
Chalk line reel 9
Cheeks, dormer 184
Chevron bracing 195
Chimney trimming 194, 195, 200
Chisels
    bevelled-edge 17
    bolster 23, 24
    cold 23, 24
    firmer 17
    impact-resistant 17
Chord
    bottom 200, 272
    top 195, 272
Cladding 184, 272
Claw hammer 15
Clearance angles (drills) 22
Cleats 56, 272
Clench-nailing 231–2, 272
Coach bolts 60, 118, 272
Cold deck (flat roof) 183
Collars
    arch-centre 232–3
    roofing 156

Collimation height 266
Concave shaping (easing) 122
Cone 220
Connectors, timber 61, 95, 195
Convex shaping (easing) 122
Coping saw 14
Corbels 194, 272
Cordless
    Angle Finish Nailer 31
    circular saw 24
    drill 25
    jigsaws 27
    nail guns 30
    planers 27
    screwdriver 25
    SDS Rotary hammer drills 28
Cornice blocks (architrave) 239
Counter-battens 187
Counterboring 68, 69
Countersink bits 21
Cradling brackets 96, 157
Cramp 272
Cramping method (floors) 90
Cross bearers (tank supports) 201
Crosscut saw 12
Crowbar 23
Crown 218, 219
    rafter 153–5
Crown wool 186
C-studs 213–7
Cupping 272
Curb (skylights) 188
Cut clasp nails 69
Cutting angles (drills) 22

Dado rails 242
Damp-proof
    course 89
    material 64
    membrane 92
Datum 115, 272
Datum level 265
Decking material 181
Deemed length 129, 130
Delaminating (trussed rafters) 196
Diamond
    Abrasive Lapping Fluid 19
    whetstones 18
Dimensioning sequence 2
Diminish, jack rafter 162, 167
Door
    indication 7
    Knobs 250
    linings 68–71
Doorsets 75
Doors, fire-resisting 76

Dormer windows 184
Dot-and-dab technique 49
Double
    architraves 238, 239
    floors 86
    pitch lines (stairs) 135–6
Dovetailed housing 95
Dowel
    metal 68, 118, 272
    wooden 119, 272
Draw-bore holes/pins 119, 272
Drills
    cordless 25, 26
    hammer 22, 26
    masonry 22
    percussion 26
    powered 26
    SDS 22, 26
    twist 22
Dry lining 212

Easing (strings) 122, 123, 272
Eaves
    closed 156–7
    concealed (in wall) 159, 170–4
    diagonal (on hip rafters) 173, 174
    drop 173–5
    open 156
    visible (projecting) 156, 159, 170–82
Electronic Detectors 24
Ellipse 220
Escutcheon plates 250
'E' staff 266
European projection 3
Expanded metal lath (EML) 72
Expansion gap 89–93
External grounds 113
Extrados 218, 219
Eyebrow window 188

Fall-arrest equipment 202
Fanlight 72
Fan truss 195
Fascia boards 157
Feather-grained 29
Febond adhesive 91
Felt
    built-up (3-layer) 181–3
    mineral 181–3
Fender wall 87–90
Fillets 272
Finished floor level (ffl) 65, 70, 115
Fink truss 195
Fire-resisting doors 75, 76
Firring 180, 181, 272

First-angle projection 3, 4
Firtree gasket 76
Fittings (hardware) 250, 251
Five-twelve-thirteen (5:12:13)
    method 268
Fixing band 81, 98
Fixings:
    cavity fixings 39
    Helical warm-roof batten-fixings 187
    Inskew 600 fixings
        (Helifix) 187
Fixing ties 65, 66
Flange 272
Flashings 181–9, 272
Flat Bits 20
Flight, stair 129
Floor
    boarding 90, 91
    brads, cut and lost-head 34, 35, 90
    clips 92
    cramps 90, 91
    expansion 90–3
    joists 86–110
    joist spacings 86
    joist Span Tables (TRADA) 95
    span 87, 93
    traps 90, 91
    ventilation 87–93
Flooring
    hardwood strip 108
    laminate 108
    plastic-laminate 108
    real-wood laminate 108
Flooring-grade chipboard 92
Floors
    double 86
    embedded-fillet 92
    floating 91–3
    ground 87–93
    single 86
    surface-battened 92
    suspended 87
    upper 94
Foresight (FS) 266
Frame
    cramps 65
    Holdfasts 66
    screw-ties, Owlett's 66
    ties 65
Framed grounds 73
Frame-fix screws 38, 39, 67
Frames
    door 63–8
    fire-resisting 75, 76
    internal 71
    storey 72

    sub 72, 73
    window 77
Framing anchors 95, 158
Free-flighting 31

Gable 152
    ladders 199
Gablet 152, 272
Galley kitchen 256
Gang-Nail Systems Ltd 195
Gap-filling adhesives 241
Gas fuel cells 30
General Access Stair (category 3) 134
Geometrical
    shapes 220
    stair 130
Geometry
    arch centre 218
    roof 160
Girder
    trusses 197
    truss shoes 199
Glazing safety issues 79
Glue blocks 121, 124, 273
Gluing 120, 121
Going 130–7, 273
Graphical representation 6
    symbols 6
Grinding angles (chisels) 17
Gripfill adhesive 241
Grounds
    architrave 112, 273
    lining 73, 273
    skirting 111, 121, 273
    wall-panelling 112, 273
    external 113, 273
Gusset plate (saddle board) 153, 154, 273

Hacksaw 23
Hammer
    drill 25
    -fix screws 38, 67
Handboard 58
Handrail regulations 132, 136, 138
Handrails
    fixing 120, 125
    protecting 122
Handsaws 12–14
Hangers 156
Hardcore 88, 273
Hardpoint saws 13
Hardwood, definition of 273
Hatch 156, 200
Hawk 58
Headroom 132
Hearth, concrete 87, 90

Helical stair 130
Herringbone
    noggings 206
    strutting 98
High-density fibreboard (HDF) 108
High-velocity principle 31
Hinges
    butt 246
    rising butt 246
Hip
    flat-top trusses 198
    girder trusses 197, 198
    mono (or monopitch) trusses 198
    rafters 153–5
Honeycombed sleeper walls 87–90
Hop-ups 58, 59
Horns 68
Housing joints 56
Hybrid (flat roofs) 183
Hyperbola 220
Hypotenuse 269
H-studs 213–7

Infill-strip (baluster-spacers) 127–8
Inner
    skin (wall) 273
    string (stairs) 114–7, 273
In situ concrete 273
    utility stair (category 2) 133
Insulating material 88–94, 181–7
Intermediate Sight (IS) 266
Intersection girder trusses 199
Intrados 218, 219
Intumescent
    paint 76
    paste 76
    strip 75, 76
Isometric projection 5
Isosceles roof shape 159
I-studs 213–7

Jabfloor Type 70 (insulation) 91
Jack plane 19
Jambs 68, 273
Jig
    lagging 231
    pencil 222
Jigs
    hinge 29, 248
    kitchen worktop 29, 262
    letter plate 29, 252
    lock 29, 252
    worktop 29, 262
Joints
    bare-faced tenon 116
    bevelled-housing 95

butt 195, 208
blind tenon 95
comb 68
dovetailed housing 95
framing 94, 95
half-lap 89
housing 95
oblique tenon 119
splayed-heading 90
splay-housed, mortice and tenoned 207
square-heading 90
stopped housing 95
stub-tenoned 207
tongued and grooved 81, 90
tongued-housing 70, 124
tusk-tenon 95
Joist hangers 96, 97
Joists
    bridging (common) 94, 273
    ceiling 153, 273
    sectional-size of 87, 94
    table for size of 94, 95
    trimmed 94, 273
    trimmer 94, 273
    trimming 94, 273
    return 180, 182, 184

KD
    cover caps 264
    Fixit blocks 263
Key, arch 219
Kicker strut 120
Kicker(s) 149, 150
Knots 273
Knotting 273

Lagging
    close 230
    open 230
Lagging jig 231
Laminate:
    flooring 108
    worktops 261
Landings 130, 132–3
Latches
    cylinder 251
    mortice 250
    tubular 250
Lay board 156
'L' cleats 107, 108
Letter plate 251–2
Levelling
    instruments 265
    staff 266
Levelling-boards, concrete 61

Levels
  boat 9
  torpedo 9
Lever furniture 250
Lining leg 64
Linings, door 63
Lintels 218, 273
Load-bearing walls 151, 155
Locating blocks (uPVC windows) 78, 79
Location drawings 7
Locks
  dead 251
  mortice 249, 250
  rim 251
Louvres 72
Low-velocity principle 31, 32

Mallet 15
Margin 235
  template 236
Marking gauges 16
Masonry drills 22
Mastic seal 67
Method Statements 202
Metres 2
Metric rafter square 11, 163
Microllam® 94
Millimetres 2
Mineral wool 76
  -wool slab 89
Mirror plates 47, 48
Mitre 235, 236, 273
  block 237
  box 237
  saw 15, 237
  square 10
Moisture absorption 73, 121
Moisture vapour 186
Mono (or monopitch) trusses 199
Mortar boards 59, 60
Mounting plates, metal 258, 263

Nail
  bar 23
  punches 23
Nailers
  Framing 30
  Finish 30
  Multi T 31
  Spit Pulsa 700P 31
Nailing guns 30
Nails
  annular-ring shank 34
  brad-head oval 33, 34
  cut clasp 34
  grooved-shank 34, 35

lost-head oval 33, 34
lost-head wire 33, 34
masonry 34, 35
round-head wire 33, 34
T headed 33
New Build Policy
  Guidance Notes 98, 180, 181
Newel
  caps 128
  drops (pendants) 128
  posts 116–21, 128, 273
Node points 195
Noggings
  fixing 206
  floor-joist 98, 99, 273
  herringbone 206, 273
  perimeter 104, 273
  roof-joist 179, 273
  rafter 179, 273
  staggered 206, 273
  straight 206, 273
  vertical 209, 273
Nosing
  landing 126, 273
  tread 130, 273
  window board 80, 81
Notching 99, 120

Oblique projections 5
Obtuse angles (on stair-strings) 122
Oilstone boxes 17, 18
Oilstones 17, 18
Open-rise steps 131, 135, 273
Optical squaring instruments 267
Ordnance
  Bench Mark (OBM) 265, 266
  datum 265
Oriented strand board (OSB) 86, 90, 94
Orthographic projection 3
Outer
  skin 273
  string 114, 273
Oversite, concrete 89–92, 273

Panel
  adhesives 241, 242
  bolt connectors 28–30, 262
  pins 34, 35
Panel saw 12
Parabola 220
Parallam® PSL 101, 102
Parallel perspective 6
Partition
  braced (traditional) 203
  common (modern) 204–12
  trussed (traditional) 203

Pellets 69, 273
Pencil gauge 10
    jig 222
Pendants (newels) 128
Percussion drill 25
Perspective projections 5, 6
Pictorial projections 5
Picture rails 242
Pilot holes 22, 274
Pincers 23
Pinch
    bar 23
    rod 124, 205, 274
Pins, wooden 119
Pitch angle
    roof 151, 159, 274
    stair 134, 274
Pitch Board 144–9
Pitching details (roofs) 176–180, 197–200
Pitch lines
    roof 159, 274
    stair 131, 135–6
Plane
    jack 19
    power 26
    smoothing 19
Planometric projection 5
Planted
    moulding 122, 274
    stops 68, 274
Plasterboard
    foil-backed 182
    sizes of 212
Plasterer's
    board and stand 59, 60
    folding stand 60
    handboard 58
    hop-up 58, 59
Plate connectors 195
Plates
    gusset 154
    head 204, 274
    sill (floor) 204, 274
    wall 89, 176–7, 274
Plinth blocks (architrave) 239
Plugging 37, 38
Plugs
    colour-coded 37
    plastic wall 37
Plumb 274
    bob 10
    cut 159, 274
    rule 62
Pneumatic nail guns 30
Pocket screwing 121, 274
Polystyrene underlay 91

Polythene sheet vapour-check/DPM 91, 92
Polyvinyl acetate (PVA) adhesive 91
Posi-Joist™ Steel-web system 105
Powered drills 25, 26
Powered planer 26
Powered router 28
Private stair (category 1) 133
Private Stair Regulations 133
Profile 274
    frame 74, 274
    lining 75, 274
    pattern pair 196
Profiles 267
Props, timber 123, 234
Protection strips 64, 122
Protractor facility
    metric rafter-square 163–6
    Roofmaster square 167–70
Pullsaws 14
Punch, nail 23
Purlins 156, 178

Quarter-turn stair 123
Quilt
    mineral wool 76

Radius rod 228
Rafters
    common 154, 170
    cripple 162, 166
    crown 154
    hip 154, 155
    jack 162–168
    pattern 170, 171
    pin (crown) 154
    trussed 195
    valley 158, 177
Ranging lines 267
Ratchet brace 20
Reduced Level (RL) 266
Reflex angles (stair strings) 122
Resin-bonded 274
Restraint straps 153, 179, 180
Reveals 274
Rib formation
    semi-hexagon 233
    semi-octagon 233
Ribs 228, 233
Ricochets 31
Ridge board 153–6
Rip saw 13
Rise
    arch geometry 218, 219
    roof 159
    stair 130, 131, 133, 134
Riser board 114, 121

Risk Assessments 202
Roof
    design 152
    falls 157, 181
    hatch (trap) 156, 200
    windows 184, 188
Roof
    flat 151, 180–3
    gable 152
    gambrel 152
    hipped 152
    jerkin-head 152
    lean-to 151
    mansard 152
    pitched 151
    trussed-rafter 195–201
    valley 152
Roofing
    bevels 159–77
    felt, built-up 181–3
    formulas (bevels) 168
    Ready-Reckoner 161
    square 11, 163
Roofing
    modern 195–202
    traditional 151–94
Roofmaster square 11, 167
Rose plates 249
Rounded step 136
Rule
    folding 9
    scale 2
    steel tape 8
Run
    common rafter 159–70
    hip 160–70
    scaled 165
Runners 274

Saddle and block (door-hanging device) 244
Saddle board 153, 154
Safety
    glass 79
    nets 202
Safety bracing (floor joists) 104
Sarking 274
Saw
    pitch-angle 12, 13
    stool 50
Saws
    electric circular 24
    hand 12
Scale rules 2
Scales 2
Screw gauges 35
Screwdriver bits 21, 22

Screwdrivers 16
Screwing and pelleting 69
Screwing, pocket 121
Screws
    Allen recessed head 35
    bugle-shaped head 33, 35
    collated floorboard 33
    countersunk head (CSK) 35, 36
    Frame-fix 38
    Frame 38, 39
    gauge of 35
    Hammer-fix 38
    nylon-sleeved 38
    Phillips' head 35
    Pozidriv head 35
    raised head (RSD) 35
    round head (RND) 35
    single-thread 36
    spaced-thread 36
    square recessed head 35
    Supadriv head 36
    Torx or T Star-slotted 35, 38
    twinfast-thread 38
    Window-fix 38
Scribing 238, 240, 274
    technique 238, 240, 241
SDS
    chuck systems 21
    drills 22
    -Max chuck 28
    -Plus chuck 28
Seat cut 159, 274
Setting blocks 78, 79
Setting out
    arch shapes 218–227
    boards 232, 233
    common rafters 170
    crown rafters 171
    hipped ends 179
    hip rafters 172
    jack rafters 175
    partitions 203, 204
Setting-out pegs 266
    tapered treads 130–6
    terms (roofing) 158, 159
    wall plates and ridge board 177
Shank holes 121, 274
Sharpening
    angle 17, 18
    oil 17, 18
    procedure 18, 19
Sheathing 184, 190, 274
Shims, plastic 78
Shrinkage 74, 122
Silent Floor® System 94
Sill 65, 67, 203

Silicone sealant 78
Single
    floors 86
    pitch-line 131–6
Site plans 7
Skew-nailing 121, 205, 274
Skin
    inner (wall) 273
    outer (wall) 273
    plywood or hardboard 230
Skirting 240–2
    bullnose 242
    chamfered 242
    dual pattern 242
    Grecian ogee 242
    grounds 111, 122
    ovolo 242
    torus 242
Skylights 188
Slatted shelves 43
Sleeper walls 87–9
Smoothing plane 19
Soffit 63, 274
    boards 157
Softwood, definition of 274
Soldier pieces 112
Spacing (infill) strip 126–8
Spalling 32, 275
Span 159, 218, 275
Spandrel 114
Spindle (lock) 249, 251
Spindles (stair) 114, 275
Spiral stair 131
Spirit level 9
Splayed step 123, 136–7
Splicing 238, 241
Spot boards (see *mortar boards*)
Spotting 71, 275
Spreader beam 201
Springing line 218
Sprocket pieces 157
Sprung edges 178, 275
Square
    combination mitre 10
    roofing 11, 163
    Roofmaster 11, 167
Squaring technique 269
Stability/integrity rating 75
Stair
    formula 133, 134
    gauge fittings 164, 165
    width 132
Staircase 114
Stairway (see *flight*)
Stairwell 94
Stakes 267

Steel square 11
Stepped measurements 270
Stepped soldiers 112
Steps
    parallel 123
    tapered (winding) 123, 131
Storey frames 72–3
    height 116
Straightedges 9, 61
Straining pieces 156
Stretchers 69, 70, 275
Striking plate 250, 253
String-line technique 111
Strings
    inner 114–7, 273
    outer 114, 273
    return 124
    wall (see *inner* strings)
Strongbacks (in Posi-web floors) 107
Structurally-graded timber 87
Struts 120, 156, 275
Strutting 97
Stuck
    moulding 122, 275
    rebate 68, 275
Stud
    corner 209
    door 205
    door-head 205
    intermediate 205
    wall 204
Studding 275
    fixing problems 211
    junctions 209
    spacings 212
Sub frames 72
Sunken rebate 68, 274
Surveying/measuring tapes 267
Suspension brackets 258, 263
Swollen timber 73, 121

Tangential cuts 229
Tank supports 201
Tapered treads 123, 129, 131, 135–6
T-Bars 109
Templates
    arch centre 232
    margin 236
    saw stool 52, 53
Temporary Bench Mark (TBM) 266
Tenon saws 13
Third-angle projection 4, 5
Three–four–five (3–4–5)
    method 177, 268
Threshold 67, 74
Tie beams 231

Ties
    angle (corner) 155, 173
    fixing 65, 66
    roof (see *ceiling joists*)
Tilting fillets 157, 158
Timber connectors 95
Timberstrand™ LSL 101, 102
TJ-Beam® software 102
TJI® joists 101
TJ-Xpert® software 102
Tonk strips 42
Toothed timber-connectors 61, 211
Trammel
    beam compass 228
    frame 229
    rod 228, 229
Transom 72
Trap, roof 156, 200
Tread 114, 130, 275
Trim-It™ floors 105
Trusses
    Fan 195
    Fink 195
    flat-top 198
    girder 197, 198
    hip girder 198
    hip mono 198
    mono or monopitch 199
    W 195
Trus Joist 93, 94
Turning piece 228
Tusk tenon joint 95
Twist 275
    bits 20
    drills 22
Tyvec®
    Butyl Tape 187
    Eaves carriers 186
    Solid membrane 187
    Supro membrane 184
    Supro Plus membrane 187

Uni-Screw heads 33, 35
Upper floors 94
uPVC windows 77
Utility Stair (category 2) 133
U-track 213–7

Valley
    boards 157, 158
    frames, diminishing 199

junctions 199
    rafters 158, 177
Vapour check (barrier) 91–3, 181–6
Vee-ended stool 58
Ventilation
    floor 87–94
    roof 157, 183–6
Verge 155, 158
    projection 158
Viability graphs 134, 135
Vials 9
Voussoirs 218–9, 275

W truss 195
Waist-thickness 147–9
Wall
    bearings 86, 97
    plates 89, 176–7
    strings 114–7, 273
    studs 204
Waling(s) 149, 150
Warm deck (flat roof) 183
Warp 275
Water bars 67
Waterproof adhesive tape 91
Weatherboard 67
Web 275
Wedge gasket 78
Wedges 234
Wet trades 63
Winder
    half kite 123, 124
    kite 123
    skew 123, 124
    square 123, 124
Winding 275
    sticks 57
Window
    boards 80
    indication 6
Windows
    casement 77
    dormer 184
    eyebrow 188
    fixing 77
    roof 188
    skylight 188
    uPVC 77
Work triangle, the 256
Wrecking bar 23

9781138296008